T0371995

Enabling Technologies for the Internet of Things: Wireless Circuits, Systems and Networks

EDITOR

Sergio Saponara
University of Pisa, Italy

Tutorials in Circuits and Systems

For a list of other books in this series, visit www.riverpublishers.com

Published, sold and distributed by:
River Publishers
Niels Jernes Vej 10
9220 Aalborg Ø Denmark

River Publishers
Lange Geer 44
2611 PW Delft
The Netherlands

Tel.: +45369953197 www.riverpublishers.com

ISBN: 978-87-93609-74-7 (Print)
978-87-93609-73-0 (Ebook)

Published by River Publishers.

Library of Congress Cataloging-in-Publication Data: July 2018
Editor: Sergio Saponara
Title: Enabling Technologies for the Internet of Things: Wireless Circuits, Systems and Networks

Table of contents

Introduction 7

1. Wireless Power Transfer for RFID Systems 11
   By Smail Tedjini

2. Computational Electromagnetics and Electromagnetic Compatibility for the Internet of Things 41
   By Ludger Klinkenbusch

3. Integrated Circuits & Systems for mm-wave/RF Wireless Transceivers in IoT Applications 61
   By Sergio Saponara

4. ICs and VLSI Architectures for mm-wave/RF Wireless Transceivers in IoT Applications 89
   By Sergio Saponara

5. Chipless RFID for Identification and Sensing 109
   By David Girbau, Antonio Lazaro, Simone Genovesi, and Filippo Costa

6. Near-Field Focused Antennas for Short-Range Identification and Communication Systems 159
   By Giuliano Manara, Andrea Michel, Alice Buffi, and Paolo Nepa

7. Ultra-low-power Devices, and Application of New Materials to mm-wave Antennas and Circuits 181
   By Massimo Macucci

8. A Functionally Safe SW Defined Autonomous and Connected IoT 215
   By Riccardo Mariani

About the Editor 249

About the Authors 253

Introduction

This book collects slides and notes from the lectures given during the 2017 Seasonal School Enabling Technologies for the Internet-of-Things.

Supported by IEEE CAS Society and by INTEL funding, the Seasonal School was held at University of Pisa, Italy, from 17th to 28th July 2017 [1], see Fig. 1. The school, organized by **Sergio Saponara**, Full Professor of Electronics at University of Pisa, and **Giuliano Manara**, Full Professor of Electromagnetism at University of Pisa, involved in two weeks lecturers from academia (e.g. Universities of Pisa in Italy, University of Kiel in Germany, Rovira i Virgili University of Tarragona in Spain, University of Grenoble-Alpes in France) and industry (e.g. Intel). Moreover, the school hosted two special events, chaired by Prof. S. Saponara, the "2nd Workshop IoT Industry Day" [2] and the "1st Workshop INTEL Functional Safety Day" [3], with the participation of about 70 people per event. As social event, a dinner was held in a typical Italian restaurant with all students (coming from 17 different countries from Europe, Asia, Africa and South America).

Lecturers discussed new trends in Internet-of-Things (IoT) technologies, considering technological and training aspects, with special focus on electronic and electromagnetic circuits and systems. IoT involves research and design activities both in analog and in digital circuit/signal domains, with issues to be solved for sensors interfacing and conditioning, energy harvesting, low-power signal processing, wireless connectivity, and networking. The Summer School is recognized as an official exam from University of Pisa. A written exam was held by students at the end of the Summer School.

As special lecturers, we had also the honor to host Prof. **Franco Maloberti**, IEEE CASS President, see Fig. 1, and Dr. **Federico Faggin**, the "father" of the microprocessor, and Dr. **Riccardo Mariani**, INTEL Senior Research Fellow, see Fig. 2. Functional Safety will be one of the key issue in emerging IoT applications in safety critical domain like industry 4.0, autonomous and connected vehicles and e-health. The world is becoming more and more interconnected. We currently estimate that two hundred billion of smart objects will be part of the IoT by 2020. This new scenario will pave the way to innovative business models and will bring new experiences in everyday life. The challenge is offering products, services and comprehensive solutions for the IoT, from technology to intelligent and connected objects and devices to connectivity and data centers, enhancing smart home, smart factory, Autonomous Driving Cars and much more, while at the same time ensuring the highest safety standards. In safety-critical contexts, where a fault could jeopardize the human life, safety becomes a key aspect. That's why, Functional Security is among the most important challenges for the future of IoT.

The main material presented during the school is now available through IEEE CASS & River Publishers with this slide and notes book. After this Introduction, the teaching material is organized in the following chapters:

Chapter 1, by Prof. S. Tedjini, University Grenoble-Alpes, Valence, France, "Wireless Power Transfer for RFID Systems"

Chapter 2, by Prof. L. Klinkenbusch, University of Kiel, Germany, "Computational Electromagnetics and Electromagnetic Compatibility for the Internet of Things"

Chapter 3, by Prof. S. Saponara, University of Pisa, Italy, "Integrated Circuits and Systems for mm-wave/RF Wireless Transceivers in IoT Applications (Communications)"

Chapter 4, by Prof. S. Saponara, University of Pisa, Italy, "ICs and VLSI Architectures for mm-wave/RF Wireless Transceivers in IoT Applications (Remote Sensing)"

Chapter 5, by Prof. D. Girbau Sala and Prof. A. R. Lázaro Guillén, Rovira I Virgili University of Tarragona, Spain, and by Dr. S. Genovesi and Dr. F. Costa, University of Pisa, Italy, "Chipless RFID for Identification and Sensing"

Chapter 6, by Prof. G. Manara, Dr. A. Michel, Dr. A. Buffi, and Prof. P. Nepa, University of Pisa, Italy, "Near-Field Focused Antennas for Short-Range Identification and Communication Systems"

Chapter 7, by Prof. M. Macucci, University of Pisa, Italy "Ultra-low-power Devices, and Application of New Materials to mm-wave Antennas and Circuits"

Chapter 8, by Dr. R. Mariani, INTEL, "A Functionally Safe SW Defined Autonomous and Connected IoT"

Prof. Sergio Saponara
IEEE CASS Senior Member

Fig. 1: IEEE CASS Seasonal School group with Prof. F. Maloberti

Fig. 2: IEEE CASS Seasonal School. Lecturers G Iannaccone, L. Fanucci, M. Schaecher, B. Neri, F. Faggin, S. Saponara, R. Mariani, and the Director of DII-University of Pisa G. Anastasi

Web References

[1] https://www.dii.unipi.it/didattica/summer-school-on-enabling-technologies-for-iot

[2] https://www.dii.unipi.it/didattica/summer-school-on-enabling-technologies-for-iot/item/1278.html

[3] https://www.dii.unipi.it/didattica/summer-school-on-enabling-technologies-for-iot/item/1279.html

Summer Schools at University of Pisa

The Summer/Winter school programme has been activated at the University of Pisa in 2012/2013 to capture, on one hand, the desire of some researchers and professors to offer short courses taught in English on specific subjects to an international audience, and on the other hand, to promote the internationalization of the University of Pisa (https://www.unipi.it/summerschool).

The program has been a success above the most optimistic forecasts. From the 6 initial Summer Schools, we have moved on to the current 25, with an increase in the number of participants from 70 to more than 600 in the last edition. Moreover, while in the first edition the international participants were around 50%, in the last edition they reached 60%, thus witnessing the international dimension of the programme.

In the framework of the Summer School Programmes, the University of Pisa collaborates with several international Institutions both in Europe and other countries (e.g. Universiteit Leiden, Universitetet Agder, Université Lille 2, Bergische Universität Wuppertal, Univerzita Karlova, Universität Mannheim, Universitet i Oslo, Universidad Complutense Madrid, University of Economics of Prague, University College London, Universidada Politecnica de Cataluña, The University of Edinburgh, Chang Gung University, Aristotle University of Thessaloniki, Instituto Superior de Agronomia Lisboa) , and it has also been chosen as Study Abroad by prestigious English universities, such as Sheffield, and by the Istituto Tecnológico de Monterrey (Mexico), which both offer scholarships to their students to attend the Summer schools of the University of Pisa. Furthermore, this year has been made an agreement with the King's College for a Summer School Students Exchange in the two universities.

This enormous success of the summer school programme has been only possible thanks to the great enthusiasm and commitment of the researchers and professors of the University of Pisa, who have been able to propose a teaching offer of great appeal, especially to an international audience and to the constant and effective support provided by all the staff of the International Cooperation Unit. The summer school "Enabling Technologies for industrial Internet of Things", coordinated by Prof. Sergio Saponara, with more than 30 participants per year from 20 different nations in the last three years is certainly one of the brightest representatives of this success.

Prof. Francesco Marcelloni
Vice-Rector for International Cooperation and Relations,
University of Pisa, Italy

Education, Research and Technology Transfer Activities on IoT at the Department of Information Engineering (DII), University of Pisa

The Summer School on *Enabling Technologies for the Internet of Things* is one of the key educational activities run by the Department of Information Engineering (DII) of the University of Pisa.

DII is an International Center of Excellence for research and higher education in the field of Information and Communication Technology (ICT), Robotics and Bioengineering, and among the top leading Universities in Italy for research about Enabling Technologies for Industry 4.0 (http://www.dii.unipi.it/en).

The Department has been promoting technological transfer since its foundation, by means of spin-off projects, cooperation with private and public institutions, actions meant to provide innovative solutions to key issues in different ICT sectors, and to bridge the gap between academic and industrial research. Currently, DII is involved in about 20 research projects funded by the European Commission (49 in the last three years), 2 projects funded through ERC Grants, and 22 research projects funded by the regional government of Tuscany.

At the beginning of 2018, DII was selected as "Department of Excellence" by the Italian Ministry of Education and Research (MIUR) with the "CrossLab" project, which aims at structuring six interdisciplinary and integrated laboratories (CrossLabs) for Industry 4.0. CrossLabs will try to put together Research and Industry, since they will be open to enterprises willing to find innovative solutions for their productive chains. In particular, Small and Medium Enterprises (SMEs) will be able to use advanced technological devices and receive support and assistance by the academic researchers. CrossLabs will cover all the key areas of Industry 4.0, namely *Additive Manufacturing, Advanced Manufacturing, Cloud Computing, Big Data, Cybersecurity, Augmented Reality, and Industrial Internet of Things* (http://crosslab.dii.unipi.it/).

The CrossLab focused on Industrial Internet of Things (IIoT) is the biggest one, in terms of involved researchers and available facilities. It aims at developing solutions to connect sensors, actuators and machines in an industrial context, in order to improve efficiency, to create opportunities for new customized products and services, and to increase workplace safety. The main applications domains, in addition to Industry 4.0, are autonomous vehicles, digital applications for sanitation, smart grids, smart agrifood, and logistics.

CrossLabs are an incredible opportunity for the Italian SMEs to access the technology and know-how needed to move toward the new industrial revolution.

Prof. Giuseppe Anastasi
Head of the Department

CHAPTER 01

Wireless Power Transfer for RFID Systems

Smail Tedjini

University Grenoble Alpes
Valence, France

INTRODUCTION. CONTEXT & HISTORICAL FACTS 14
1. Agenda 14
2. Iot Is Here Now – And Growing! 14
3. Back to the 19th Century 15
4. Marconi Application 1901 : Communication 15
5. Tesla Application 1900 Power Transfer 16
6. The Spirit of RFID : The Thing Designed by Léon Theremin 16

WPT RFID SYSTEM COMPONENTS 17
7. The Thing : 1945 The Great Seal Bug 17
8. Enabling Technology for Iot 17
9. Available Energy Sources 18
10. Harvesting System 18

NEAR-FIELD WIRELESS POWER TRANSMISSION 19
11. HF RFID System 19

MAGNETIC WIRELESS SYSTEM 19
12. Magnetic Loop Characteristics 19
13. Coupling Two Coils 20
14. Improving the Coupling by Resonance 20
15. Typical Voltage at Transponder Coil 21
16. Quality Factor 21

APPLICATION TO HF RFID AT 13.56MHZ 22
17. Protecting the Processing Circuit 22
18. HF RFID Reader 22
19. Voltage at Reader Coil/Cap 23
20. Signal Analysis at Reader Coil (I) 23
21. Signal Analysis at Reader Coil (II) 24
22. Effect of Z_l Variation 24
23. Effect of Z_l Variation : Load Modulation 25
24. Subcarrier Modulation 25
25. Load Modulation for HF Tags 26
26. Generation of Subcarrier (I) 26

FAR-FIELD WIRELESS POWER TRANSMISSION 27
27. Generation of Subcarrier (II) 27
28. Characteristics of Far-Field 27
29. Antenna Radiation Regions 28
30. Antenna Radiation Parameters (I) 28
31. Antenna Radiation Parameters (II) 29
32. The Dipole $\lambda/2$ (Usually for Tag) 29
33. The Patch (Usually for Reader) 30
34. Backscatter Communication 30

RECTIFIERS 31
35. Basic Rectifier Parameters 31
36. Schottky Diode As Rectifier 31
37. Cockcroft-Walton Voltage Multipliers. 32
38. CMOS Rectifiers 32
39. Voltage Multipliers Comparison 33
40. Efficiency of Voltage Multiplier 33
41. Rectenna Device 34
42. Loaded Rectifier Structure 34
43. Harmonic Balance (HB) Analysis 35
44. Input Impedance of Rectifier 35
45. Matching Options 36
46. RFID Tag Antenna Case 36

ADVANCED APPLICATION EXPLOITING NONLINEAR CHARACTERISTICS OF RFID CHIPS 37
47. UHF Tag Read-Range Evolution 37
48. Non-Linearities in Passive UHF RFID 37
49. Powering a Sensor 38
50. Concluding Remarks 38
51. Some References of Interest 39

Backscatter technique is becoming a topic of great interest, in particular for the implementation of the rapidly emergent internet of things. Indeed, due to the huge number of objects to be connected (some tens of billions), we need battery-less or passive devices for performing the identification of the object to be connected and information on its immediate environment such as temperature, humidity, pressure, and speed. For battery-less devices, RadioFrequency IDentification (RFID) tags seem to be a major player currently used for identification purposes, but some advanced researches demonstrated its ability to perform additional duties, in particular sensing. To perform these actions, an ASIC is used, which requires a tiny power to operate. Such a power is delivered thanks to Wireless Power Transmission (WPT)...

The lecture is dedicated to WPT and its application for RFID systems. The course is organized in several sections as follows

Introduction: This section introduces the context and some historical facts for the development of wireless systems and their first exploitation for communication and power transmission. Also, some examples of early devices exploiting wireless techniques such as RFID tags are highlighted.

WPT RFID System Components: This section describes the basic architecture of all WPT systems together with the main components. It also introduces the two classes of WPT systems i.e. Near-Field and Far-Field. These two classes differ with respect to the transmission distance.

Near-Field WPT: This section focuses on Near-Field systems, which are exploiting the magnetic coupling between two coils correctly located close to each other. The principle of operation and main characteristics of such systems are discussed, as well as their application to RFID at HF frequency.

Far-Field WPT: This section discusses the Far-Field configuration, which is based on the use of a couple of antennas one as transmitter and the other as receiver. The transmission characteristics are governed by Friis equation. The impact of the antenna characteristics, in particular their radiation patterns, on the transmission performance are discussed. Then, a comparison between Near-Field and Far-Field is reported.

Rectifier: This section focuses on the rectifier device that is necessary to convert the transmitted RF signal into DC in order to empower the circuitry of the tag. The main rectifier architectures like Cockcroft-Walton voltage multipliers are discussed. Also, the matching between antenna and rectifier is highlighted and some matching methods are reported and compared.

Concluding remarks: In this section, several examples of tags and up-to-date performance are presented. In particular, UHF RFID tags having a read-range of tens of meters are on the market. Finally, we report some advanced examples and hot topics in terms of research and development in the field of RFID and WPT.

1  Agenda

- INTRODUCTION. CONTEXT & HISTORICAL FACTS
- WPT RFID SYSTEM COMPONENTS
- NEAR-FIELD WPT
    - MAGNETIC WIRELESS SYSTEM
    - APPLICATION TO HF RFID
- AR-FIELD WPT
- RECTIFIER
- ADVANCED APPLICATION
    - EXPLOITING NONLINEAR
- CHARACTERISTICS OF RFD CHIP
- CONCLUDING REMARKS

INTRODUCTION. CONTEXT & HISTORICAL FACTS

2 

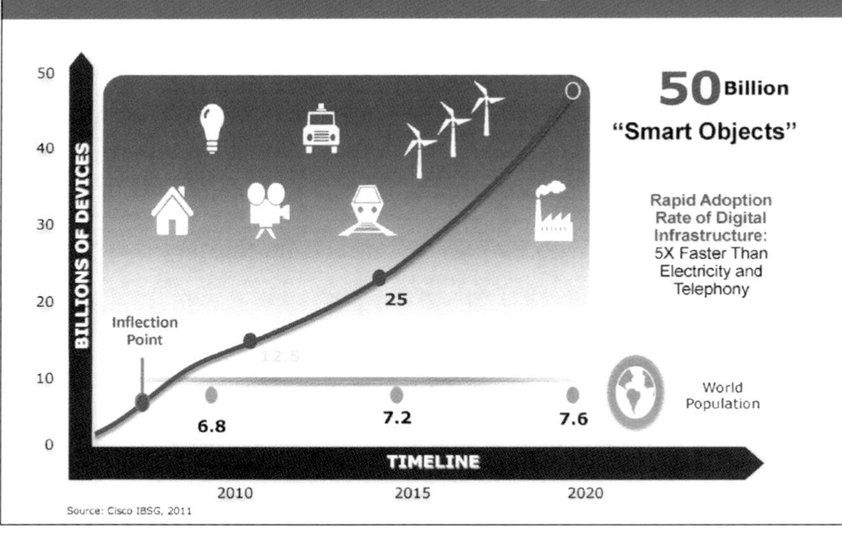

IoT isn't "science fiction" or something that will happen in the future – it's real, and it's here now. Cisco believes that the inflection point – the point at which the number of connected devices began outnumbering the number of men, women, and children on the planet – happened about five years ago; others in the industry believe that it happened about a year and a half ago. Similarly, Cisco believes that the number of connected objects will grow to about 50 billion over the next several years, while other estimates put that number at 25, 30, or even as high as 200 billion!

Who's right doesn't really matter ... the point is that we all universally agree on two things: 1) the point of inflection is in the past; and 2) gap is expected to widen exponentially over the next several years. *So, IoT is here today, and will continue to grow*!

3

The precursors of WPT : Maxwell for theory and Hertz for some experimental demonstration. However, many other scientists have contributed to the birth of Wireless Technology

Back to the 19th Century

All electromagnetic behaviors can ultimately be explained by Maxwell's four basic equations:

$$\nabla \cdot \mathbf{D} = \rho \qquad \nabla \times \mathbf{E} = -\frac{\partial \mathbf{B}}{\partial t}$$

$$\nabla \cdot \mathbf{B} = 0 \qquad \nabla \times \mathbf{H} = \mathbf{j} + \frac{\partial \mathbf{D}}{\partial t}$$

Maxwell 1864

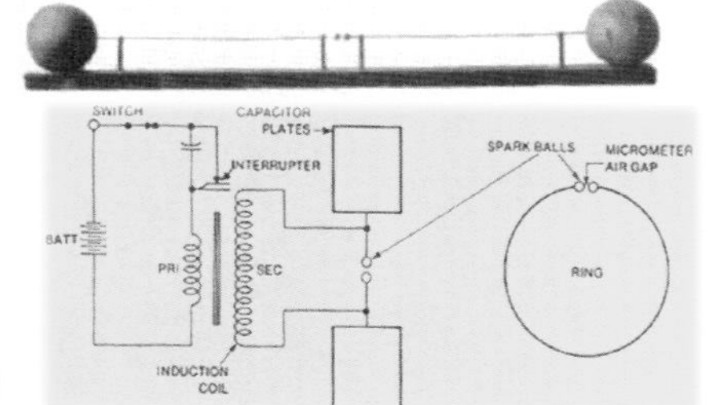

Hertz 1886

4

First transatlantic Wireless transmission by Marconi. Russian POPOV have also made some demonstrations. In a March 24, 1896 demonstration, he used radio waves to transmit a message between different campus buildings in St Petersburg. His work was based on the work of other physicists such as Oliver Lodge and contemporaneous with the work of radio pioneer Guglielmo Marconi.

Marconi Application 1901 : Communication

5

Tesla's Wardenclyffe plant on Long Island in 1904. From this facility, Tesla hoped to demonstrate wireless transmission of electrical energy across the Atlantic.

Tesla Application 1900 Power Transfer

The Famous Tesla Tower erected in Shoreham, Long Island, New York, was 187 feet high, the spherical top was 68 feet in diameter. The Tower, which was to be used by Nikola Tesla is his "World Wireless" was never finished.

Tesla demonstrating wireless lighting by electrostatic induction during an 1891 lecture at Columbia College via two long Geissler tubes (similar neon tubes) in his hands.

6

This is the spirit of RFID. Simple technology applied to basic concept leads to magic device. The beauty of the "thing" is it really works !

The Spirit of RFID : The Thing Designed by Léon Theremin

Tiny capacitive membrane (microphone) connected to a small λ/4 antenna

Passive cavity resonator, became active @ 330 MHz.

Sound waves caused the microphone to vibrate, turn modulated reflected radio waves

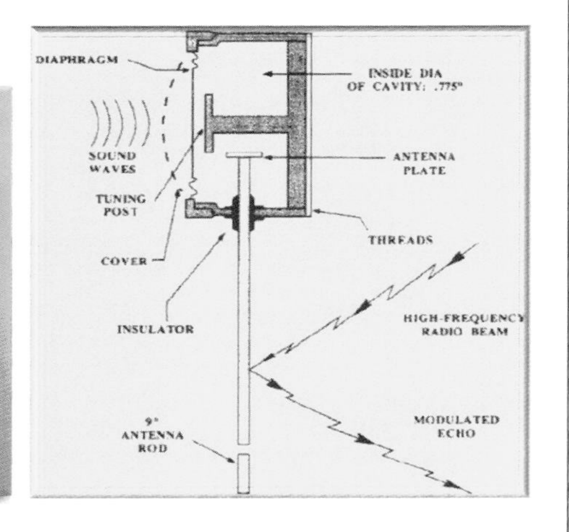

Some historical facts demonstrating the powerf of the future RFID.

The Thing : 1945 The Great Seal Bug

The device was embedded in a carved wooden plaque of the US Great Seal . On August 4, 1945, Soviet school children presented it to U.S. Ambassador A. Harriman, as a « gesture of friendship ». It hung in the ambassador' s Moscow residential office until it was exposed in 1952 during the tenure of Ambassador G. F. Kennan. The existence of the bug was accidentally discovered by a British radio operator who overheard American conversations on an open radio channel as the Russians were beaming radio waves at the ambassador's office. The CIA found it after an exhaustive search of the American Embassy, and P. Wright, a British scientist and former MI5 counterintelligence officer, eventually discovered how it worked.

WPT RFID SYSTEM COMPONENTS

8

How to enable IOT ? Only wireless and passive devices with certain features can be realistically envisaged. RFID is surely a serious option if "Augmented tags" are developed.

Enabling Technology for Iot

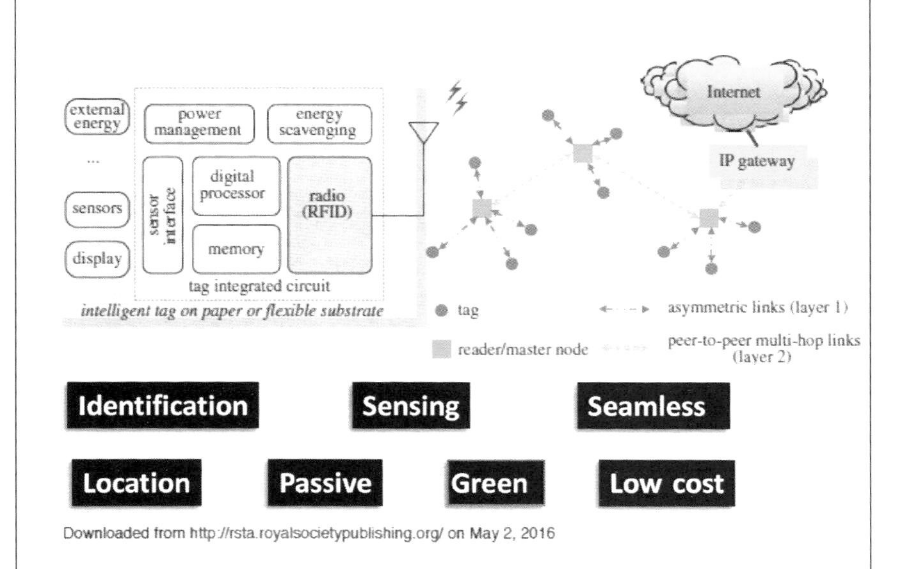

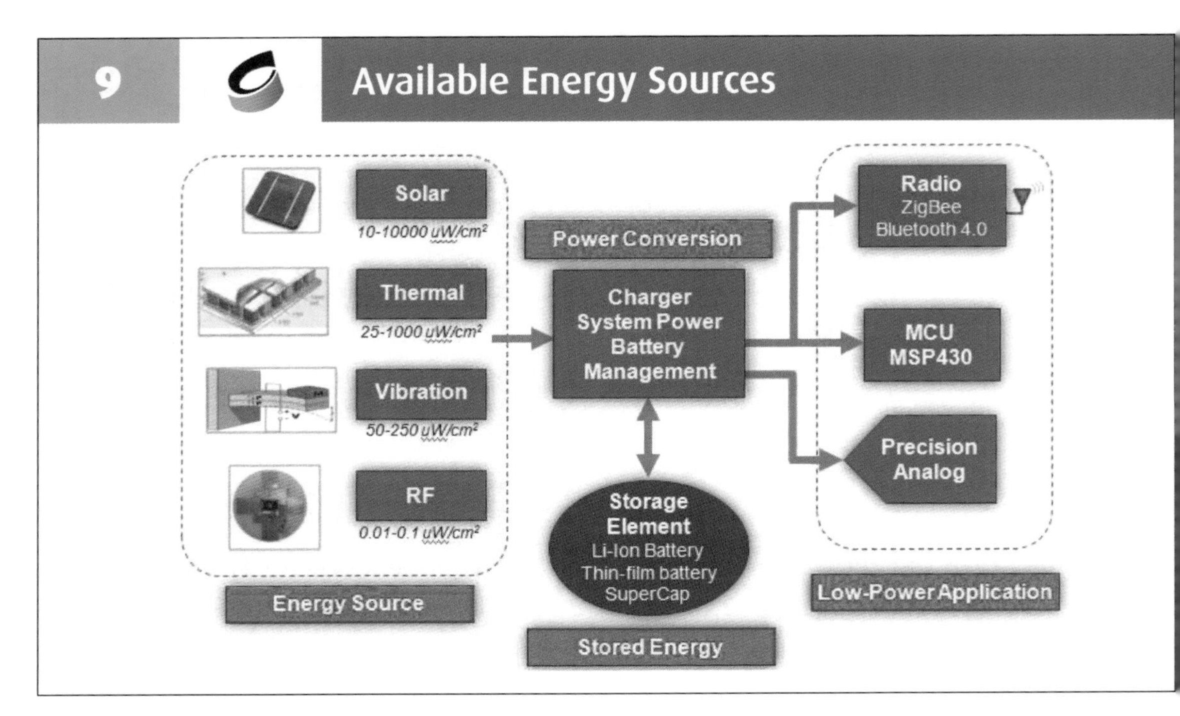

Without energy, nothing is possible. Some ambient energy sources and potential available power.

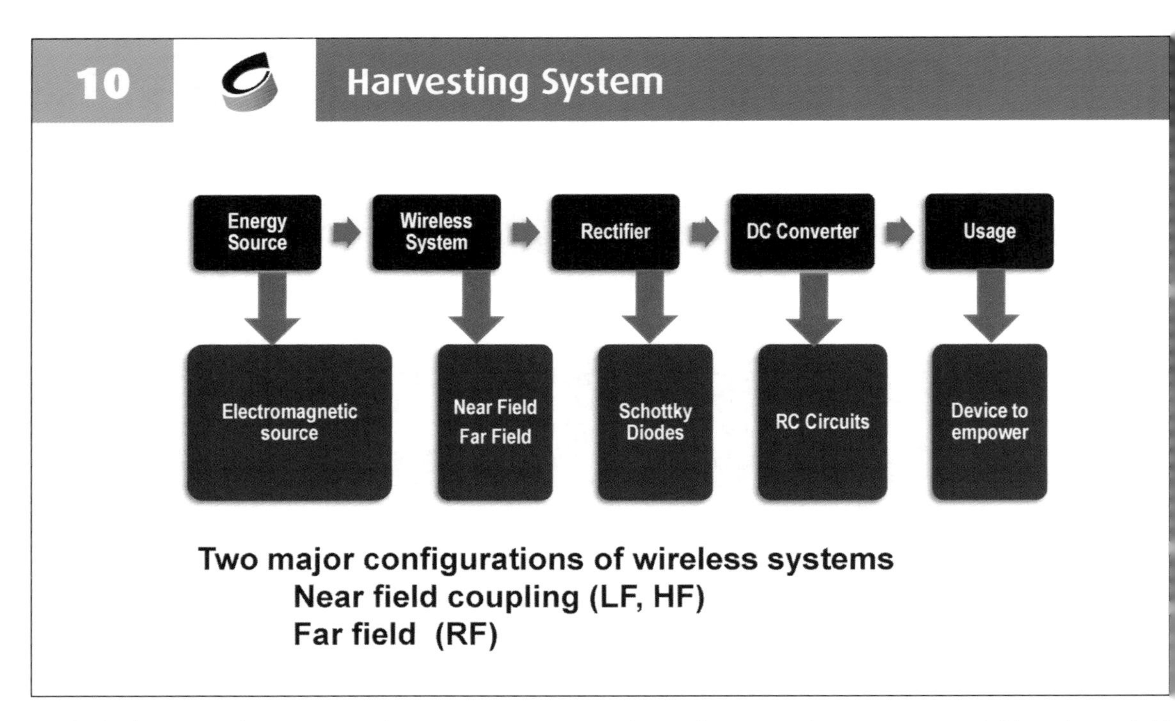

Block diagram of any power harvesting system. Make a distinction between Near-Field and Far-Field configurations.

NEAR-FIELD WIRELESS POWER TRANSMISSION

11

The example of HF RFID that uses the same signal to harvest power from the reader, but also ensures the communication thanks to load modulation technique (also known as Backscatter). This is an example of Near-Field system exploiting magnetic coupling between inductive loops.

HF RFID System

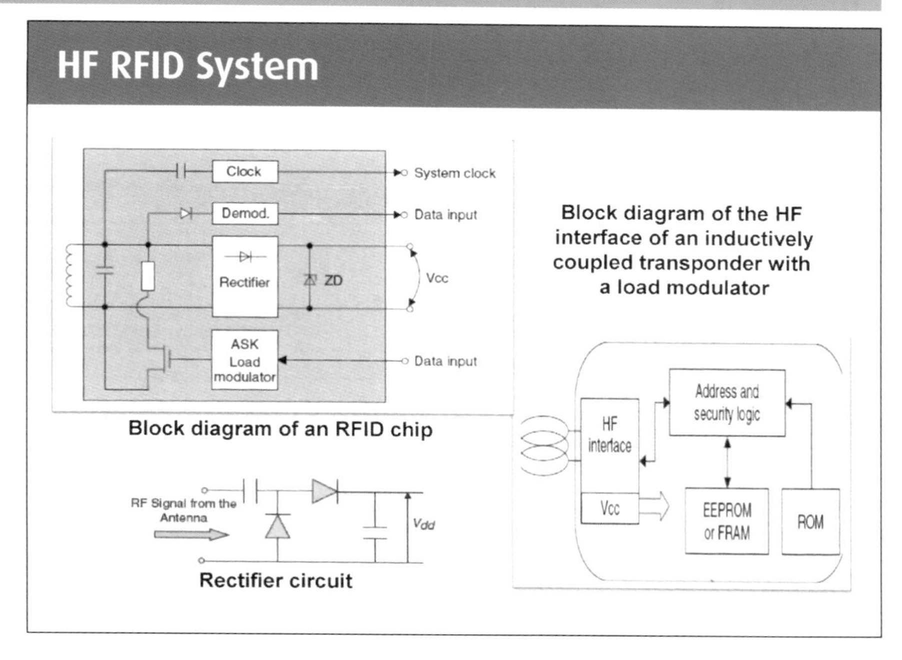

MAGNETIC WIRELESS SYSTEM

12

There is an optimal geometry for maximum coupling between two loops.

Magnetic Loop Characteristics

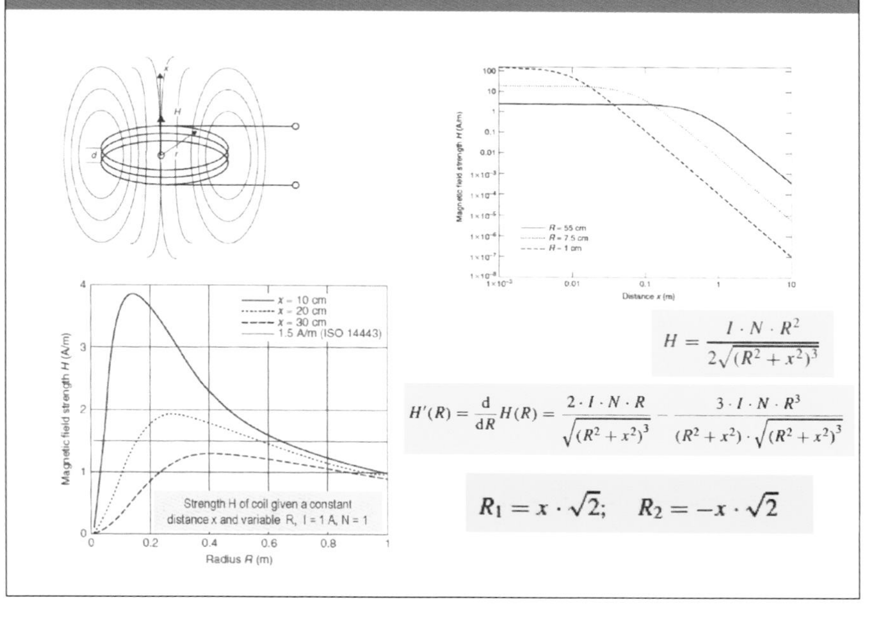

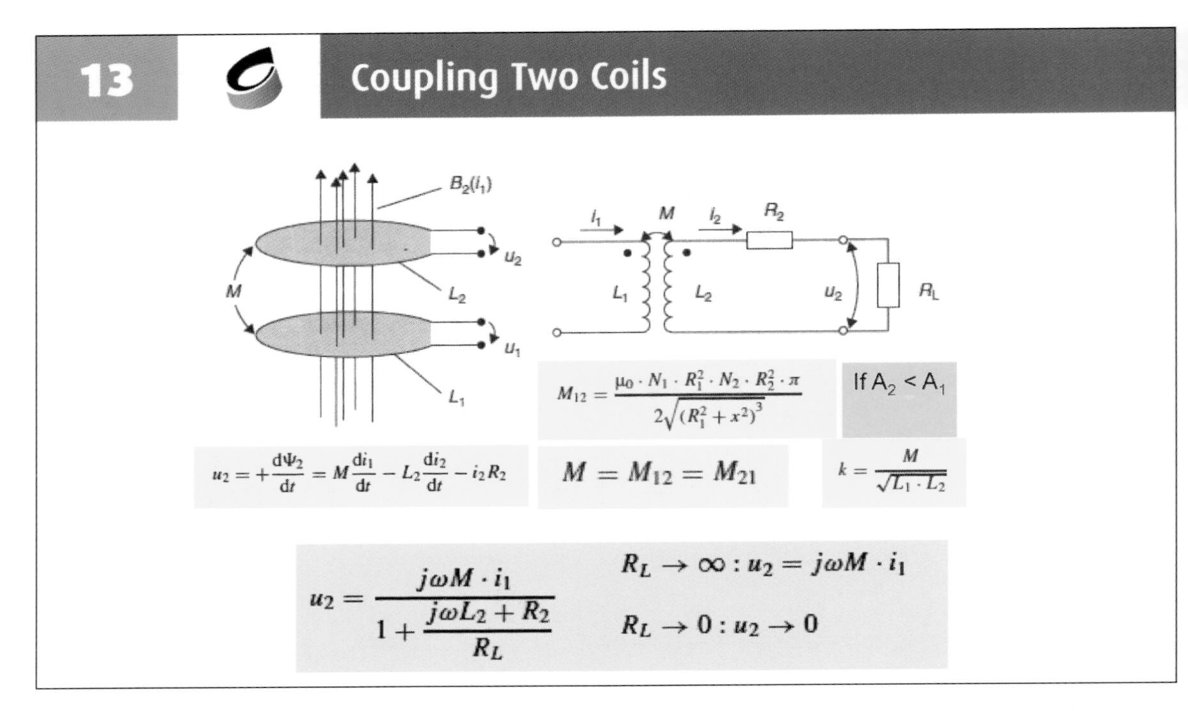

To increase the magnetic coupling, use resonance technique, by adding a series capacitor to the loop.

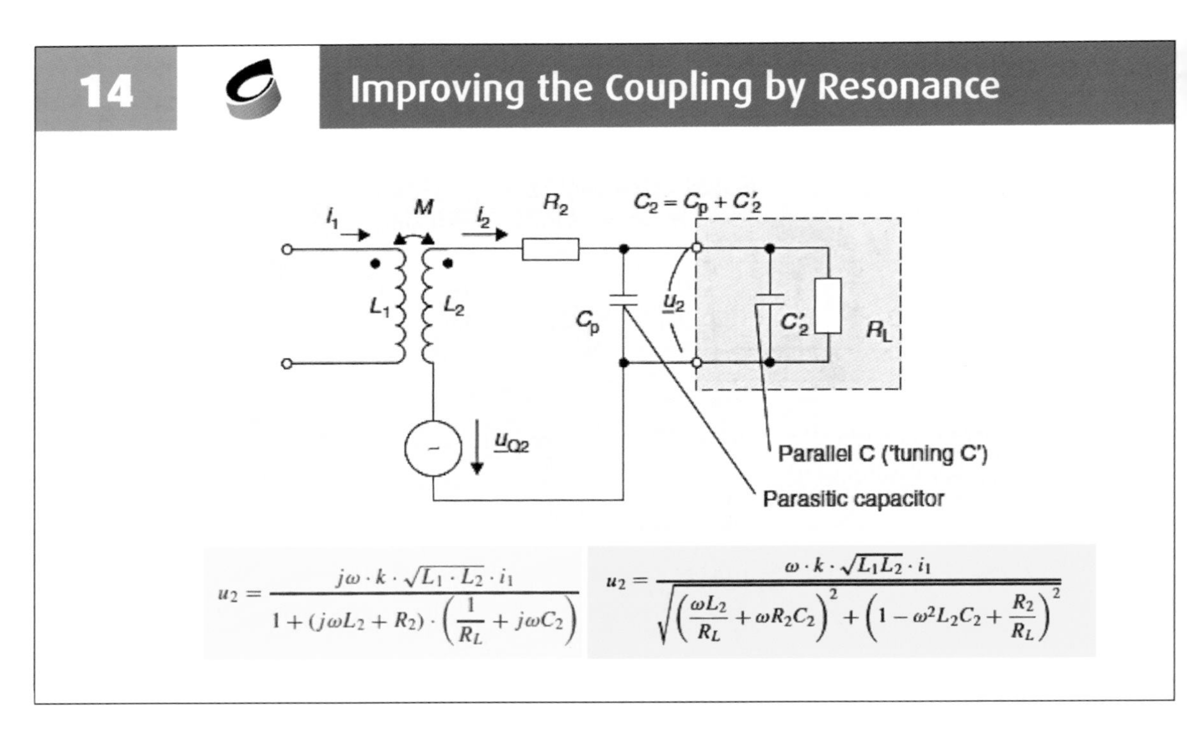

Huge improvement can be obtained by tuning the circuits Q factors.

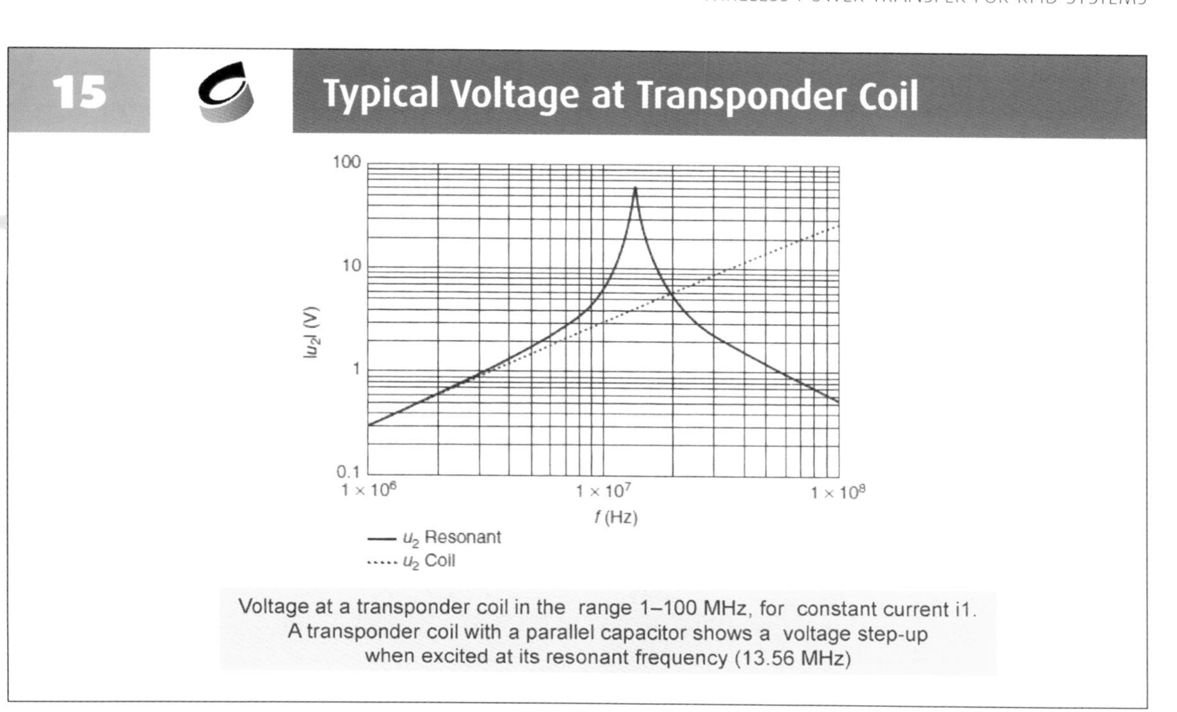

Voltage at a transponder coil in the range 1–100 MHz, for constant current i1. A transponder coil with a parallel capacitor shows a voltage step-up when excited at its resonant frequency (13.56 MHz)

Voltage at a transponder coil in the range 1–100 MHz, for constant current i1. Comparison between resonant and non-resonant coupling.

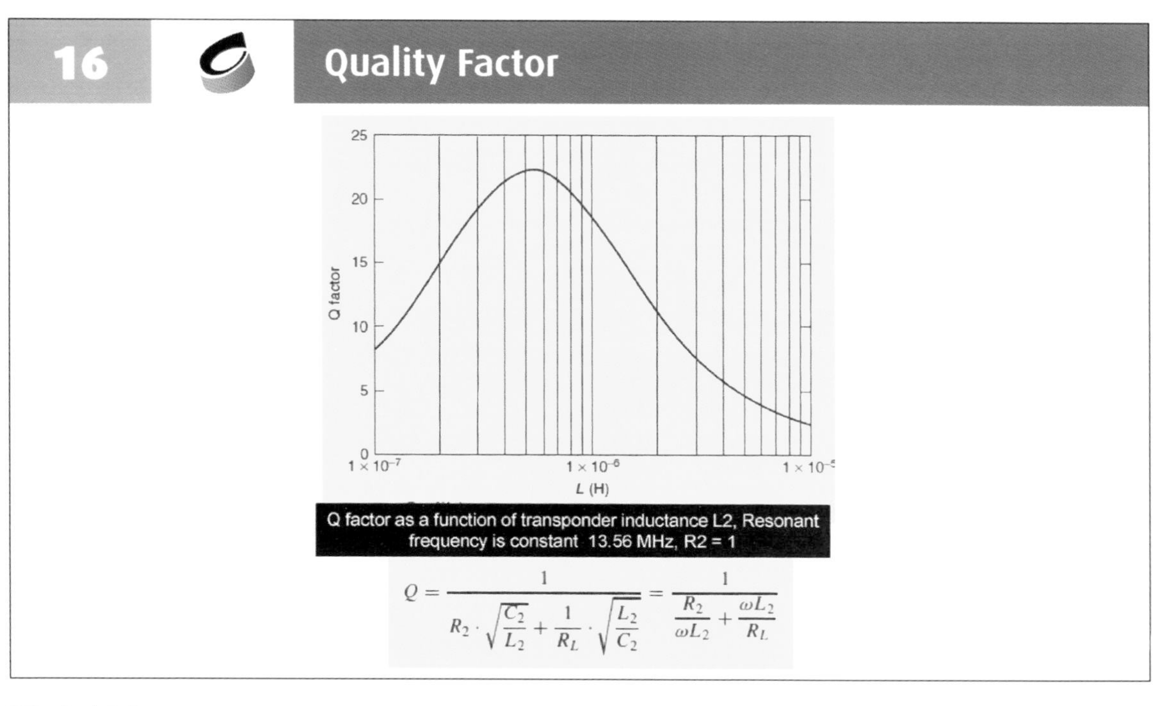

Q factor as a function of transponder inductance L2, Resonant frequency is constant 13.56 MHz, R2 = 1

$$Q = \frac{1}{R_2 \cdot \sqrt{\frac{C_2}{L_2}} + \frac{1}{R_L} \cdot \sqrt{\frac{L_2}{C_2}}} = \frac{1}{\frac{R_2}{\omega L_2} + \frac{\omega L_2}{R_L}}$$

Typical Q factor : some tens.

17  Protecting the Processing Circuit

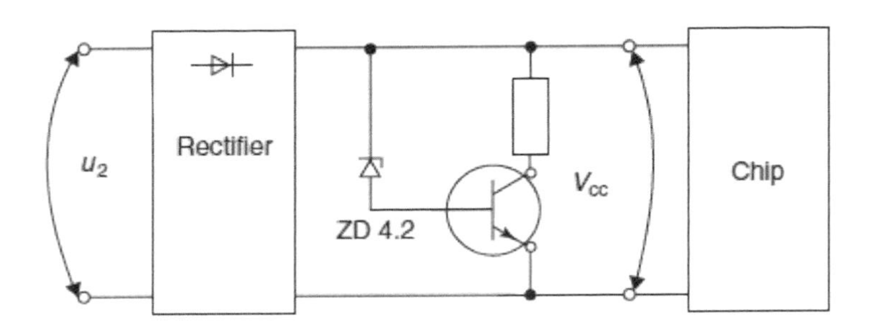

A simple shunt regulator based upon a zener diode

How to protect against overvoltage ?

APPLICATION TO HF RFID AT 13.56MHZ

18  HF RFID Reader

Equivalent circuit of reader. The effect of coupling with tag is modeled by a series impedance Z'T.

Schematic diagram of reader

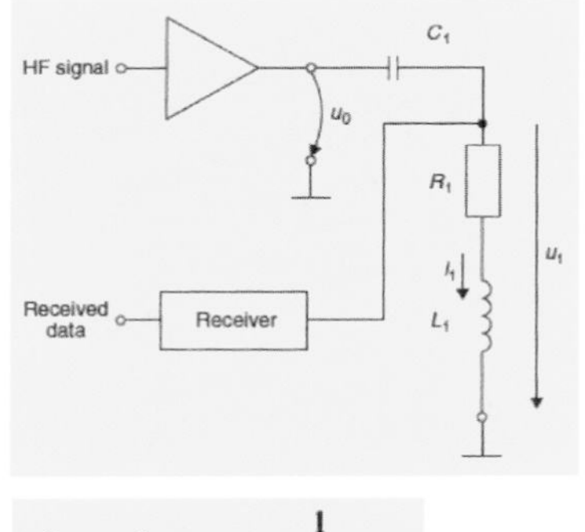

Equivalent circuit of reader

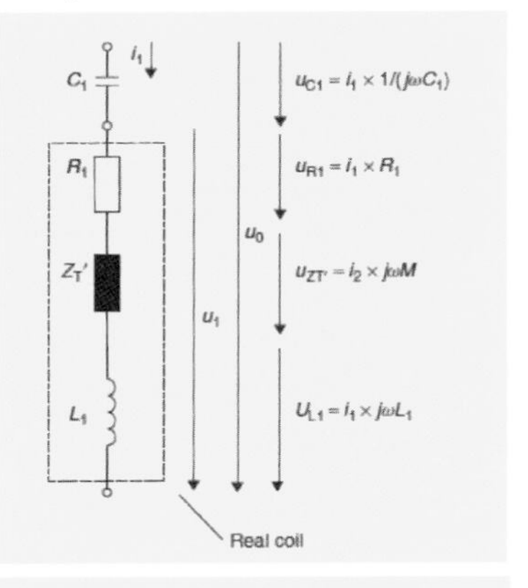

$$f_{TX} = f_{RES} = \frac{1}{2\pi\sqrt{L_1 \cdot C_1}}$$

$$Z_1 = R_1 + j\omega L_1 + \frac{1}{j\omega C_1}$$

$$j\omega L_1 + \frac{1}{j\omega C_1} = 0\Big|_{\omega = 2\pi \cdot f_{RES}} \Rightarrow Z_1(f_{RES}) = R_1$$

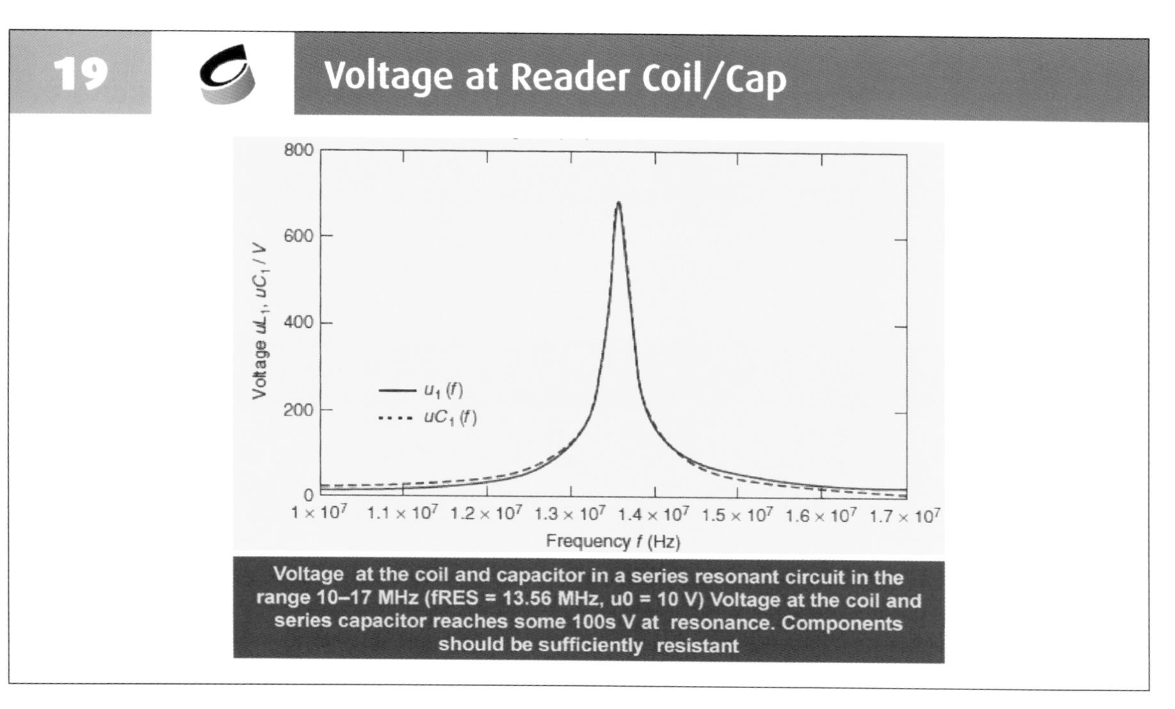

Overvoltages can reach hundreds of volts; therefore, capacitors must be compliant with huge voltage.

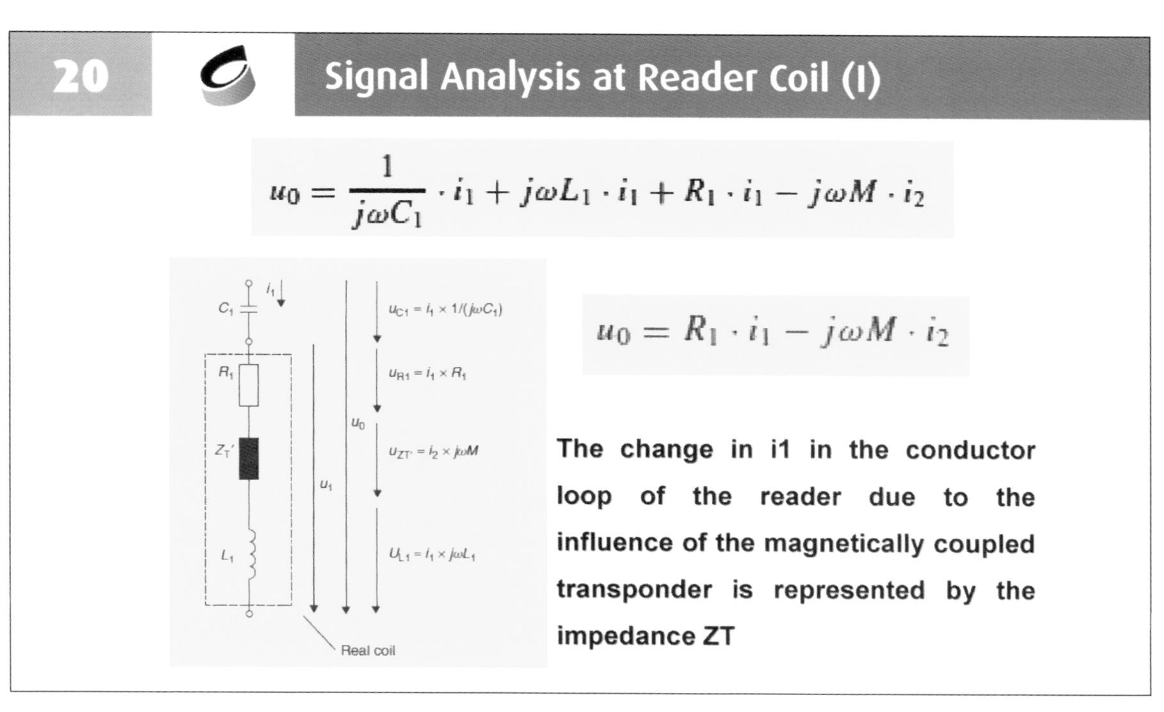

Signal analysis model at reader coil.

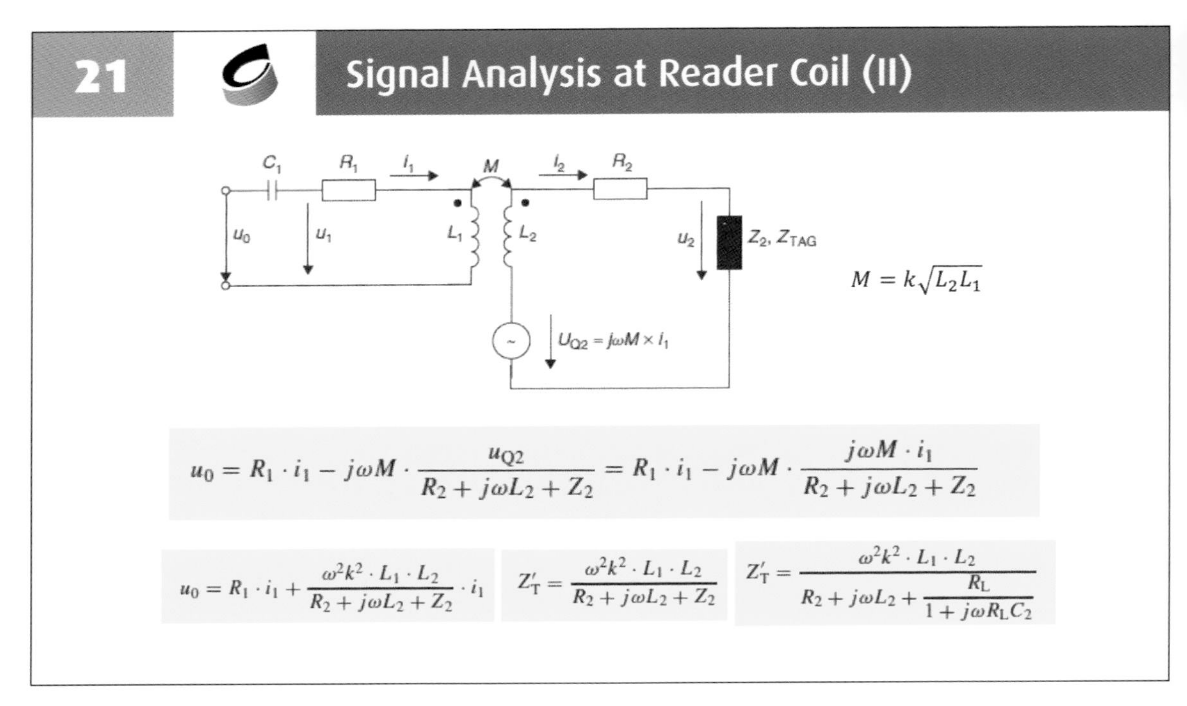

Calculation of the "perturbation" impedance seen at the coil reader.

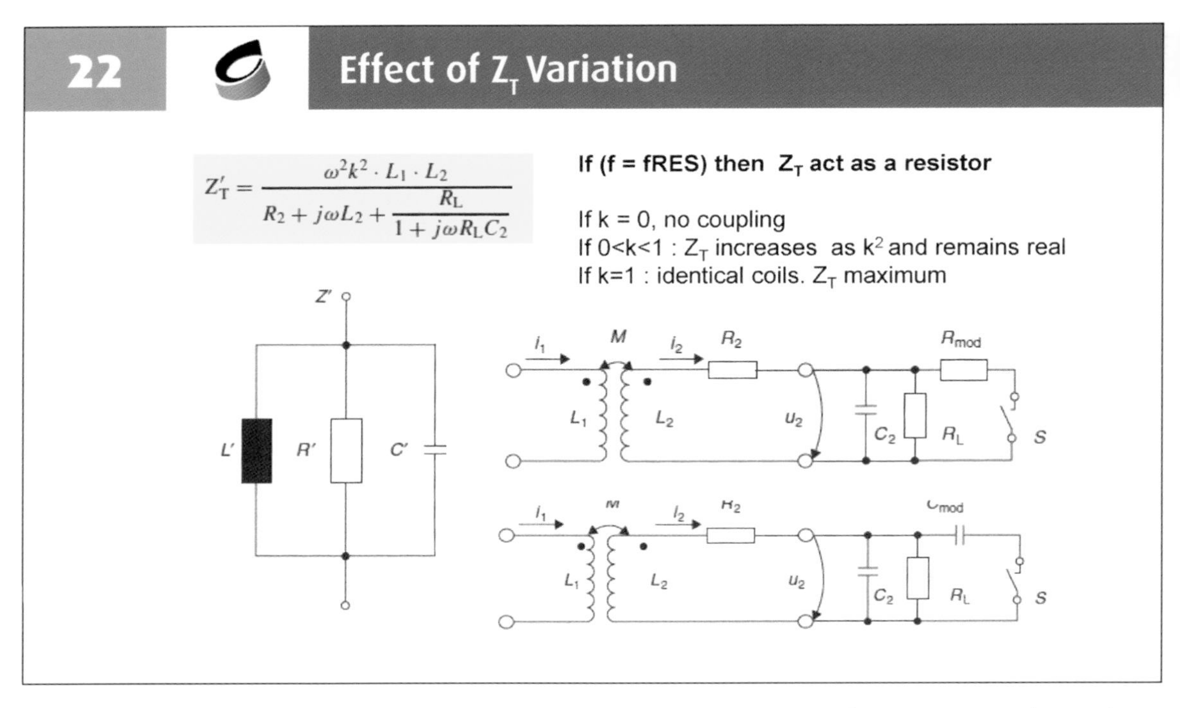

Modulating the perturbation impedance allows the tag to generate specific responses to the reader. It consists in controlling the load impedance (assured by RFID chip circuitry).

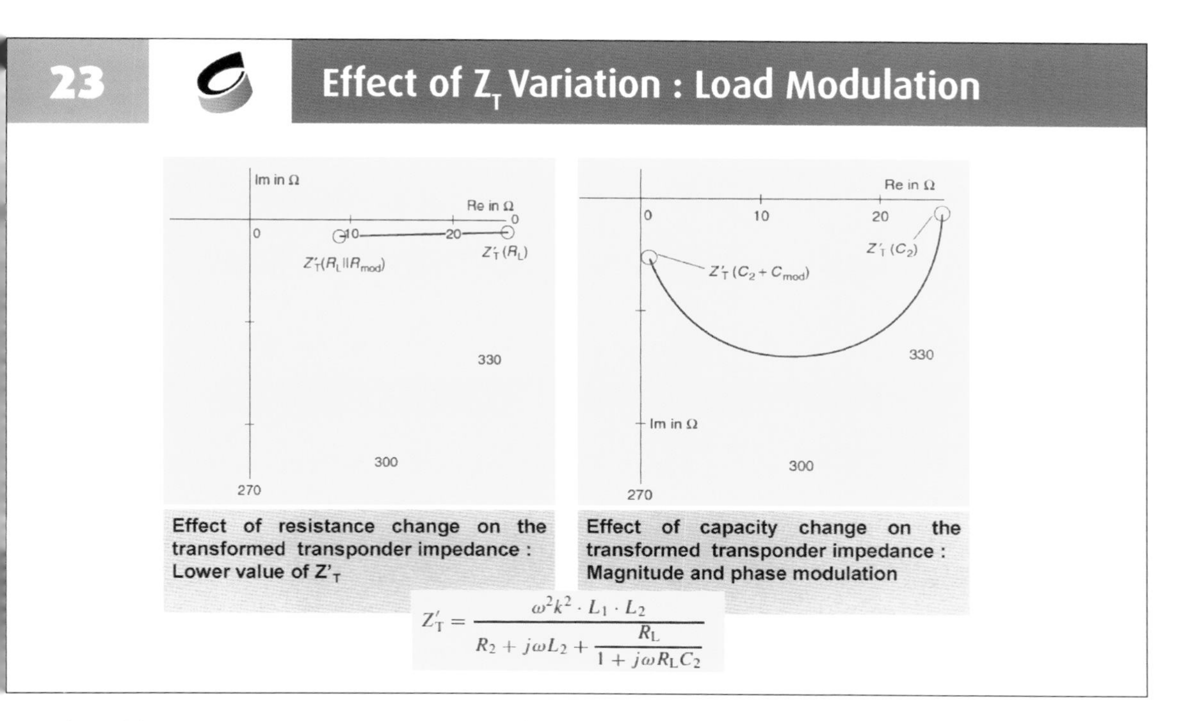

Load modulation option : Resistance variation of Capacitive variation.

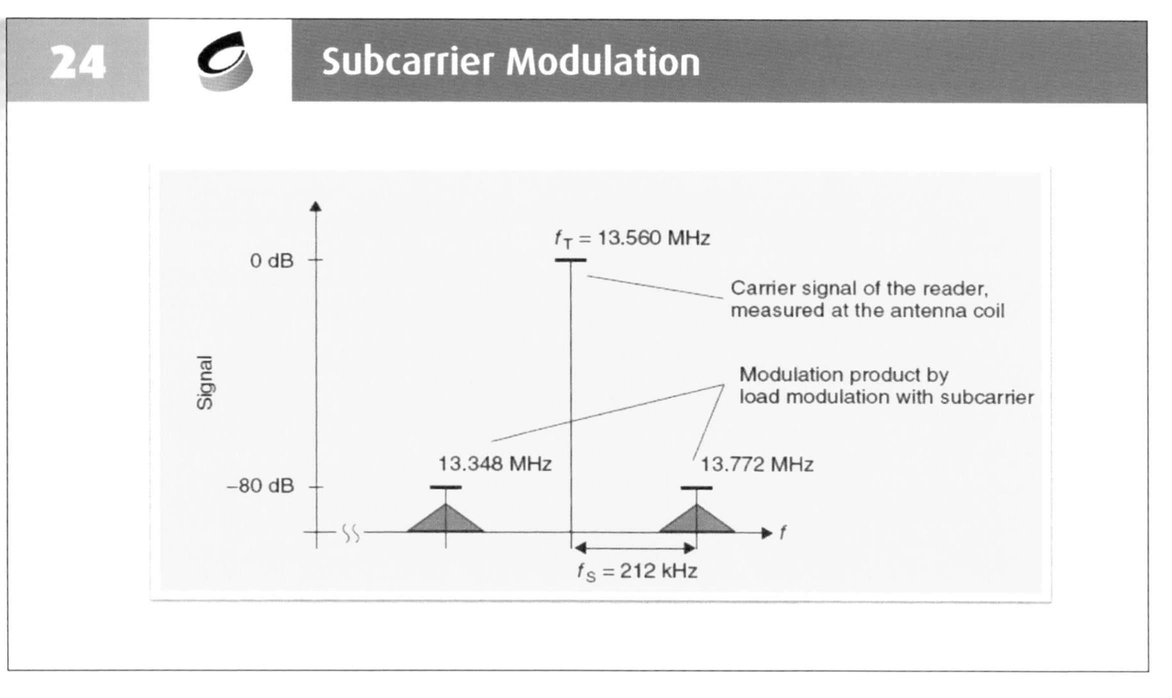

Subcarrier technique to avoid huge SNR level.

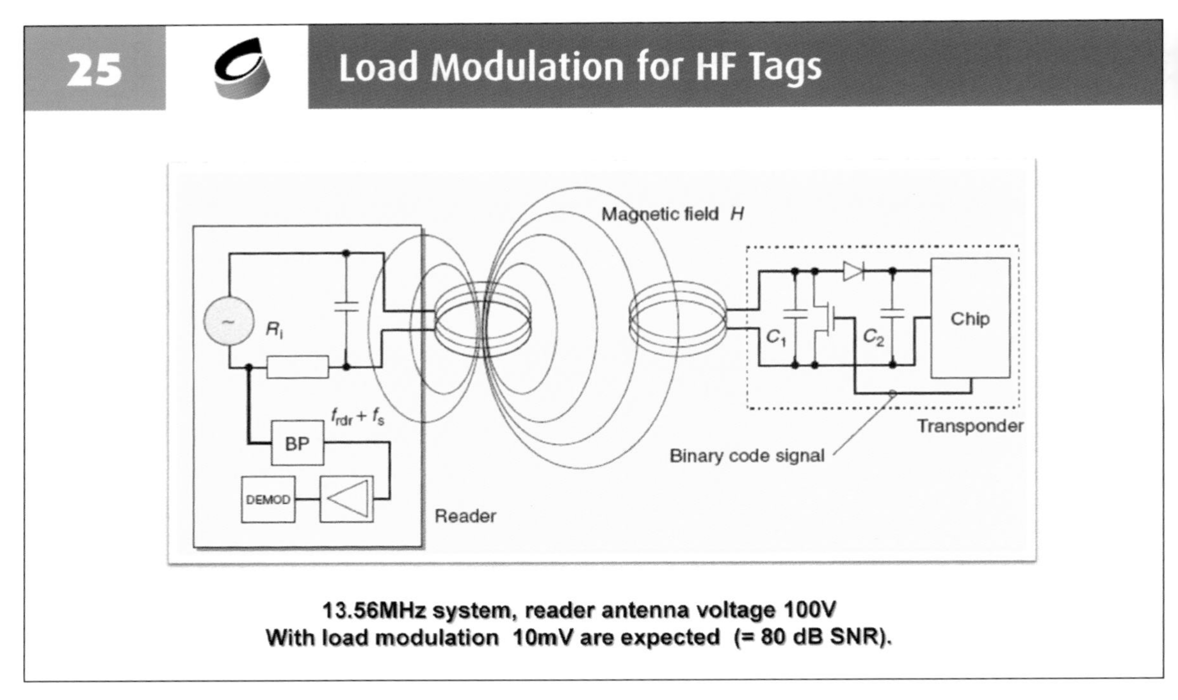

Architecture for subcarrier option.

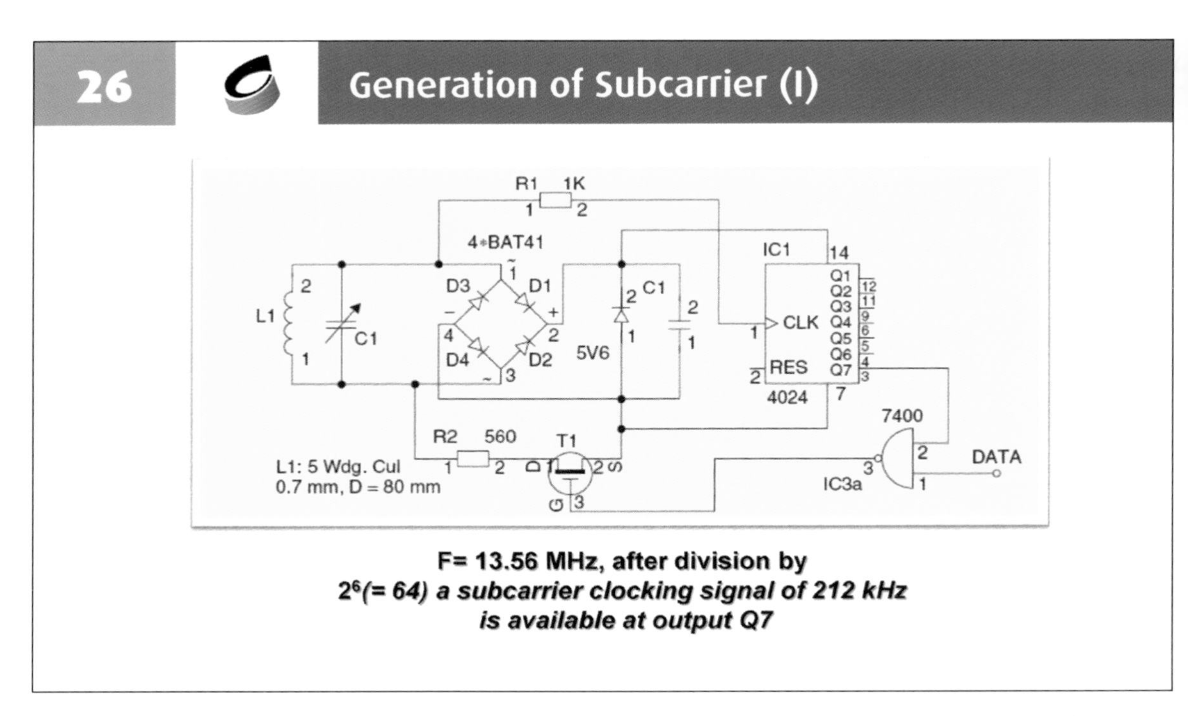

A simple frequency divider can be used to generate different subcarriers from the communication frequency generated by the reader.

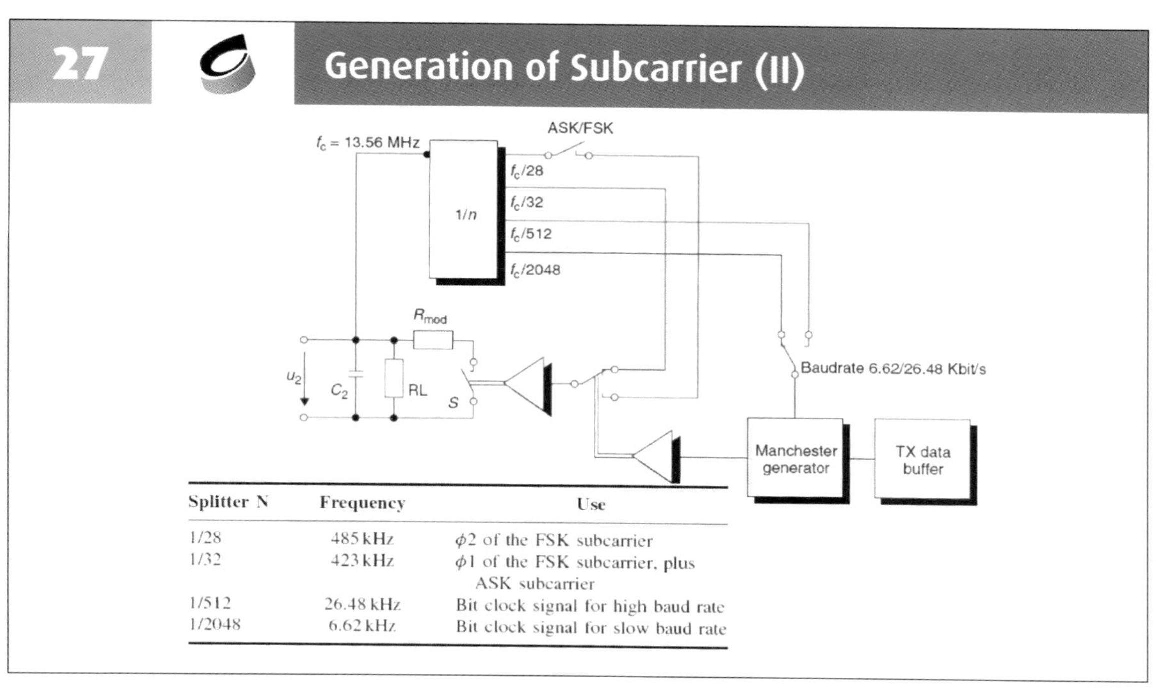

Splitter N	Frequency	Use
1/28	485 kHz	ϕ2 of the FSK subcarrier
1/32	423 kHz	ϕ1 of the FSK subcarrier, plus ASK subcarrier
1/512	26.48 kHz	Bit clock signal for high baud rate
1/2048	6.62 kHz	Bit clock signal for slow baud rate

Some standard values for subcarrier and modulations performed by the RFID chip.

FAR-FIELD WIRELESS POWER TRANSMISSION

28 Characteristics of Far-Field

Far-Field WPT exploits the antenna characteristics and Friis law. Observe the free space attenuation in the order of 30dB/m around 1 GHz.

The transmission process is based on the propagation of an EM wave between two antennas. So the transfer function is dependent on the distance between the two antennas, the parameters of the antennas, and the structure of the electromagnetic wave. Finally, the transmission is governed by the so-called FRIIS equation

$$P_r = P_t \cdot \frac{G_t G_r \lambda_0^2}{(4\pi R)^2}$$

Z_L

TX Antenna Gt

RX Antenna Gr

Freq MHz	860	915	960	2450
Atten dB/m	31,1	31,7	32	40

Antenna Radiation Regions

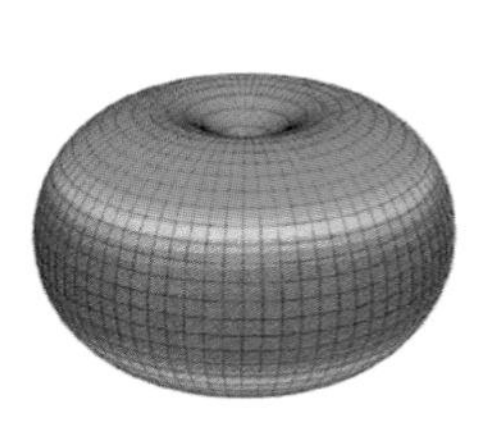

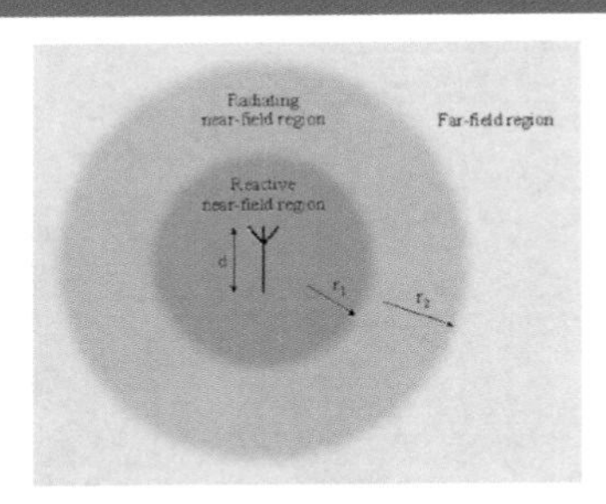

Rayleigh	Fresnel	Fraunhofer
$d < 0.62\sqrt{D^3/\lambda}$	$0.62\sqrt{D^3/\lambda} < d < 2D^2/\lambda$	$d > 2D^2/\lambda$
Near Field Stationnary	Stationary&Propagation	Far-field Propagation

Freq (MHZ)	13,56	900	2450	5800
Rayleigh (cm)	484,96	7,31	2,68	1,13
Fresnel (cm)	1106,19	16,67	6,12	2,59

Antenna size: D, Communication Distance : d, Wavelength : λ

Main properties of antenna device.

30

Antenna Radiation Parameters (I)

> ***Radiation pattern***
  + ***Directivity***
  + ***Aperture***
  + ***Gain***
> ***Polarization***
  + ***Circular***
  + ***Linear***
> ***TX mode***
  + ***Gain***
  + ***Impedance***
> ***RX mode***
  + ***Reception surface***
  + ***Impedance***
> ***Reflection mode***
  + ***Radar Cross Section***

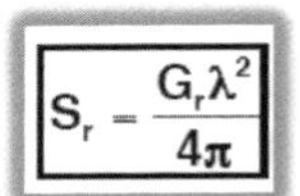

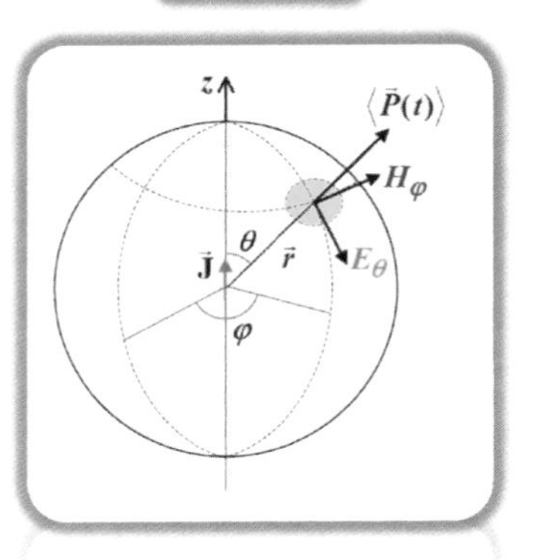

Antenna Parameters. Radiation Parameters.

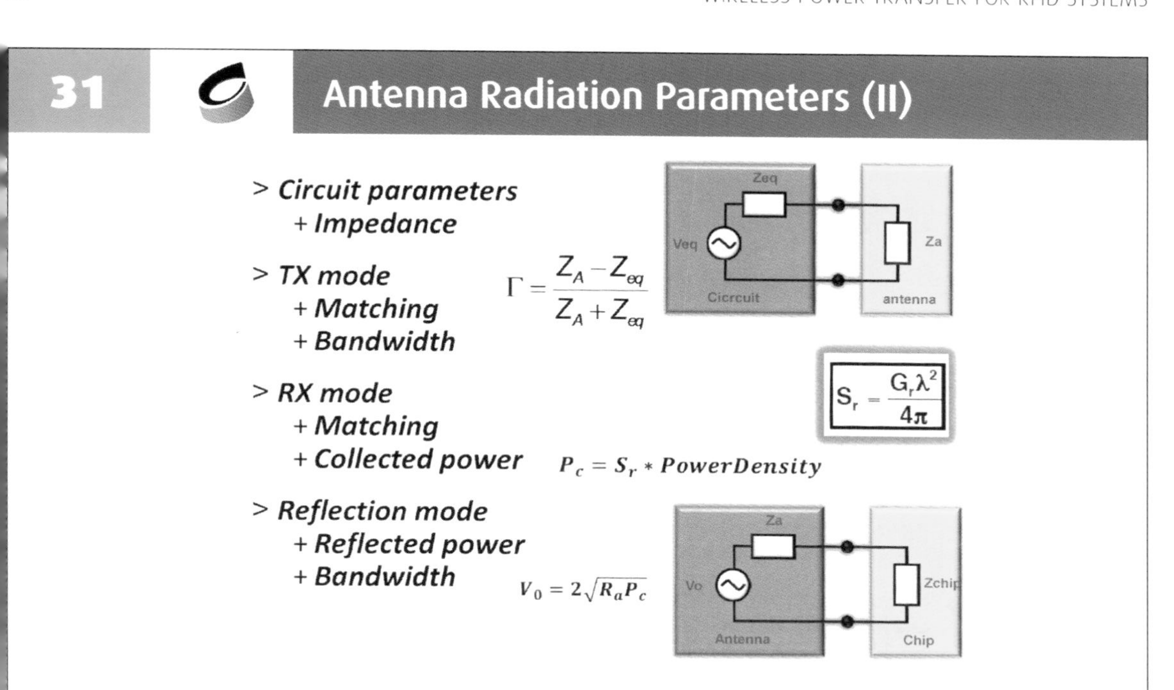

Antenna Parameters. Circuit Parameters.

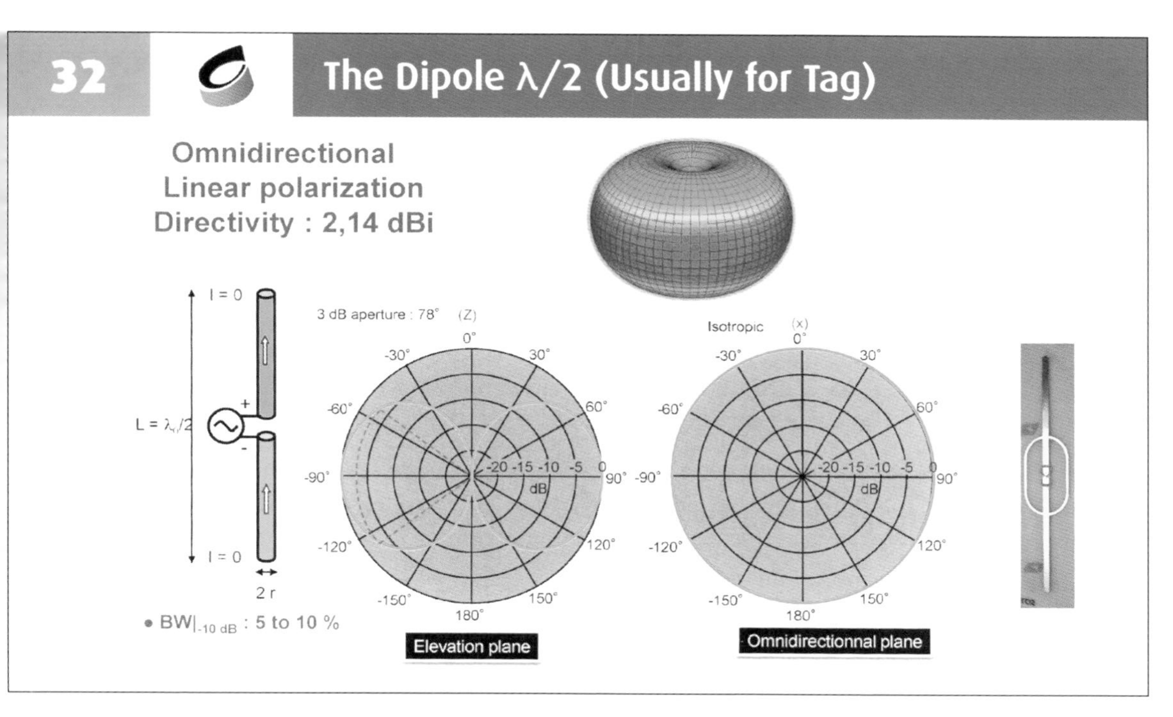

Dipole Case. Most of the UHF RFID tags are based on Dipole antenna. The advantage is the omnidirectional operation, but length must be reduced. This is why meander configurations are very popular.

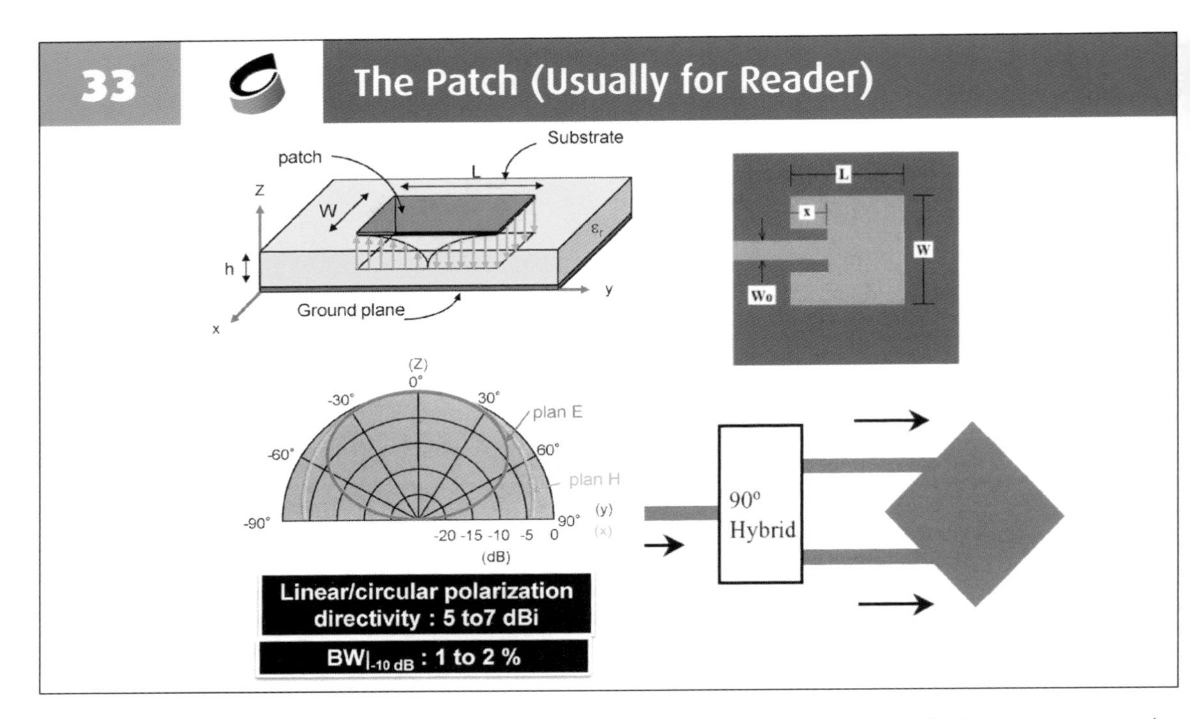

The patch antenna is used at the reader side. The major advantage is the easy method to generate circular polarization.

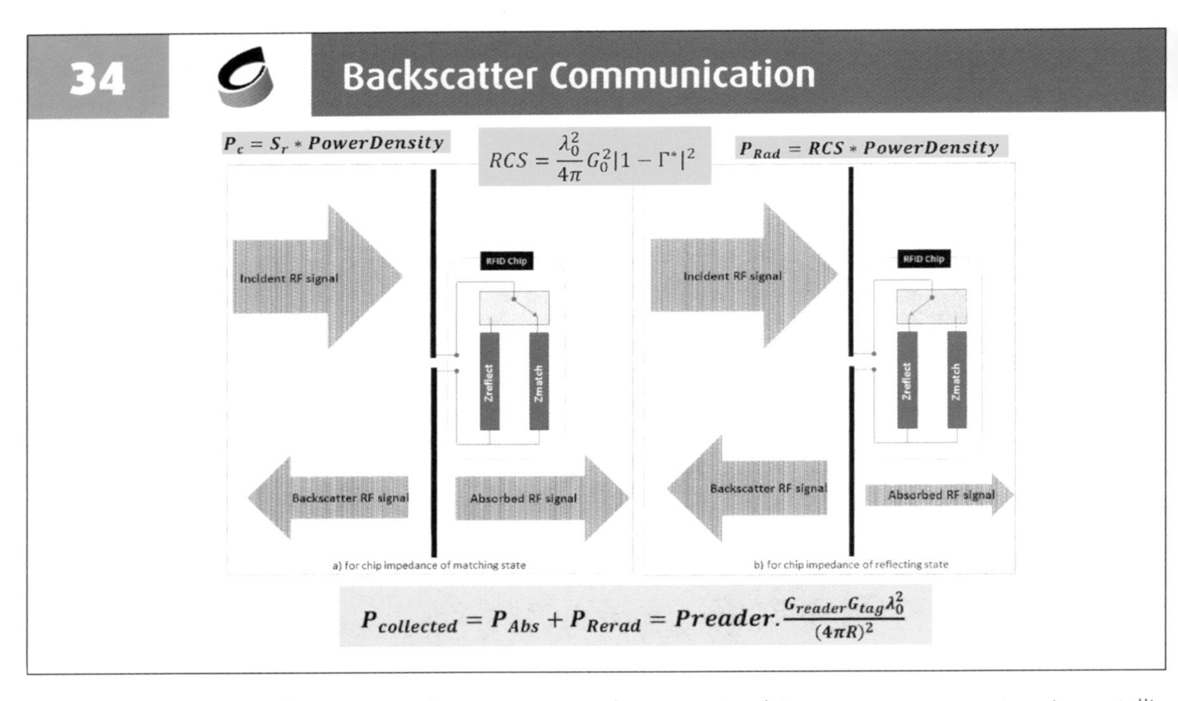

Re-radiation process of antennas. When an antenna detects a signal, it generates a current on its metallic part, which turns in new radiation signal. This is the backscatter process.

When an antenna is loaded by a match load, it backscatters the same power as the one absorbed by the matched load.

RECTIFIERS

35

Basic Rectifier Parameters

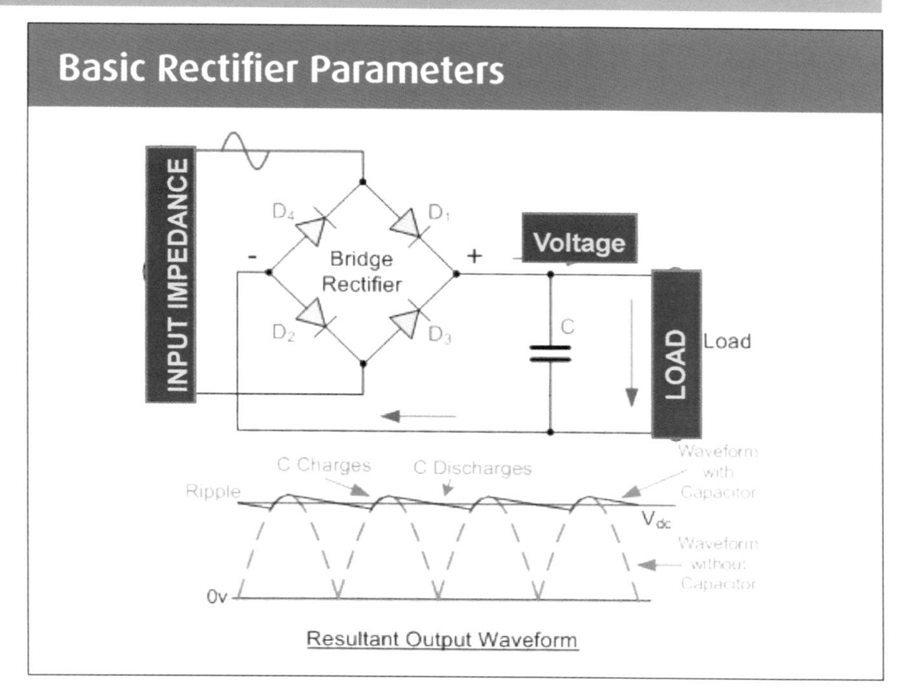

Rectifiers are using nonlinear characteristics of diodes. Different architectures are possible. This architecture is based on a 4 diode bridge to rectify positive and negative alternations of the feeding input signal.

36

Schottky Diode As Rectifier

- **Low forward voltage drop between 0.15 and 0.45 volts**
- **Very fast switching action.**
- **Less energy is wasted as heat**

Table 1. Schottky Diode Packaging and Marking

Single	Single	Series Pair	Reverse Series Pair	Unconnected Pair	Single
SC-79 Green™	SOT-23	SOT-23	SOT-23	MIS Green™	SOD-882 Green™
				SMS7621-517 Marking: H Pb-Free	
♦ SMS7621-079LF Marking: Cathode and SA	SMS7621-001LF Green™ Marking: XH1	♦ SMS7621-005LF Green™ Marking: XH2	♦ SMS7621-006LF Green™ Marking: XH8		SMS7621-040LF Marking: E
♦ SMS7630-079LF Marking: Anode and SC		SMS7630-005LF Green™ Marking: XD2	♦ SMS7630-006LF Green™ Marking: XD8		SMS7630-040LF Marking: P
L_S = 0.7 nH	L_S = 1.5 nH	L_S = 1.5 nH	L_S = 1.5 nH	L_S = 0.6 nH	L_S = 0.45 nH

Schottky diodes have low forward voltage and fast-switching capability. They are very popular as rectifiers at GHz frequencies.

Many products are commercially available.

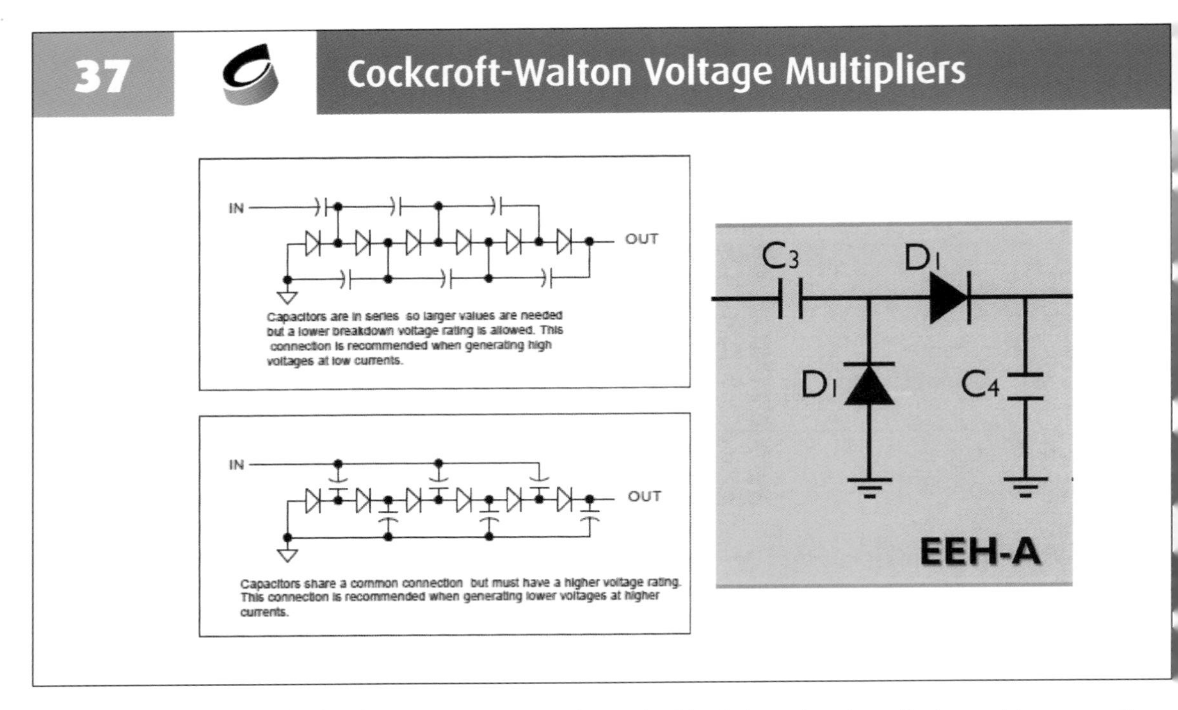

Circuit to multiply rectified voltages or currents. Voltage doubler is based on two diodes and two capacitors.

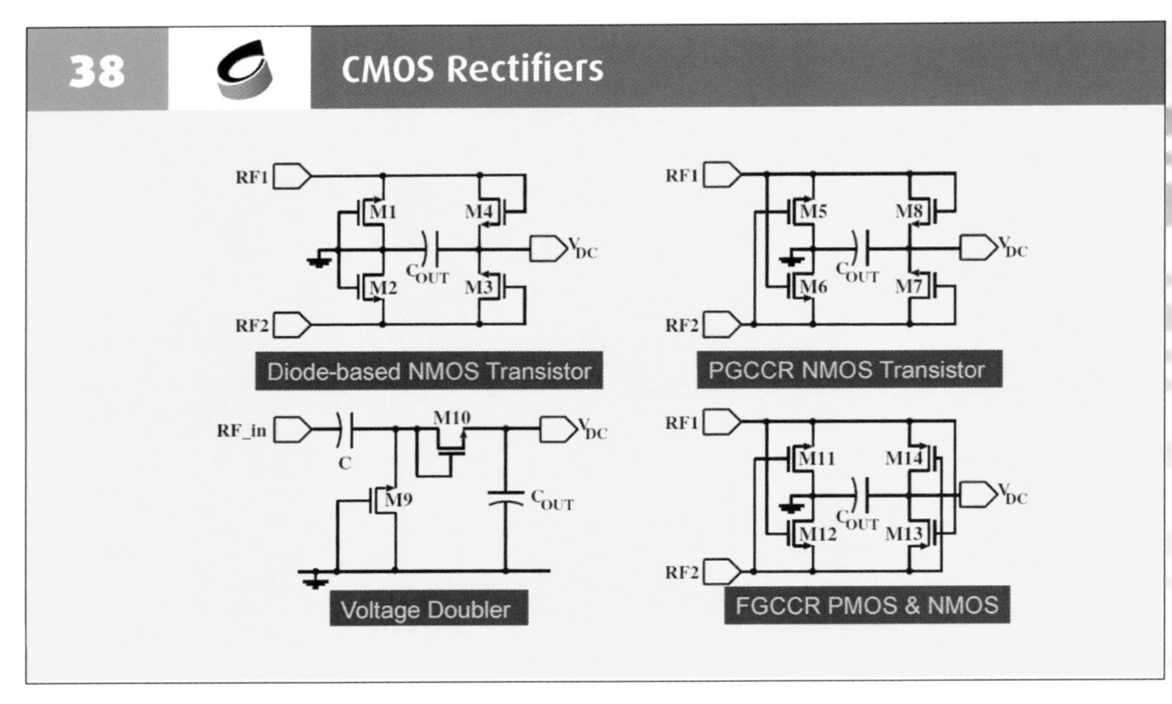

Topology of CMOS rectifiers.

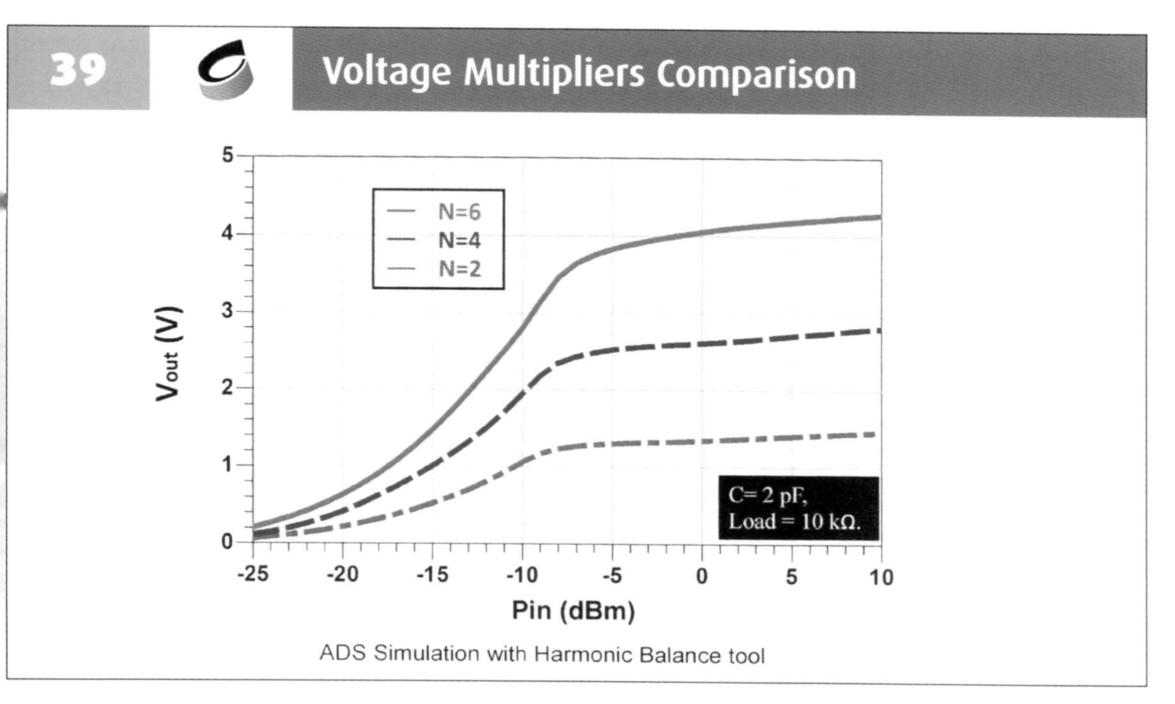

Voltage level depends on the number of cells (Diode and Capacitor). Calculation is made for a specific load.

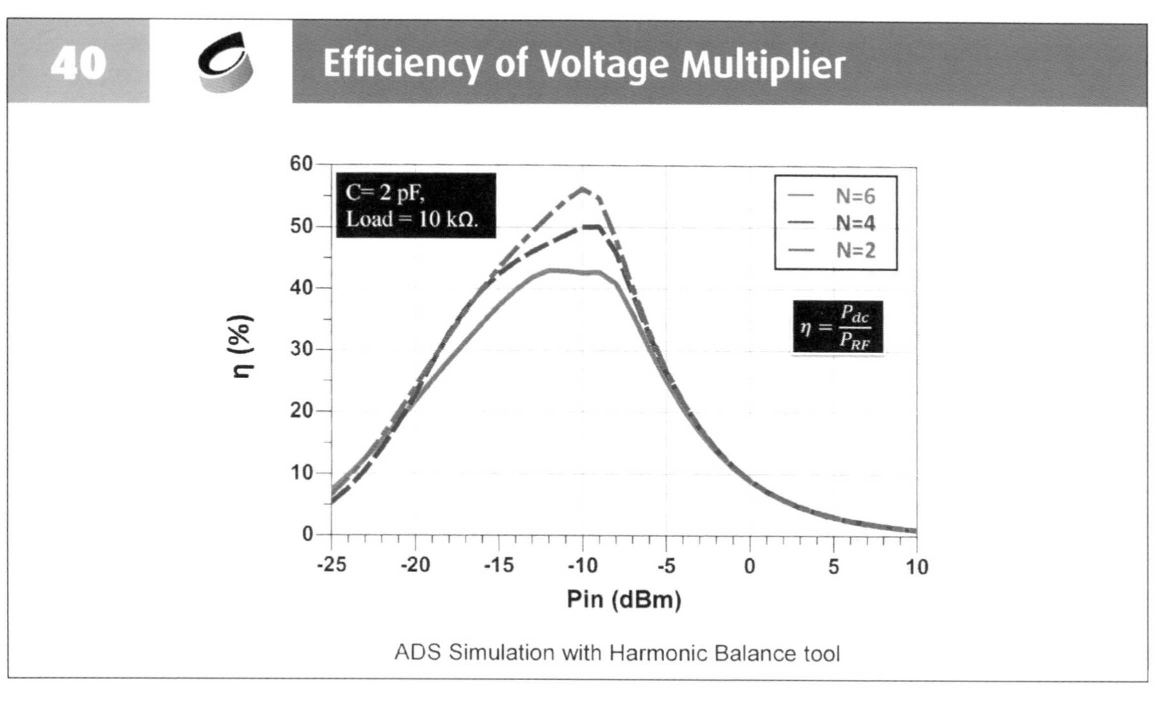

Trade-off between efficiency and voltage level.

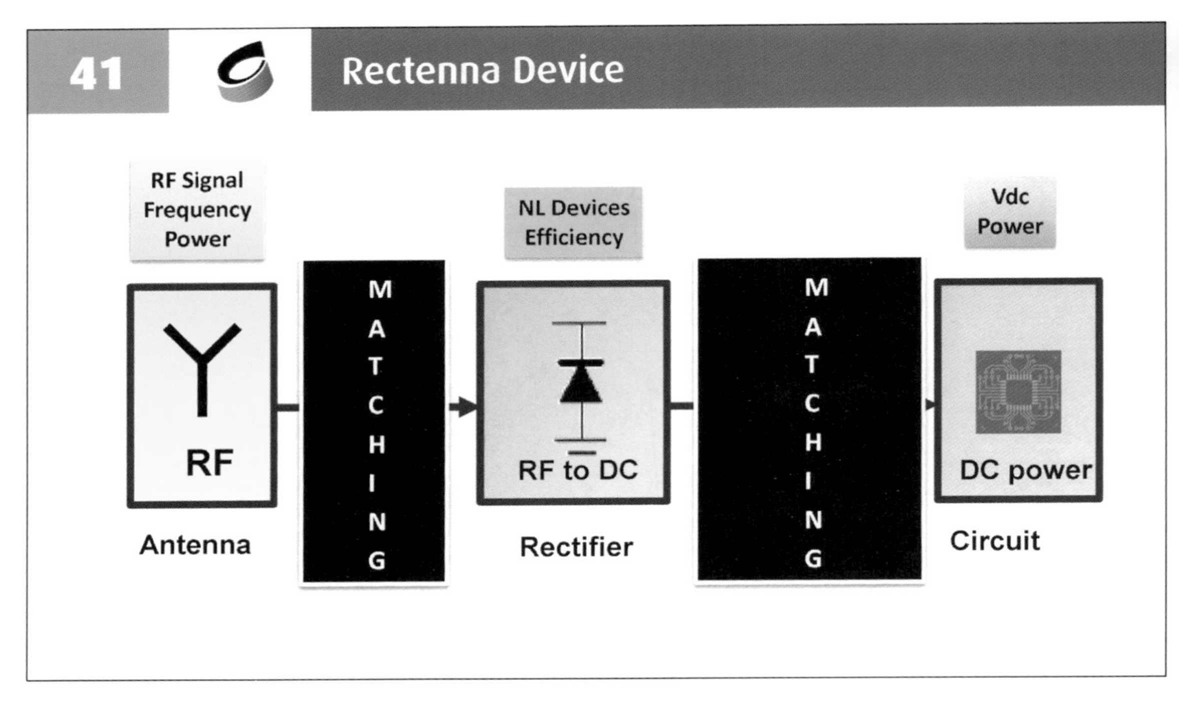

Rectenna = RECTIFIER + ANTENNA. But you need some matching circuits.

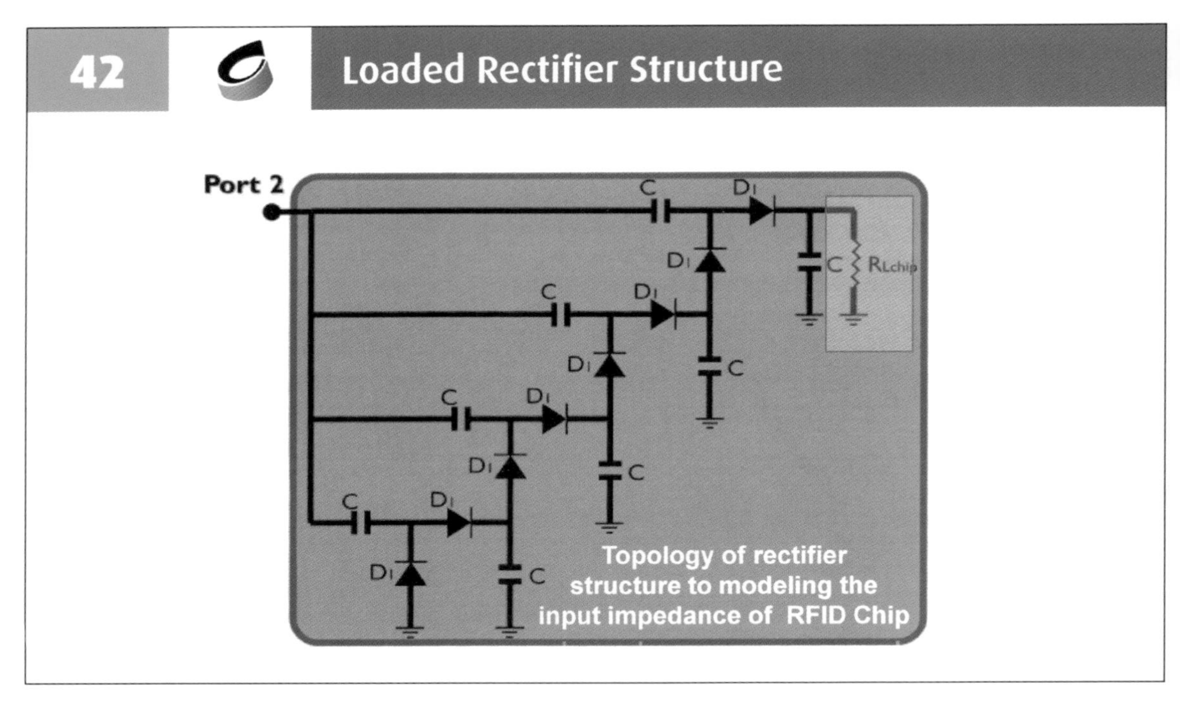

Matching requires an estimation of the input impedance of the rectifier. This is a simplified configuration.

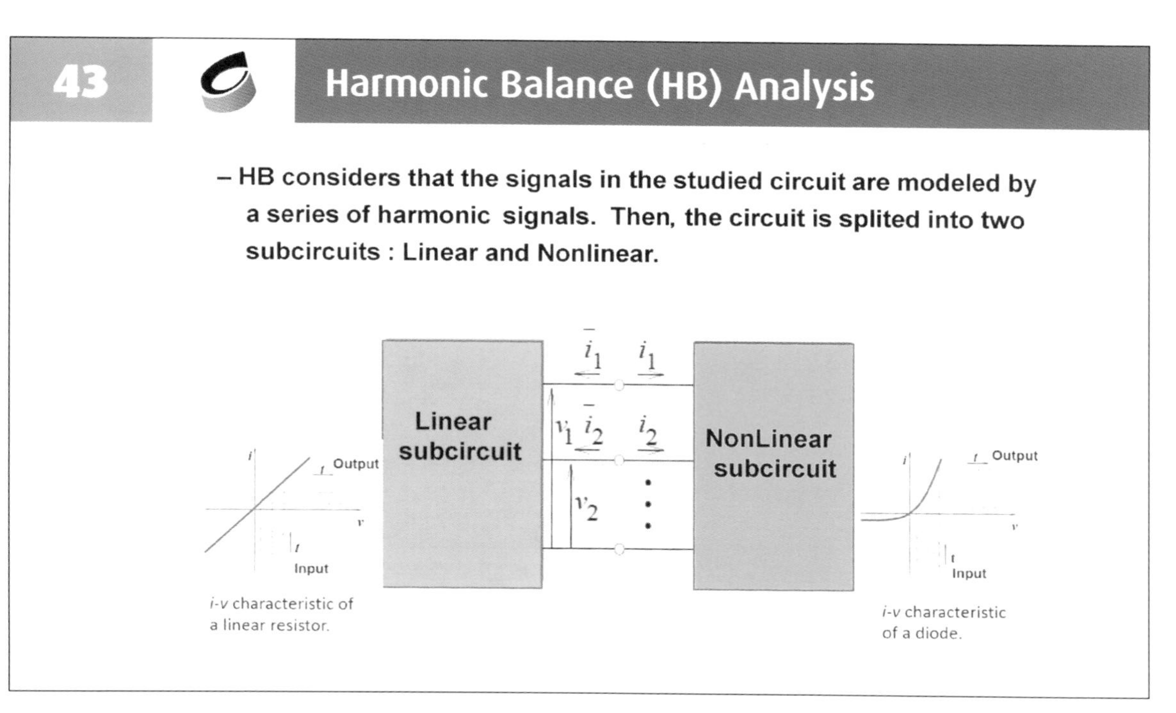

Simulation should take into account the NL behavior. So Harmonic Balance simulation is suitable.

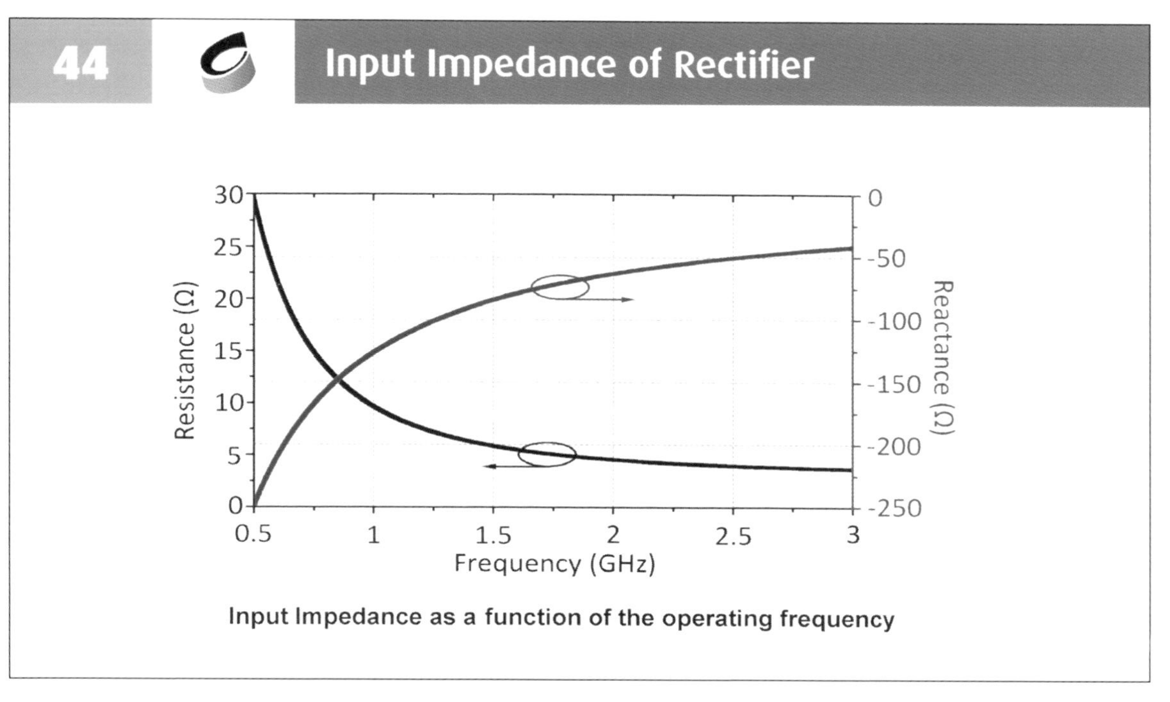

Example of simulation. Estimated equivalent impedance of the rectifier as a function of frequency. In fact, only the value of the impedance at harmonic frequency is meaningful.

45  Matching Options

- ❑ **LUMPED COMPONENTS**
  - ➢ **L, C elements**
  - ➢ **Possible values**
  - ➢ **Losses : Q factors**
- ❑ **DISTRIBUTED CIRCUIT**
  - ➢ **Transmission lines**
  - ➢ **Propagation Losses**
  - ➢ **Size**
- ❑ **ANTENNA IMPEDANCE**
  - ➢ **Design for given impedance**

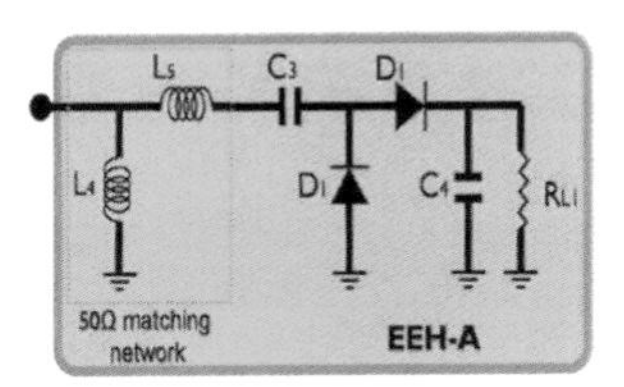

Different options are possible for matching. The most effective method is the design of antenna with specific impedance.

46  RFID Tag Antenna Case

What are the constraints ?

- Matched to chip.
  - Complex impedance Zchip=A-jB
  - (10<A<30, -600<B<-100)
  - Zant=Zchip*
- Low-cost substrate AND conductors AND process
- Nonconventional substrate, conductors
- Size must be as small as possible.
- Meet regulations and standards
- Robustness to Environment
- Modulation type (BASK / BPSK)
  - BASK (Matching High impedance)
  - BPSK (Matching Mean impedance)

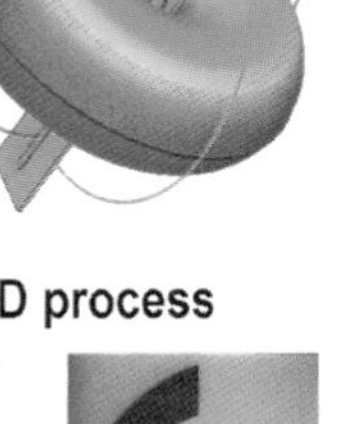

Main characteristics of tag antennas.

47 UHF Tag Read-Range Evolution

Year	IC sensitivity	Read Range*
1997	-8 dBm	5.2 m
1999	-10 dBm	6.5 m
2005	-12 dBm	8.2 m
2007	-13 dBm	9.3 m
2008	-15 dBm	11.7 m
2010	-18 dBm	16.5 m
2011	-20 dBm	20.7 m
2014	-22 dBm	26.1 m

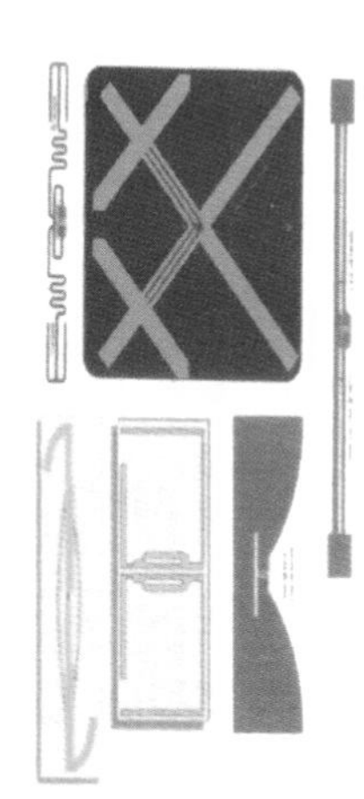

* (FS, 36 dBm EIRP, 2 dBi tag antenna) $r_{tag} = \frac{\lambda}{4\pi}\sqrt{\frac{EIRP}{P_{tag}}}$ $P_{chip} = P_{tag}\, p\, G\, \tau$

Read range of UHF RFID tags depends on the matching between RFID chip and antenna. It also depends on the sensitivity of the RFID chip. In the last decade, the read range was double. It exceeds 25 m !

ADVANCED APPLICATION EXPLOITING NONLINEAR CHARACTERISTICS OF RFID CHIPS

48 Non-Linearities in Passive UHF RFID

Some advanced techniques to exploit the NL characteristics of Rectifier.

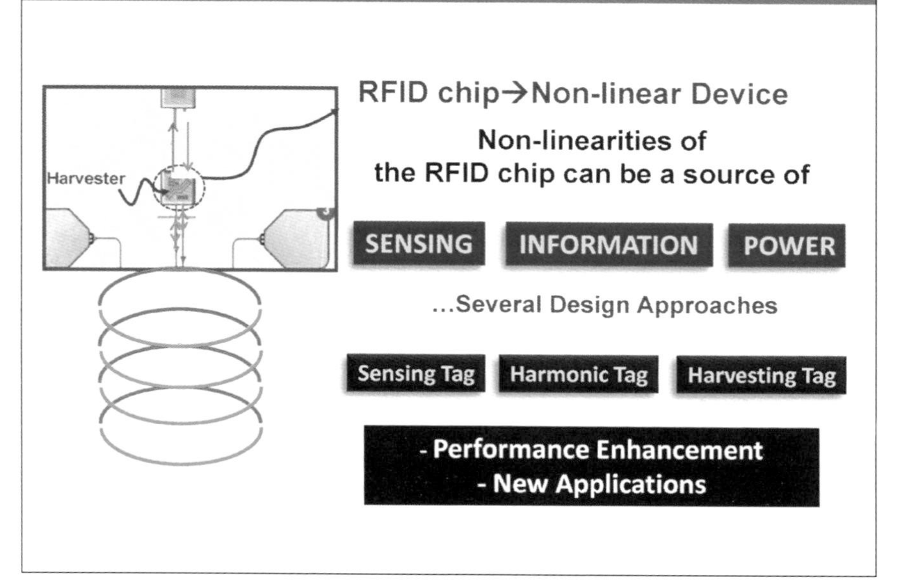

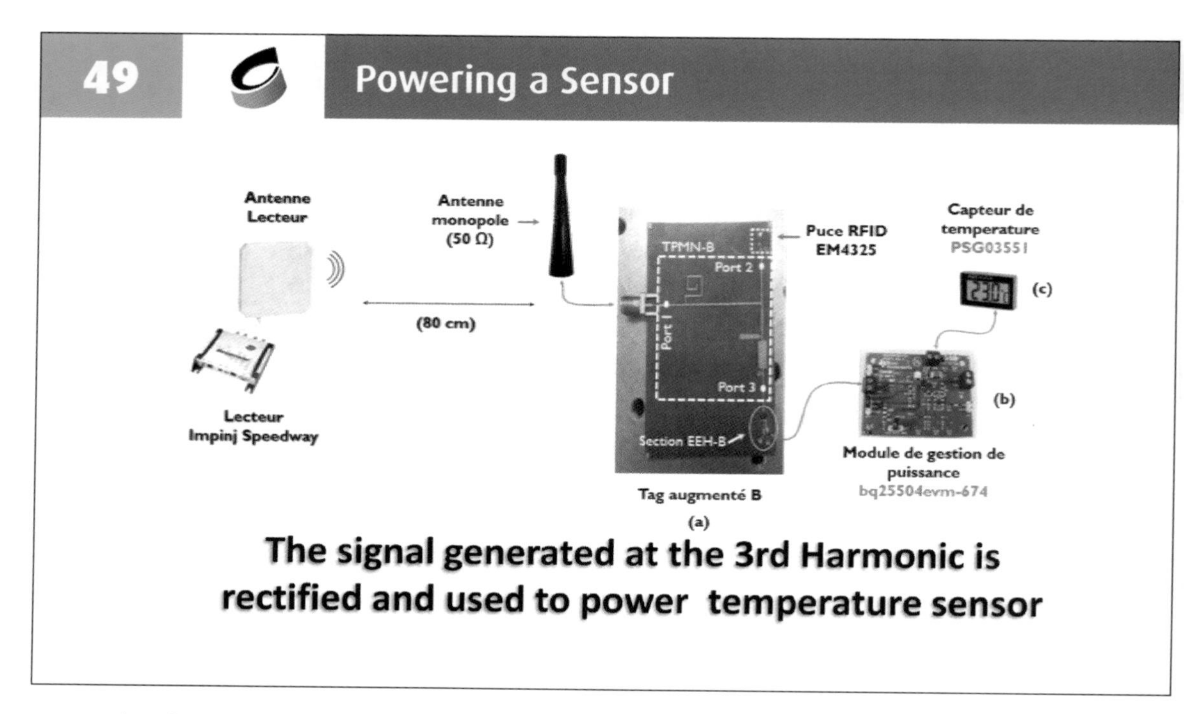

Example of powering a sensor by exploiting the rectifier nonlinearity.

50 Concluding Remarks

- NOTHING IS POSSIBLE WITHOUT ENERGY !!!
- WIRELESS TECHNIQUES DEMONSTRATED A CENTURY AGO
- TWO WIRELESS VISIONS : COMMUNICATION, POWERING
- ADVANCES IN MICROELECTRONICS ALLOW WIRELESS COMMUNICATION & POWER TRANSMISSION (WCPT)
- AMBIENT ENERGY COULD DELIVER ENOUGH POWER TO ACTIVATE RFID TECHNOLOGY

Exploitation of the NL in passive RFID means a boost of existing applications and the birth of other new ones.
CHANNEL DIVERSITY, SENSOR HARVESTING, AUTHENTICATION, LOCATION

These are some of the lessons we can learn from the matter in the previous slides. Remember that nothing is possible without energy, but sustainable applications should avoid using batteries and prefer harvesting techniques. When exploited effectively, backscatter signal can lead to unsuspected applications and concepts.

51  Some References of Interest

[1] P. Nikitin and K. Rao, «Harmonic scattering from passive UHF RFID tags», *Antennas and Propagation Society International Symposium*, APURSI 09 IEEE, pp. 1–4 June. 2009.

[2] L. Mayer and A. Scholtz, «Sensitivity and impedance measurements on UHF RFID transponder chips», in *EURASIP Workshop on RFID Tech*, pages 1–10, 2007

[3] Gomes, H.G. and Carvalho, N.B.C., «RFID for Location Proposes Based on the Intermodulation Distortion», *Sensors & Transducers Magazine*, vol.106, n.7, pp. 85-96, July, 2009.

[4] G. De Vita and G. Iannacone, «Design criteria for the RF section of UHF and microwave passive RFID transponders», *Microwave Theory and Techniques, IEEE Transactions on 53*, (9), 2978–2990, 2005.

[5] S. Tedjini, Y. Duroc, G. Andia-Vera, C.Loussert, «RFID communication system» WO2014072812, International Application No.:PCT/IB2013/002628 (2014).

[6] G. Andia Vera, Y. Duroc and S. Tedjini, «Analysis of harmonics in UHF RFID signals», *Microwave Theory and Techniques, IEEE Transactions* on 61, (6), 2481–2490, 2013.

[7] G. Andia Vera, Y. Duroc, and S. Tedjini, «RFID Test Platform: Nonlinear Characterization», *Instrumentation and Measurement, IEEE Transactions,* on vol.63, no.9, pp. 2299-2305, Sept. 2014

[8] G. Andia Vera, Y. Duroc and S. Tedjini, «Analysis and exploitation of harmonics in wireless power transfer (H-WPT): passive UHF RFID case», *Wireless Power Transfer*, Vol. 1, N° 2, , pp 65-74, Cambridge Univ. Press, 2014.

[9] D Allane, Y Duroc, GA Vera, R Touhami, S Tedjini «On energy harvesting for augmented tags», *Comptes Rendus de l'Académie des Sciences, Physique*, 18, N°2, pp 710-718, Fev. 2017.

[10] D Allane, GA Vera, Y Duroc, R Touhami, S Tedjini «Harmonic power harvesting system for passive RFID sensor tags», *IEEE Transactions on Microwave Theory and Techniques,* 64 (7), 2347-2356, 2016.

[11] G Andia-Vera, S Nawale, Y Duroc, S Tedjini «Exploitation of the nonlinearities in electromagnetic energy harvesting and passive UHF RFID», *Wireless Power Transfer*, 3 (01), 43-52, 2016.

[12] Andia Vera, G.; Nawale, S.D.; Duroc, Y.; Tedjini, S., «Read Range Enhancement by Harmonic Energy Harvesting in Passive UHF RFID», in *Micro. & Wire. Comp. Letters, IEEE*, vol.25, no.9, pp.627-629, Sept. 2015.

[13] S.Tedjini, GA Vera, Z Marcos, RCS Freire, Y Duroc, «Augmented RFID tags», *IEEE Radio and Wireless Week*, (WiSNet) Austin, 23-27 January 2016.

[14] S. Tedjini, GA Vera, Y Duroc, «Energy paradigms of augmented tags for the internet of things deployment», *Applied Computational Electromagnetics Society*, Symposium-Italy (IEEE-ACES), April 2017.

CHAPTER 02

Computational Electromagnetics and Electromagnetic Compatibility for the Internet of Things

Ludger Klinkenbusch

Kiel University, Germany

1. Agenda 44
2. Overview (I) 44
3. Overview (II) 45
4. Computational Electromagnetics (I) 45
5. Computational Electromagnetics (II) 46
6. Computational Electromagnetics (III) 46
7. The Finite-Difference Time-Domain Method (I) 47
8. The Finite-Difference Time-Domain Method (II) 47
9. The Finite-Difference Time-Domain Method (III) 48
10. The Finite-Difference Time-Domain Method (IV) 48
11. The Finite-Difference Time-Domain Method (V) 49
12. The Finite-Difference Time-Domain Method (VI) 49
13. The Finite-Difference Time-Domain Method (VII) 50
14. The Finite-Difference Time-Domain Method (VIII) 50
15. The Finite-Difference Time-Domain Method (IX) 51
16. The Finite-Difference Time-Domain Method (X) 51
17. The Finite-Difference Time-Domain Method (XI) 52
18. The Finite-Difference Time-Domain Method (XII) 52
19. The Finite-Difference Time-Domain Method (XIII) 53
20. Electromagnetic Compatibility Issues for the IoT (I) 53
21. Electromagnetic Compatibility Issues for the IoT (II) 54
22. Electromagnetic Compatibility Issues for the IoT (III) 54
23. Electromagnetic Compatibility Issues for the IoT (IV) 55
24. Electromagnetic Compatibility Issues for the IoT (V) 55
25. Electromagnetic Compatibility Issues for the IoT (VI) 56
26. Shielding (I) 56
27. Shielding (II) 57
28. Shielding (III) 57
29. Shielding (IV) 58
30. Shielding (V) 58
31. Shielding (VI) 59
32. Shielding (VII) 59

In this chapter, we will explain some electromagnetic basics and tools that are necessary to understand the physics of wireless propagation in the context of the Internet of Things. We will introduce a computational tool suitable for the evaluation of electromagnetic fields transmitted from antenna to antenna. These fields are the carrier for the signals controlling the IoT devices and possibly they also transport the energy necessary to run them. On the other hand, all electromagnetic systems are susceptible to receiving unwanted electromagnetic radiation coming in from the environment. The corresponding scientific area is referred to as Electromagnetic Compatibility (EMC), which also will be addressed in this chapter.

1  **Agenda**

- **Overview**
- **Computational Electromagnetics**
- **The FDTD Method**
- **Electromagnetic Compatibility Issues for the IoT**
- **Shielding**

The chapter is organized as follows. After a brief introduction in Maxwell's equations, we will introduce the Finite-Difference Time-Domain (FDTD) Method as an example for a computational electromagnetics (CEM) tool. The chapter also includes a brief overview on the area of Electromagnetic Compatibility (EMC) as an important issue for a reliable and safe function of the Internet of Things. In particular, shielding as one of the most important measures to ensure the EMC of devices will be discussed in more detail. A 2D FDTD MATLAB© code will be discussed and delivered to the students to numerically compare different designs of a shielding structure.

2 **Overview (I)**

- Need for a systematic design and layout of wireless systems by electromagnetic **modeling and simulation:**

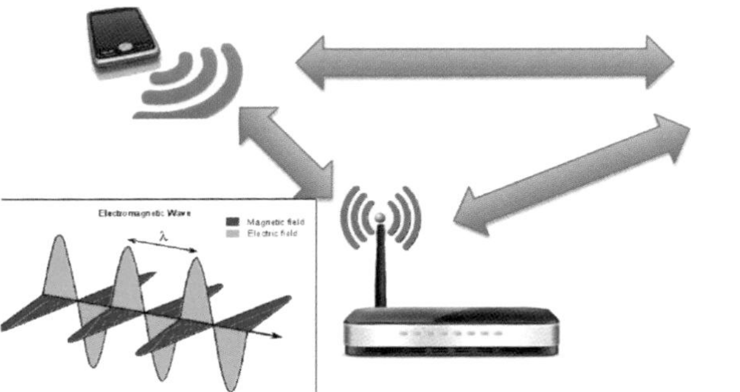

The idea of the Internet of Things generally requires reliable electromagnetic connections between quite different "things", i.e. devices. Such a connection can be realized by means of cables ("wired") or — as shown in this slide — using wireless techniques such as Bluetooth and WLAN. To investigate and design the physical behavior of these electromagnetic channels, first a field analysis by means of a suitable solution of Maxwell's equations is necessary. For simple (canonical) structures, the principles can be studied using analytical methods, but for realistic geometries, the use of a suitable numerical method will be necessary. Depending on the typical minimum wavelength λ of the electromagnetic wave, different numerical methods are in use.

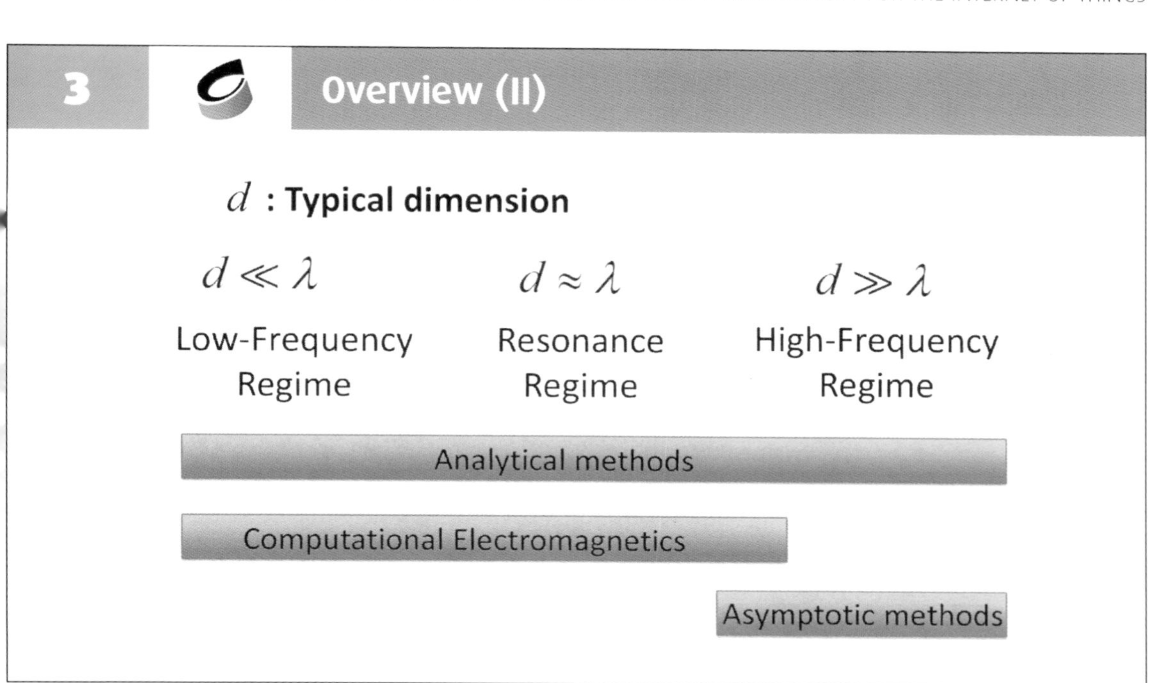

The minimum wavelength (corresponding to the highest frequency in the relevant band) is compared to the typical dimension d of the device of interest. For $d \approx \lambda$, isolated resonances may occur, thus, the problem is defined to be in the "resonance regime". Computational electromagnetics (CEM) software, i.e. codes which numerically solve the full Maxwell's equations, is capable of handling structure dimensions up to the resonance regime and — to some extent — up to the high-frequency regime. CEM methods include finite element methods, finite difference methods, and integral equation techniques, among others. If CEM methods are not applicable or too inefficient, asymptotic methods like geometrical optics and physical optics, the geometrical theory of diffraction, and the uniform theory of diffraction may be applied to tackle high-frequency problems. Finally, analytical methods like modal analysis typically are not limited to a specific frequency range, but they cannot handle arbitrary geometries.

4 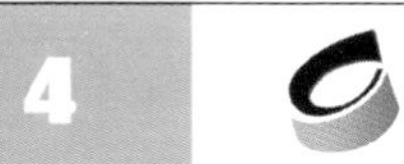 Computational Electromagnetics (I)

- Solution of Maxwell's equations by means of numerically exact methods
- Delivers numbers (not formulas)
- Usually requires some discretization
- Results obtained for a certain limited accuracy (depending on the available hardware)
- Visualization is usually easily obtained.

5

Computational Electromagnetics (II)

Domain discretization

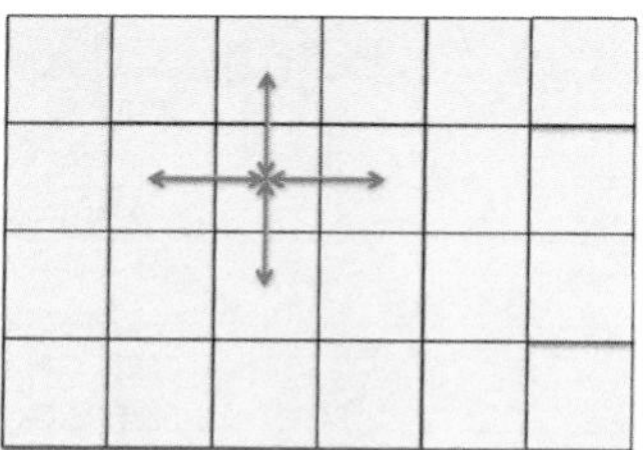

Local numerical method:
Each element is directly interacting with the nearest neighbors.
(Finite Element Method, Finite Difference Method, ...)

Within the CEM techniques, we usually distinguish between local and global numerical methods. First, we suppose a discretization of the solution domain into elements (e.g. triangles, rectangles, tetrahedrons, or rectangular parallelepipeds). In local numerical methods, just the interaction between neighboring elements is considered. Depending on the type of the field expansion in each element, the neighboring elements may include the direct neighbors only or those ones behind the next neighbors as well. Corresponding CEM codes include the Finite Difference and Finite Element methods widely used for problems with a large variety of different and/or complex materials. They usually lead to large systems of linear equation which are populated only around the diagonal (representing the interaction with the nearest neighbors) while the other matrix entries are zero.

6

Computational Electromagnetics (III)

Domain discretization

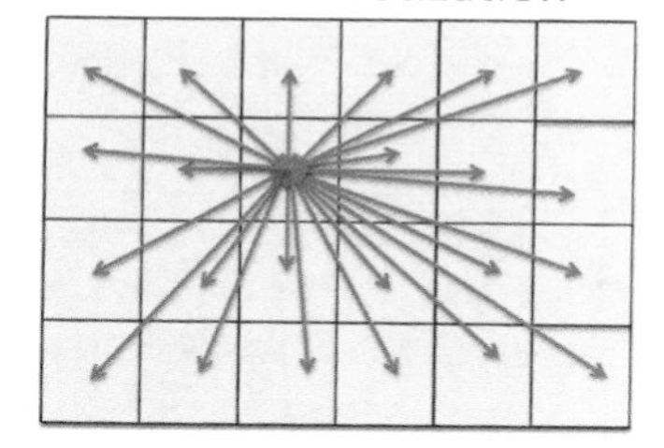

Global numerical method:
Each element is directly interacting with all other elements.
(Method of Moments, Integral Equation Method, ...)

In contrast to local methods in global numerical methods, the field expansions in each element interact with the field expansions in all other elements. They are preferred if the problem consists of only a few different materials, for instance, problems bounded by perfectly conducting structure. Usually the mathematical basis for this formulation is an electric, magnetic, or combined field integral equation with unknowns defined on the surface between the different materials. The integral equation is then transferred to a system of linear equations by applying the Method of Moments (MoM). Consequently, such codes are referred to as integral-equation solvers, MoM-solvers, or boundary element solvers, among others. They usually lead to fully populated matrices. In case of electrically large structures, these matrix equations are solved with the aid of special techniques (spectral methods, multi-level fast-multipole methods) which asymptotically allow the computation time to be in the order of $N \log N$ (instead of N^3 for a standard solver) with N being the number of unknowns.

7  The Finite-Difference Time-Domain Method (I)

Maxwell's curl-equations:

$$\frac{\partial \vec{E}(\vec{r},t)}{\partial t} = \frac{1}{\varepsilon(\vec{r})}\vec{\nabla} \times \vec{H}(\vec{r},t) - \frac{\sigma^e(\vec{r})}{\varepsilon(\vec{r})}\vec{E}(\vec{r},t)$$
$$\frac{\partial \vec{H}(\vec{r},t)}{\partial t} = -\frac{1}{\mu(\vec{r})}\vec{\nabla} \times \vec{E}(\vec{r},t) - \frac{\sigma^m(\vec{r})}{\mu(\vec{r})}\vec{H}(\vec{r},t).$$

Exemplarily, we will explain and apply the Finite-Difference Time-Domain (FDTD) method ,which is widely used in the areas of antennas, wave propagation, and EMC. As discussed before, FDTD is a local numerical method; the differential operators are discretized in time and space, i.e. they are replaced by difference operators. We start from the two Maxwell curl equations where we assume a linear, isotropic, non-dispersive, and inhomogeneous medium which also can comprise both electric and magnetic conductivities σ^e and σ^m, respectively. Note that also dispersive and/or anisotropic media can be treated by FDTD but will not be explained here.

8  The Finite-Difference Time-Domain Method (II)

$$\frac{\partial E_x(\vec{r},t)}{\partial t} = \frac{1}{\varepsilon(\vec{r})}\left(\frac{\partial H_z(\vec{r},t)}{\partial y} - \frac{\partial H_y(\vec{r},t)}{\partial z} - \sigma^e(\vec{r})E_x(\vec{r},t)\right)$$
$$\frac{\partial E_y(\vec{r},t)}{\partial t} = \frac{1}{\varepsilon(\vec{r})}\left(\frac{\partial H_x(\vec{r},t)}{\partial z} - \frac{\partial H_z(\vec{r},t)}{\partial x} - \sigma^e(\vec{r})E_y(\vec{r},t)\right)$$
$$\frac{\partial E_z(\vec{r},t)}{\partial t} = \frac{1}{\varepsilon(\vec{r})}\left(\frac{\partial H_y(\vec{r},t)}{\partial x} - \frac{\partial H_x(\vec{r},t)}{\partial y} - \sigma^e(\vec{r})E_z(\vec{r},t)\right)$$
$$\frac{\partial H_x(\vec{r},t)}{\partial t} = \frac{1}{\mu(\vec{r})}\left(\frac{\partial E_y(\vec{r},t)}{\partial z} - \frac{\partial E_z(\vec{r},t)}{\partial y} - \sigma^m(\vec{r})H_x(\vec{r},t)\right)$$
$$\frac{\partial H_y(\vec{r},t)}{\partial t} = \frac{1}{\mu(\vec{r})}\left(\frac{\partial E_z(\vec{r},t)}{\partial x} - \frac{\partial E_x(\vec{r},t)}{\partial z} - \sigma^m(\vec{r})H_y(\vec{r},t)\right)$$
$$\frac{\partial H_z(\vec{r},t)}{\partial t} = \frac{1}{\mu(\vec{r})}\left(\frac{\partial E_x(\vec{r},t)}{\partial y} - \frac{\partial E_y(\vec{r},t)}{\partial x} - \sigma^m(\vec{r})H_z(\vec{r},t)\right)$$

Next we write these two vector equations in the form of their 6 Cartesian components. It is easily recognized that the 6 field components are either directly or indirectly related to all other field components, thus we have to solve a system of coupled first-order partial differential equations. Interestingly, Maxwell wrote in his famous treatise these equations also in the form of Cartesian components since he did not know the compact form using vector analysis introduced to electromagnetics a bit later by Heaviside. We note that we have only first-order differential operators in these 6 Maxwell 's equations.

9 

The Finite-Difference Time-Domain Method (III)

Discretization in space and time:

$$\vec{r} = i\Delta x\ \hat{x}\ +\ j\Delta y\ \hat{y}\ +\ k\Delta z\ \hat{z}$$

$$(i = 0, 1, 2, ..., I;\ j = 0, 1, 2, ..., J;\ k = 0, 1, 2, ...K)$$

$$t = n\Delta t \qquad (n = 0, 1, 2, ..., N).$$

Discretized function:

$$u(\vec{r}, t) = u(i\Delta x, j\Delta y, k\Delta z, n\Delta t) = u|_{i,j,k}^{n}$$

In the standard FDTD method, time and space are uniformly discretized. That is, we consider only space points equally separated by constant distances $\Delta x, \Delta y, \Delta z$ in the three Cartesian directions. Of course, the number of steps I, J, K in each direction must be limited for a realistic simulation. That also means that a direct simulation of a free-space problem is not possible with FDTD. We will discuss that important issue later. Furthermore, also the time is uniformly discretized into a finite number N of time steps with a time step length of Δt. For the sake of a short concise writing, we use the short notation $u|_{i,j,k}^{n}$, meaning the quantity is evaluated at the location $\vec{r} = i\Delta x\ \hat{x}\ +\ j\Delta y\ \hat{y}\ +\ k\Delta z\ \hat{z}$ and time $t = n\Delta t$.

10 

The Finite-Difference Time-Domain Method (IV)

Central difference quotient:

$$\left.\frac{\partial u}{\partial x}\right|_{i,j,k}^{n} = \frac{u|_{i+1/2,j,k}^{n} - u|_{i-1/2,j,k}^{n}}{\Delta x} + O\left[(\Delta x)^2\right]$$

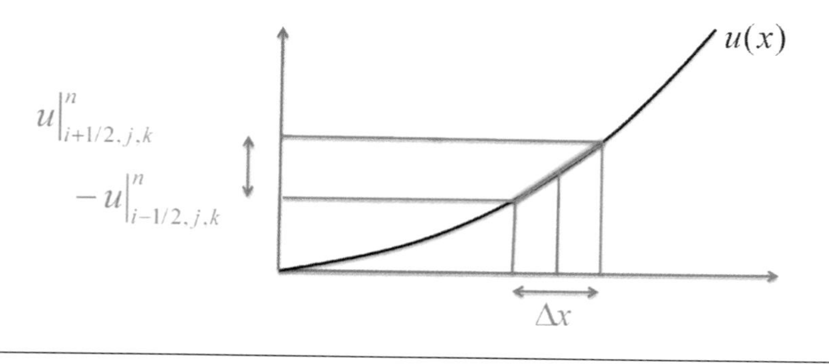

All first-order derivatives in the Maxwell's curl equations are replaced by central difference quotients. For instance, to calculate the first derivative $\partial u / \partial x$ at $x = i\Delta x$, we expand the functions $u[(i+1/2)\Delta x]$ and $u[(i-1/2)\Delta x]$ into Taylor expansions around $i\Delta x$ and subtract the results from each other. We just consider the first 3 terms of each expansion and obtain the result represented in the slide. We also may graphically interpret the central difference quotient, as indicated in the figure: The tangent at the curve at $i\Delta x$ is approximated by the straight line between $\{(i-1/2)\Delta x; u[(i-1/2)\Delta x]\}$ and $\{(i+1/2)\Delta x; u[(i+1/2)\Delta x]\}$. Note that this is a second-order approximation with respect to Δx.

11 

The Finite-Difference Time-Domain Method (V)

Discretization of Maxwell's equations:

$$\frac{\partial H_x(\vec{r},t)}{\partial t} = \frac{1}{\mu(\vec{r})}\left(\frac{\partial E_y(\vec{r},t)}{\partial z} - \frac{\partial E_z(\vec{r},t)}{\partial y} - \sigma^m(\vec{r})H_x(\vec{r},t)\right)$$

$$\frac{H_x|^{n+1/2}_{i,j,k} - H_x|^{n-1/2}_{i,j,k}}{\Delta t} = \frac{1}{\mu|_{i,j,k}}\left\{\frac{E_y|^n_{i,j,k+1/2} - E_y|^n_{i,j,k-1/2}}{\Delta z} - \frac{E_z|^n_{i,j+1/2,k} - E_z|^n_{i,j-1/2,k}}{\Delta y} - \sigma^m|_{i,j,k}\, H_x|^n_{i,j,k}\right\}.$$

Next we replace in the six Maxwell's equations all differential operators by central-difference operators as exemplarily shown here. We observe that in this equation we have the magnetic field at three different time steps $n-1/2;\, n;\, n+1/2$ while the electric field is given at one time step n. For all 6 equations, it would mean that we have all six components at three different time steps. Since FDTD is dealing with a time-stepping procedure, it follows that for the calculation of the component at the newest time step, we always had to know the values at two preceding time steps. In total, this would lead to a considerable amount of data we had to store at each time step.

12 

The Finite-Difference Time-Domain Method (VI)

Arithmetic mean:

$$H_x|^n_{i,j,k} = \frac{H_x|^{n+1/2}_{i,j,k} + H_x|^{n-1/2}_{i,j,k}}{2}$$

$$\frac{H_x|^{n+1/2}_{i,j,k} - H_x|^{n-1/2}_{i,j,k}}{\Delta t} = \frac{1}{\mu|_{i,j,k}}\left\{\frac{E_y|^n_{i,j,k+1/2} - E_y|^n_{i,j,k-1/2}}{\Delta z} - \frac{E_z|^n_{i,j+1/2,k} - E_z|^n_{i,j-1/2,k}}{\Delta y} - \sigma^m|_{i,j,k}\frac{H_x|^{n+1/2}_{i,j,k} + H_x|^{n-1/2}_{i,j,k}}{2}.\right\}$$

To reduce that amount of data, the standard FDTD uses a trick: To get rid of the third time step in the central-difference equations, we describe the value of $H_x|^n_{i,j,k}$ as an arithmetic mean of the two temporally neighbored values $H_x|^{n+1/2}_{i,j,k}$ and $H_x|^{n-1/2}_{i,j,k}$. Actually, such a linear approximation is in accordance with the general second-order FDTD scheme. Furthermore, this is needed only within conducting media. Note that in highly conducting media, this approximation can be refined taking into account the temporally exponential decay of the field. Finally with this step, we remark that we always need only one set of field values (i.e., at one time step) to evaluate the newest values.

13 

The Finite-Difference Time-Domain Method (VII)

$$H_x\Big|_{i+1/2,j,k}^{n+1/2} = +\left(\frac{1-\frac{\sigma^m|_{i,j,k}\Delta t}{2\mu|_{i,j,k}}}{1+\frac{\sigma^m|_{i,j,k}\Delta t}{2\mu|_{i,j,k}}}\right) H_x\Big|_{i+1/2,j,k}^{n-1/2} + \left(\frac{\frac{\Delta t}{\mu|_{i,j,k}}}{1+\frac{\sigma^m|_{i,j,k}\Delta t}{2\mu|_{i,j,k}}}\right)\left(\frac{E_y|^n_{i+1/2,j,k+1/2} - E_y|^n_{i+1/2,j,k-1/2}}{\Delta z} - \frac{E_z|^n_{i+1/2,j+1/2,k} - E_z|^n_{i+1/2,j-1/2,k}}{\Delta y}\right)$$

$$E_x\Big|_{i,j+1/2,k+1/2}^{n+1} = +\left(\frac{1-\frac{\sigma^e|_{i,j,k}\Delta t}{2\varepsilon|_{i,j,k}}}{1+\frac{\sigma^e|_{i,j,k}\Delta t}{2\varepsilon|_{i,j,k}}}\right) E_x\Big|_{i,j+1/2,k+1/2}^{n} + \left(\frac{\frac{\Delta t}{\varepsilon|_{i,j,k}}}{1+\frac{\sigma^e|_{i,j,k}\Delta t}{2\varepsilon|_{i,j,k}}}\right)\left(\frac{H_z|^{n+1/2}_{i,j+1,k+1/2} - H_z|^{n+1/2}_{i,j,k+1/2}}{\Delta y} - \frac{H_y|^{n+1/2}_{i,j+1/2,k+1} - H_y|^{n+1/2}_{i,j+1/2,k}}{\Delta z}\right)$$

Next, we isolate the expression with the newest time step in each of the six equations and put them on the left side of the equations as shown here for the two ***x***- components of the electric and magnetic fields. In addition, we re-order them such that the electric fields are given at "entire" time steps $(n, n+1)$, whereas the magnetic fields are given at "half" time steps $(n-1/2, n+1/2)$. We observe that the result can be interpreted as an update algorithm for the six field components. Note that the locations of the electric and magnetic field components are also shifted, the reason for which will be explained in Slide 3.15.

14 

The Finite-Difference Time-Domain Method (VIII)

´Leap-frog´-algorithm

$\vec{H}^{1/2} \rightarrow \vec{H}^{3/2} \rightarrow \vec{H}^{5/2} \rightarrow \vec{H}^{7/2}$ ...

$\vec{E}^{0} \rightarrow \vec{E}^{1} \rightarrow \vec{E}^{2} \rightarrow \vec{E}^{3}$ ...

t

$n=0 \quad n=1 \quad n=2 \quad n=3$ ...

$n=\frac{1}{2} \quad n=\frac{3}{2} \quad n=\frac{5}{2} \quad n=\frac{7}{2}$ ...

Each of the electric (magnetic) field components can be calculated from the same electric (magnetic) field component one time step before and from two of the magnetic (electric) field components one half time step before. The algorithm is performed each half time step, where — beginning from start values $\vec{E}|^0$ and $\vec{H}|^{1/2}$ of all six components at all locations — alternately the electric or the magnetic field components are updated. Therefore, this iteratively working technique is also referred to as a *leap-frog* algorithm.

15

The Finite-Difference Time-Domain Method (IX)

A look at the spatial arrangement of the field components in the six equations reveals the situation represented in this slide: The six components can be arranged on the surface of a rectangular parallelepiped with edge lengths $\Delta x, \Delta y, \Delta z$. If one corner of it is denoted as i, j, k, the magnetic field components lie on the mid of the three neighboring edges of the rectangular parallelepiped while the electric field components are located at the centers of the three neighboring surfaces. This arrangement is commonly referred to as a *Yee* cell in honor of Kane Yee who first proposed FDTD in 1966. The entire solution domain is uniformly filled with such Yee cells. The material parameters can differ from cell to cell to simulate an arbitrarily inhomogeneous structure.

16

The Finite-Difference Time-Domain Method (X)

Stability of the leap-frog algorithm:

- Available parameters: $\Delta x, \Delta y, \Delta z, \Delta t$
- Spatial discretization chosen such that $\Delta x, \Delta y, \Delta z \ll \lambda$, typically $\Delta x, \Delta y, \Delta z \leq \lambda/10$
- *Question:* How to then choose Δt ?
- *Answer:* For an absolutely stable solution, it is sufficient to choose (*Courant* condition):

$$\Delta t \leq \frac{1}{c\sqrt{\frac{1}{(\Delta x)^2} + \frac{1}{(\Delta y)^2} + \frac{1}{(\Delta z)^2}}}$$

Basically, we are now ready to discretize a given bounded domain with given boundary conditions into Yee cells and run the leap-frog algorithm. One of the remaining questions is how to choose the parameters $\Delta x, \Delta y, \Delta z$, and Δt. Usually, first the spatial parameters are set to one-tenth of the minimum wavelength. For instance, if the typical maximum frequency is 1 GHz, then the wavelength in free space is 30cm and the corresponding $\Delta x, \Delta y, \Delta z$ should be set to not more than 3cm. Once the spatial parameters are fixed, one can show that for a (absolutely) stable algorithm the time parameter Δt must be set according to the Courant condition written here. c denotes the velocity of light (phase velocity) in the medium. The time parameter should not be set much smaller than the necessary value because the computational efficiency is decreasing and — even more important — the systematic error caused by the discretization, which is called *numerical dispersion*, is increasing.

17

The Finite-Difference Time-Domain Method (XI)

Simulation of free-space problems:

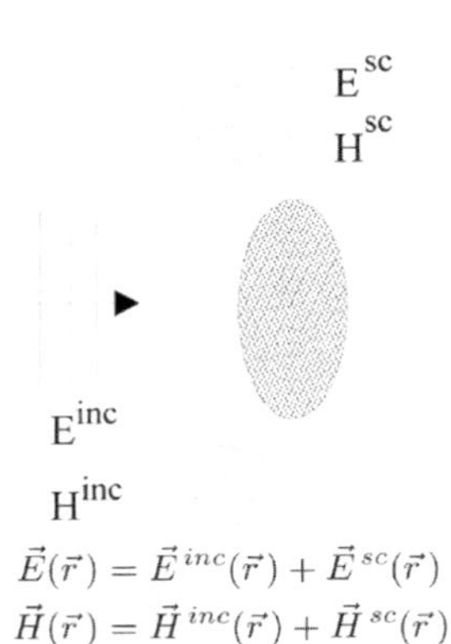

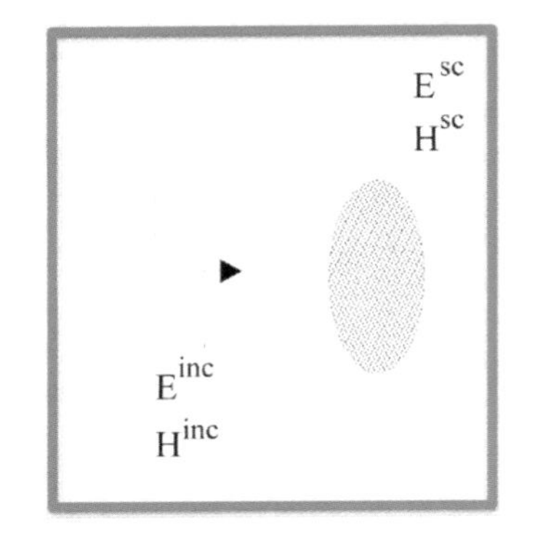

The standard FDTD algorithm requires a bounded domain with given boundary conditions, i.e. a vanishing tangential electric or magnetic field or a fixed relation between them. For the important case of an open boundary needed for instance to simulate the electromagnetic radiation from an antenna into free space, a boundary has to be defined which simulates the free space. Similar to an anechoic chamber, this boundary has to prevent any reflected field. To this end, we first separate from the total field the given incident field and the scattered field caused by any scattering objects. In case that the source is in the FDTD solution domain (e.g., an antenna), the total field is identical to the scattered field.

18

The Finite-Difference Time-Domain Method (XII)

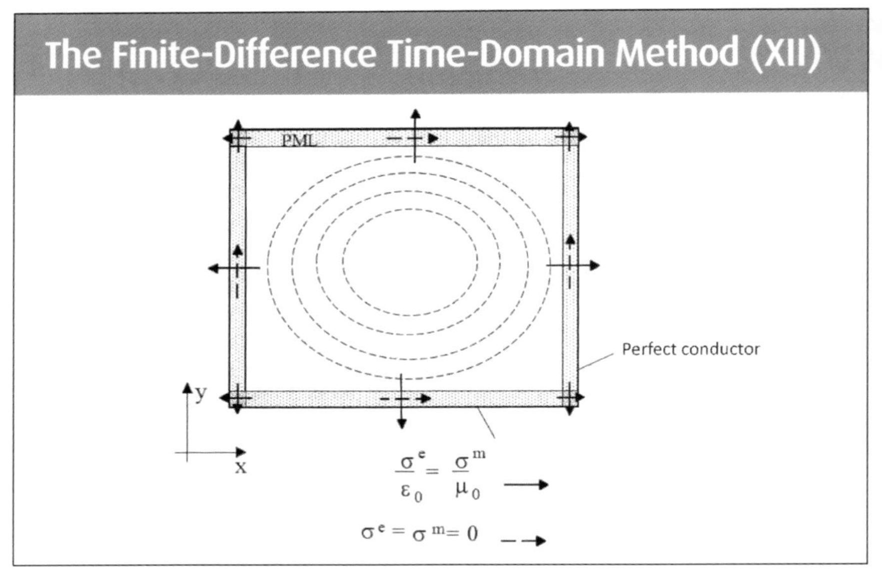

The scattered field defined in the preceding slide must not experience any reflection from the boundary simulating the open space. Such a boundary was first proposed by Berenger and is referred to as the *Perfectly Matched Layer (PML)*. Within the PML, the scattered field is split into two parts: One of these parts would - if it would be the only part - travel in the direction perpendicular to the boundary. For this part, the PML medium is perfectly matched, i.e., the ratio of the electric conductivity to the permittivity in vacuum equals the ratio of the magnetic conductivity to the permeability in vacuum. For such a medium, it is known that there is no reflection for a perpendicularly incident wave. Moreover, due to the losses, the amplitude of this part is exponentially decreasing. The other part would (if it existed alone) travel in parallel to the boundary. For this part, the PML represents vacuum and naturally not causing any reflection. However, the mentioned two parts are still coupled with the result that for any angle of incidence there is no reflection at the PML boundary, while in the PML, the field amplitude is exponentially decaying.

19

The Finite-Difference Time-Domain Method (XIII)

Total field / Scattered field formulation:

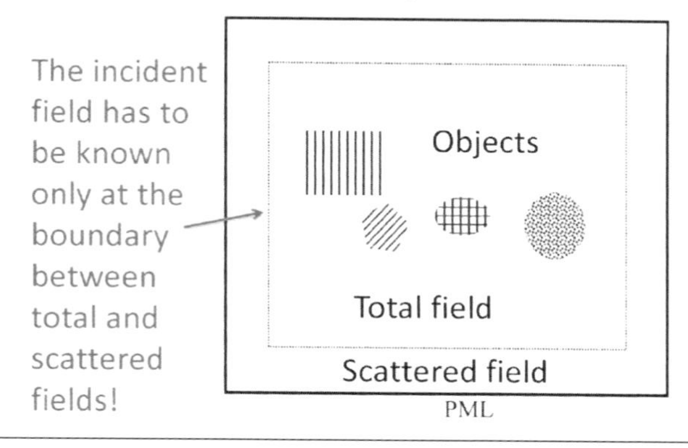

One specific problem which occurs while treating scattering problems is how to incorporate the incident field (e.g., a plane electromagnetic wave) into the FDTD scheme. A common way to do this is known as the total field/scattered field formulation. We first draw an imaginary boundary into the solution domain which encloses all objects which are different from the free space. The field values in the inner part represent the total field, and those in the outer domain are representatives of the scattered field. A simple analysis using the FDTD formulation at the boundaries reveals that the change from total to scattered fields at the imaginary boundary is accomplished by employing the known incident field values there.

20

Electromagnetic Compatibility Issues for the IoT (I)

Electromagnetic Compatibility is the ability of a device to reliably work within a well-defined electromagnetic environment without influencing the electromagnetic environment such that it would be unacceptable for other devices in the same environment.

→National and International Standards describe what is well defined and unacceptable!

The Internet of Things may naturally become susceptible to threads caused by an insufficient electromagnetic compatibility (EMC) of the systems and devices involved. Therefore, particularly devices with wireless connections must be constructed such that they become robust and compatible to any environment they were designed for. The legal definition of EMC shown here includes an active and a passive component. The level of interferences caused by an active device must not exceed well-defined values; on the other hand, any device must be able to withstand a certain level of interference without a malfunction. Of cause, these levels of interference are defined in national and international standards. For instance, any device sold in the market of the European Community has to comply with such standards. With the "CE" sign, the manufacturer does guarantee that.

21 

Electromagnetic Compatibility Issues for the IoT (II)

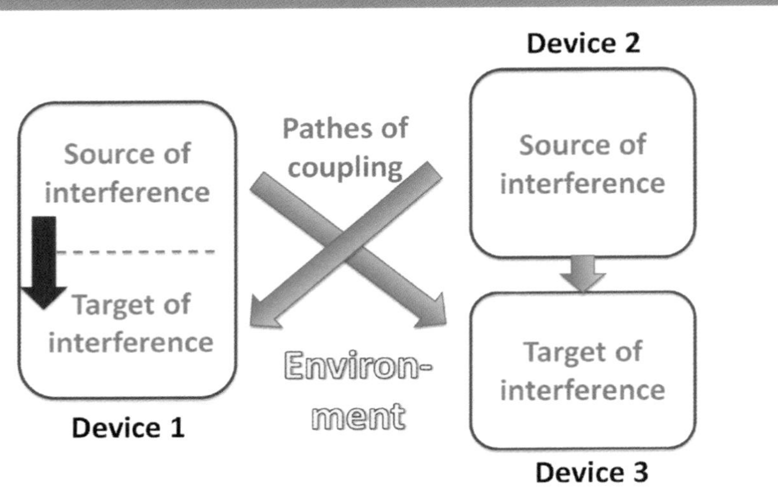

The use of the correct terminology is very important to ensure that all partners in this multi-disciplinary subject "EMC" do understand. First, all of the electric parts in a given system are denoted and numbered as "devices". There may be devices which act as sources of interference, others as targets of interference, and there also may be devices which simultaneously act as sources and targets. The EMC within the device must also be considered though it might have been ensured by the manufacturer of the device. The devices are located within a certain environment. All sources and targets are related through paths of coupling. A systematic EMC analysis includes the identification of all possible sources, devices, and paths, as well as a corresponding assessment of any possible interference.

22 

Electromagnetic Compatibility Issues for the IoT (III)

Typical sources of interference (examples):

- **Atmospheric disturbances (T-storm lightnings around the world anytime), frequencies up to MHz.**
- **Radio stations, TV stations, radar stations, mobile comm. stations and phones, up to 100 GHz**
- **Power lines, transformers, switches, power electronics**
- **Household appliences, tools, cars, trains, commutator motors, electronic devices**
- **Electrostatic discharge (ESD)**
- **...**

The most often interferences with frequencies up to several MHz are natural and originate from T-storms around the world most of them nearby the equator. Among the man-made interferences, we have to distinguish between wanted and unwanted sources. The first ones include radio and TV stations, radar, the mobile communication networks and corresponding phones. Among the unwanted sources are power lines, transformers, and power electronic devices, but also household appliances, cars, and trains. Electrostatic discharge (ESD) plays a special role as a source of interference, which is dangerous especially for sensitive high-ohmic electronic circuits.

23 

Electromagnetic Compatibility Issues for the IoT (IV)

Coupling mechanisms:

- **Galvanic coupling (e.g., using same ground line)**
- **Capacitive coupling (e.g., electrodes of the unwanted capacitor are in different devices, used at lower frequencies)**
- **Inductive coupling (e.g., circuits in different devices act as primary and secondary part of an unwanted transformer, used at lower frequencies)**
- **Electromagnetic coupling (general coupling, all effects included, mostly used for higher frequencies and radiation interference)**
- **Electrostatic discharge (ESD)**

For the coupling of interferences from the source to the targets along the paths of interference, different mechanisms are responsible: The galvanic coupling may occur if there is a galvanic connection between source and target. Sometimes these connections are not directly recognized since the non-zero resistance of a real line is not represented in the outline of the circuit. Moreover, coupling mechanisms include capacitive and inductive interactions between different devices, particularly lines. The corresponding modeling is possible only at lower frequencies where capacitors and inductors may be defined. The general case of modeling the electromagnetic coupling includes galvanic, capacitive, and inductive effects, but is only necessary if the frequency is too high for a network description, i.e., if the dimensions of the investigated structure are in the area or larger than the typical wavelength of the interference.

24 

Electromagnetic Compatibility Issues for the IoT (V)

- **Typical electrostatic discharge (ESD) modeling:**

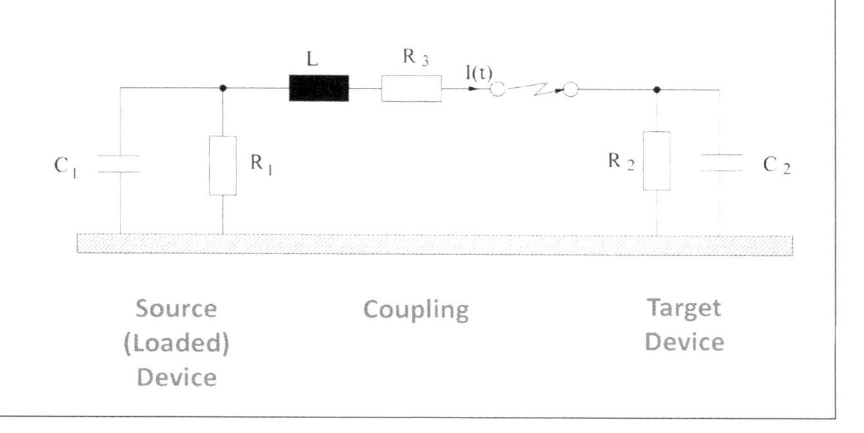

The electrostatic discharge (ESD) can generally be modeled using this circuit. The charge-loaded device (source) is modeled by a capacitor C_1 and a resistor R_1. Consequently, the source is discharged even without an ESD according to a time constant $\tau_1 = R_1 C_1$. If the distance to the target device which also is modeled by a resistor and a parallel capacitor becomes too short, there is a discharge symboled by a corresponding flash. The path of interference is modeled by a resistor R_3 and an inductor L in series.

25 

Electromagnetic Compatibility Issues for the IoT (VI)

Two most important measures to ensure EMC:

- ***Filters*: For line-connected devices**
  - **- Frequency-selective filters (passive R-L-C)**
  - **- Galvanically decoupling filters (Transformers, optically coupling devices)**
  - **- Surge-protective devices (Suppressor-diodes, varistors, surge arrestors)**
- ***Shielding*: For radiation-based interference processes**

Depending on the type of interference, mainly two different measures to prevent devices and systems from EMC problems are in use. In case of "cabled" interference, i.e., if the path of interference is built up by a line, filters are in use to secure the EMC. Here, we can distinguish between three types of filters: frequency-selective filters such as capacitors, inductors, and resistors and any combinations of them are in use to prevent interferences of certain frequencies (lowpath, highpath, and bandpass filters). Transformers and optically coupled devices (optocouplers) are in use to galvanically decouple sources from targets while surge-protective devices like suppressor diodes, varistors, and surge arrestors are employed to prevent devices and systems from being influenced by high voltages coming for instance from a lightning or from an interrupted inductor-current. It is not uncommon that sensitive devices contain several or all of the filter types. To prevent devices from "radiated" emissions of sources, the main measure used is shielding.

26 

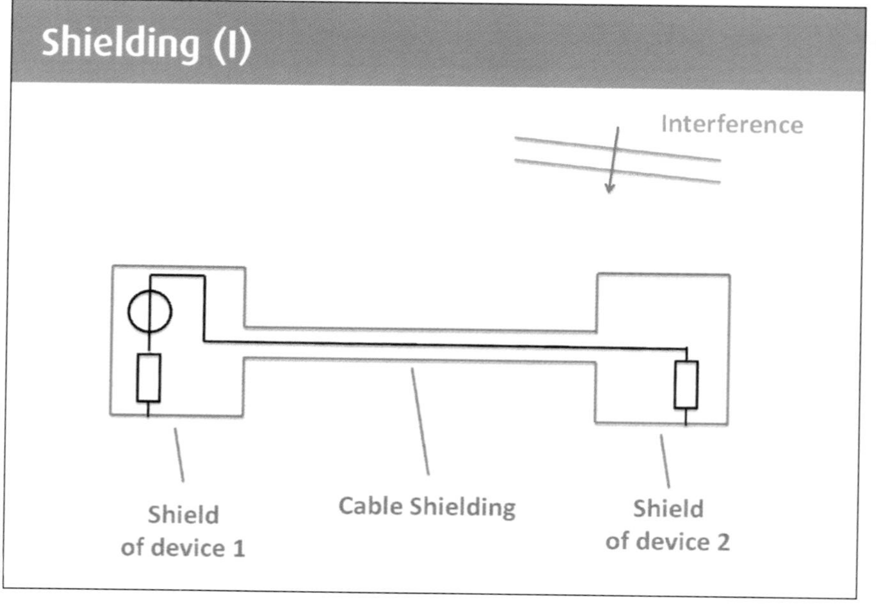

The principal design of a perfectly shielded system is shown in this slide. The two shields of the two devices are connected by a line which also is shielded. It is important to note that the cable shielding is a continuation of the device shielding. The interference caused by a radiated emission of a distant source (or any other disturbing electromagnetic field) is prevented from entering the inner system of the shielded structure. This general shielding structure is obviously present if we connect two high-frequency devices by a coaxial cable. The shielding principle depends also on the frequency regime as will be demonstrated in the next slides.

27

Shielding (II)

σ > 0

Device

$\vec{D}$

Shielding against electrostatic / low-frequency electric fields

The realization of a shielding structure against electrostatic and low-frequency electric fields can be accomplished most simply by any conducting material. Since the electrostatic lines of force end on the metallic shield, also small apertures in the shield are allowed. Because of energetic reasons, the lines of force are seeking for the shortest way to their end. Thus, they do not enter the aperture if its dimension is not too big. This principle is represented in and well known as the Faraday cage. For the proper functioning, it is important that the shielded device and the shield are on the same potential. Typically both shield and shielded device are connected and earthed.

28

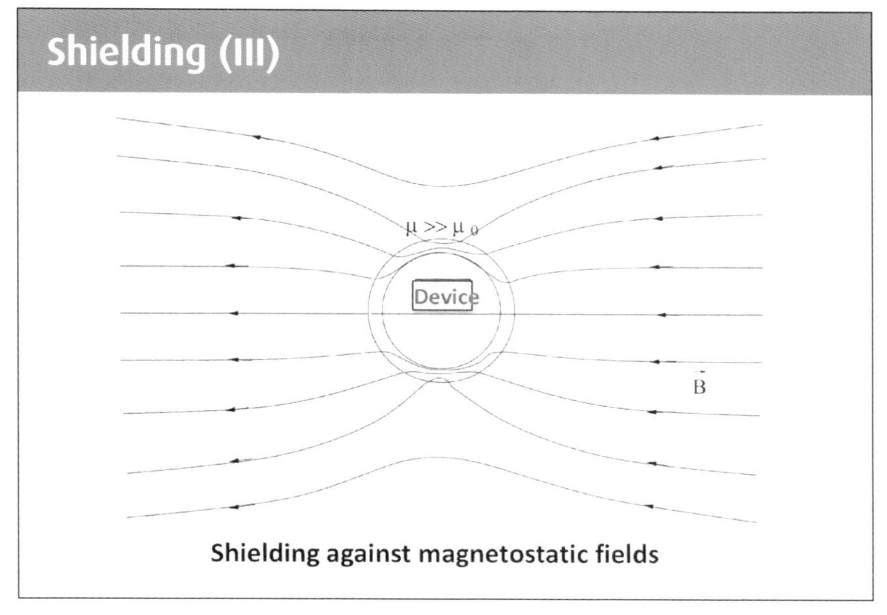

Shielding against magnetostatic fields

The electric conductivity does not have any effect in case of shielding against magnetostatic fields. Thus, and because there is no magnetically conducting material, shielding against magnetostatic fields is one of the most expensive types of shielding measures. Usually this is accomplished by employing a material with a high value of its permeability. For instance, *mu-metal* (a nickel-iron soft ferromagnetic alloy) may have a relative permeability of several hundreds of thousands. As illustrated in the slide, because of energetic reasons the lines of force of the magnetic field prefer to follow the high material instead of going through the shielded domain. However, to achieve a reduction of the magnetic field by 100 dB, tens of centimeters thickness of the shield are necessary. Moreover, apertures in the shield have to be avoided, which even increase the costs for a well-functioning shield against magnetostatic shields.

29

Shielding (IV)

Skin depth:

$$\delta = \sqrt{\frac{2}{\omega\mu\sigma}}$$

y

$\vec{H}$

x

δ

σ> 0

σ= 0

Shielding against low-frequency magnetic fields

The situation changes drastically in case of low-frequency magnetic fields. Because of the induction principle and corresponding eddy currents, the conductivity plays a crucial role in the shielding effect. A parameter to estimate the thickness necessary to obtain the desired shielding effect is the skin depth, which is related to the circle frequency ω, the permeability μ, and the electric conductivity σ of the shield material as shown in the slide. To achieve a shielding effect, the thickness must be several times smaller than the skin depth. Again, apertures in the shield should be avoided. Note that this type of shielding is also common for cable shielding.

30

Shielding (V)

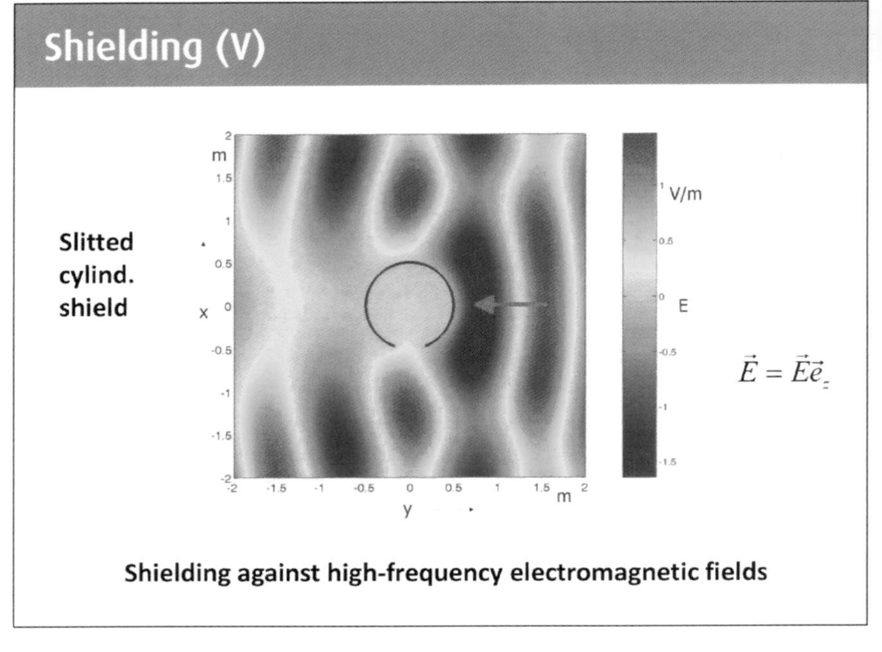

Shielding against high-frequency electromagnetic fields

For high-frequency electromagnetic fields, the full Maxwell's equations have to be solved. For arbitrary shielding structures, this can be done numerically, for instance, using the FDTD method described above. For studying the principle, also analytical methods can be applied. This slide shows a snapshot of the electric field for a plane electromagnetic wave incident on a slotted circular cylinder. The electric field is polarized in the direction of the cylinder axis (TM-case). The ratio of the diameter of the cylinder to the wavelength can be easily deduced. We observe that the amount of field penetrating through the aperture is limited.

31

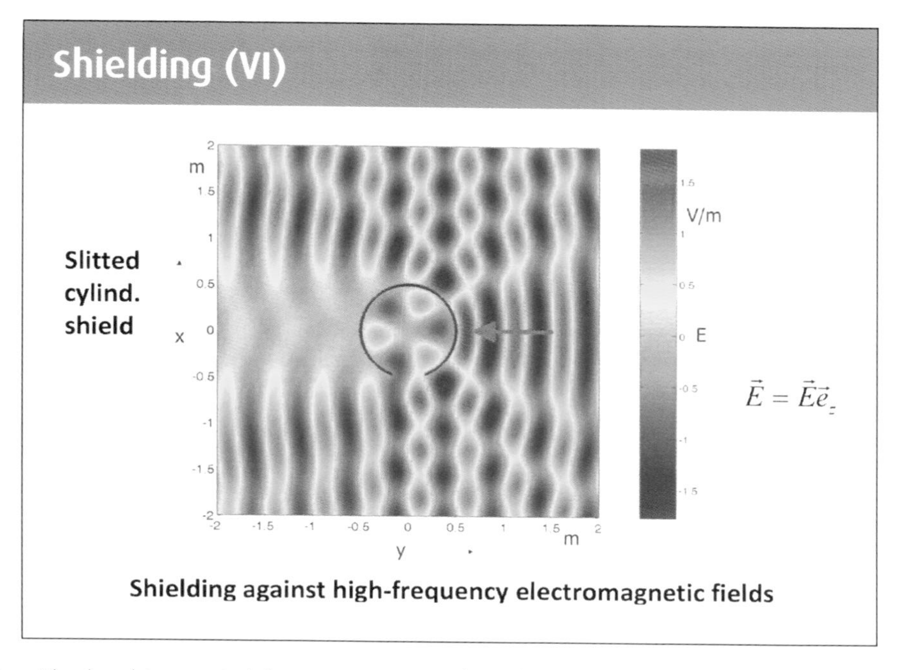

Shielding against high-frequency electromagnetic fields

The situation changes if we increase this ratio increases. We observe a resonance in the shielded domain which is obviously fed through the aperture. This effect is more or less independent of the size of the aperture and should be considered while employing shields against electromagnetic fields in the resonance regime. One way to avoid such resonances is to cover the interior walls with absorbing materials. Another way is to limit the amount of energy penetrating though the aperture exploiting the cut-off frequencies of empty waveguides.

32

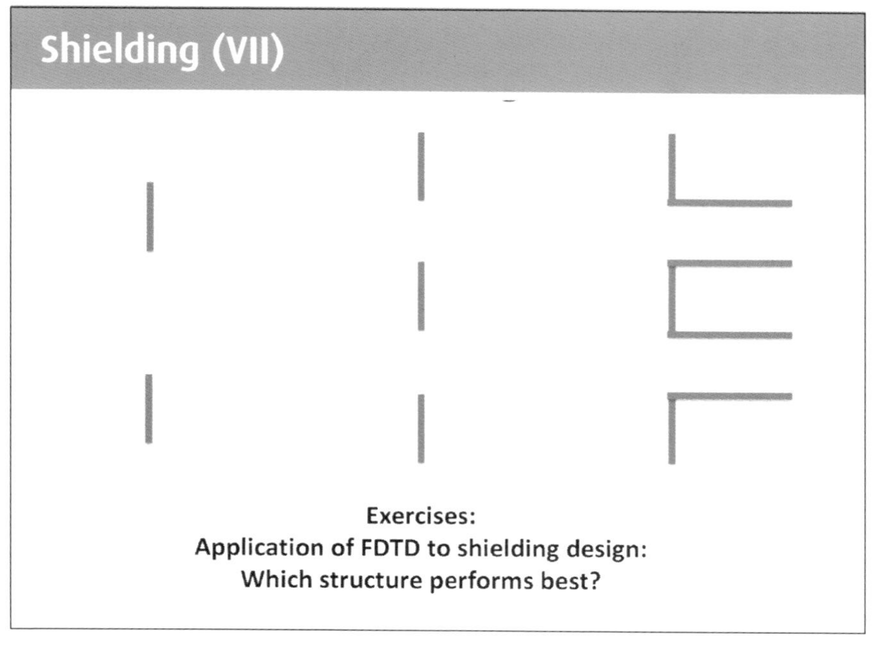

Exercises:
Application of FDTD to shielding design:
Which structure performs best?

To explore this point, we look at these three different designs of an aperture. The left one represents just one big slot while the design in the middle consists of two slots with totally the same length as the design in the left example. The right design is similar to that one in the middle; however, the apertures are extended by means of empty-waveguide, like structures. As an exercise, all three designs should be analyzed by a self-written FDTD code for different frequencies and polarizations of the incident plane wave coming from the left. The amount of energy passing through each aperture should be compared as discussed.

CHAPTER 03

Integrated Circuits & Systems for mm-wave/RF Wireless Transceivers in IoT Applications (Communications)

Sergio Saponara

University of Pisa, Italy

1. Agenda 64
2. Electromagnetic spectrum 64
3. Worldwide spectrum allocation for IoT wireless transceivers 65
4. Worldwide mm-wave frequencies for IoT wireless transceivers (communications) 65
5. Worldwide mm-wave frequencies for IoT wireless transceivers (sensing) 66
6. RF communication systems -basics 66
7. RF communication systems -basics 67
8. RF transmitter and receiver 67
9. RF transmitter and receiver 68
10. Amplitude Shift Keying (ASK) 68
11. On-Off Keying (OOK) 69
12. Frequency Shift Keying (FSK) 69
13. Phase Shift Keying (PSK) 70
14. Quadrature Amplitude Modulation (QAM) 70
15. Spectral efficiency of modulations 71
16. Spread spectrum 71
17. Direct Sequence Spread Spectrum (DSSS) 72
18. Frequency Hopping Spread Spectrum (FHSS) 72
19. Transceiver architecture with frequency hopping 73
20. Static (Top) or adaptive (Bottom) hopping 73
21. Frequency agility 74
22. Multiple access techniques 74
23. Architecture of a mobile wireless terminal 75
24. Super-heterodyne receiver 75
25. Double-conversion super-heterodyne receiver 76
26. Transceiver with super heterodyne receiver 76
27. Zero IF receiver (homodyne) 77
28. Low-IF receiver 77
29. Image rejection receiver CC1020 from TI 78
30. Example: Double conversion transceiver 78
31. Example: OOK zero-IF transceiver 79
32. Integrated power amplifier 79
33. Gain of the 60 GHz power amplifier 80
34. 60 GHz 2-stage LNA 80
35. 60 GHz 3-stage LNA 81
36. Multi-channel integrated power amplifier 81
37. RF building blocks 82
38. Balun & matching 82
39. PCB antennas 83
40. Boosting performance of RF ICs 83
41. RF ICs with off-chip PA 84
42. RF link budget analysis 84
43. BER vs. SNR 85
44. Verification and testing flow 85
45. Equipment for RF measurements 86
46. References 86

This chapter is focused on the analysis of integrated circuits and systems for wireless transceivers operating at several frequencies (from sub-GHz to mm-waves) targeting communication applications.

Basic concepts of wireless communication systems are reviewed: duplexing, modulations, multiple access techniques, VLSI transceiver architecture and main hardware building blocks like power amplifier and low noise amplifier (LNA).

Advanced communication techniques like direct sequence spread spectrum and frequency hopping spread spectrum are also detailed.

Real examples of wireless transceivers from scientific literature and from commercial products are also discussed.

1 Agenda

- Spectrum allocation for IoT services
- Basics of RF communication systems, duplexing & modulations
- VLSI transceiver architectures
- RF building blocks
- Radio link budget analysis
- Equipment for RF measurements
- Medium access techniques

Chapters 3 and 4 focus on integrated circuits (ICs) and systems for wireless transceivers operating at RF or mm-waves in communication (Chapter 3) and remote sensing (Chapter 4) IoT applications.

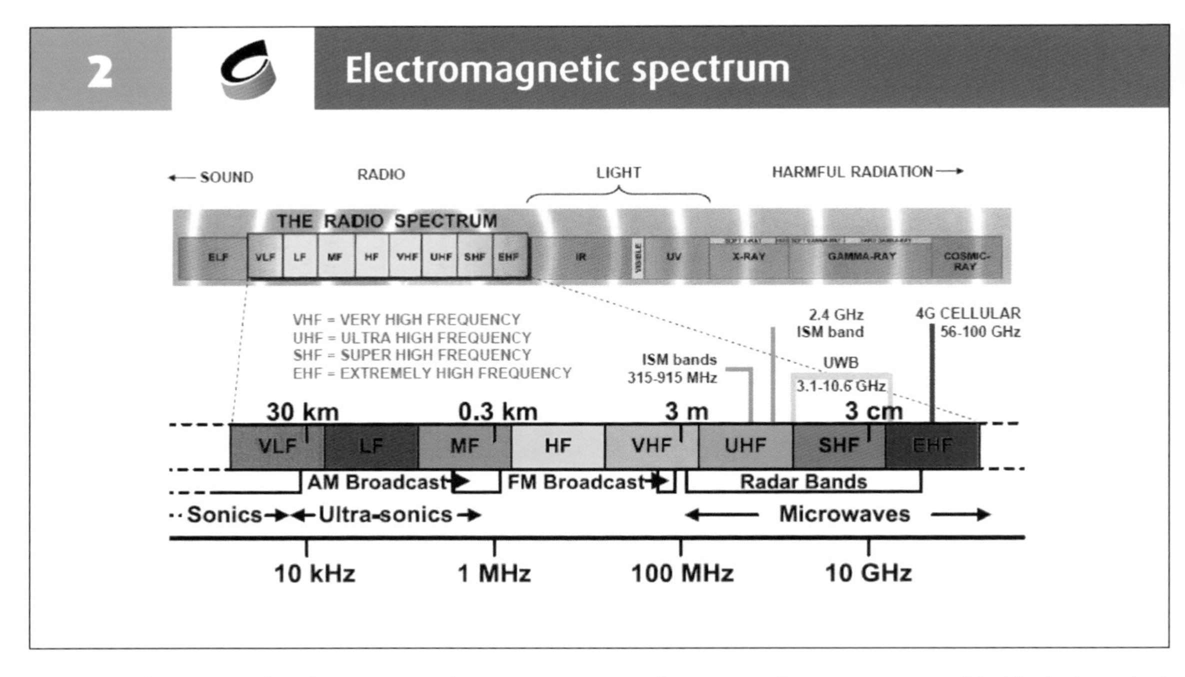

With reference to the Electromagnetic spectrum, most of IoT transceivers operate worldwide in the sub-6 GHz part, particularly in the sub-GHz portion and around 2.4 GHz, although new high-band applications are emerging at mm-wave.

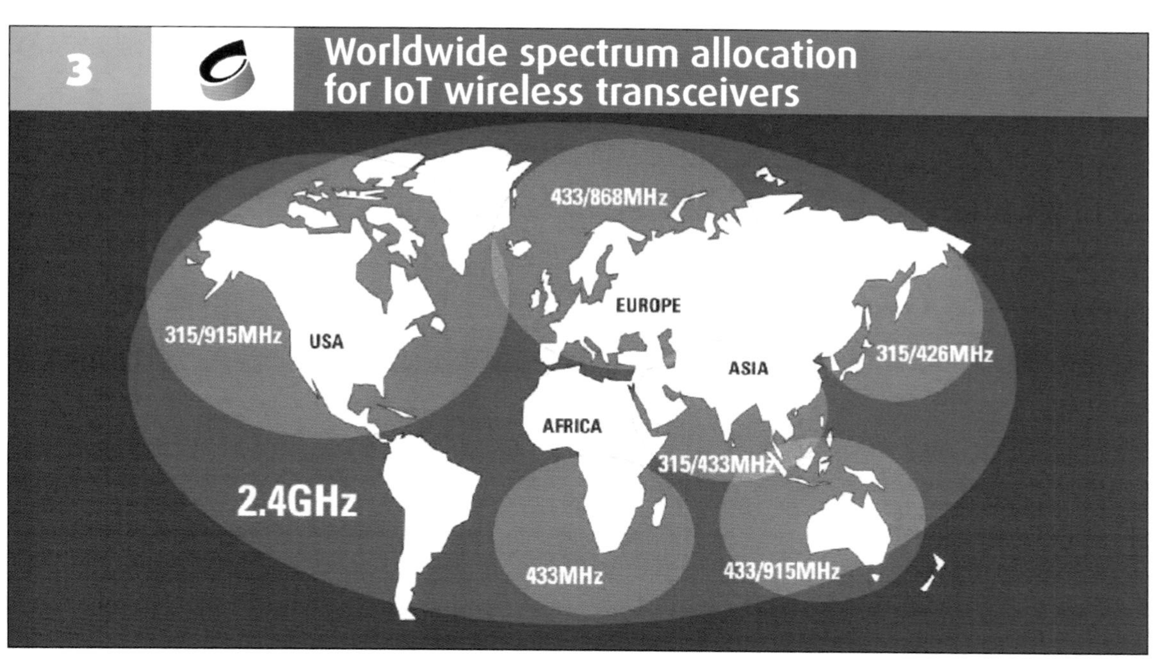

Unlicensed ISM (Industrial Scientific Medical) bands for IoT, particularly for short-range devices (SRD), are available worldwide around 315, 433, 868, and 915 MHz, plus 2.4 GHz. As an example, in Europe, the bands 433-434.79 MHz, 863-870 MHz, and 2.4 – 2.48835 GHz are used according to ETSI EN 300 220/440/328.

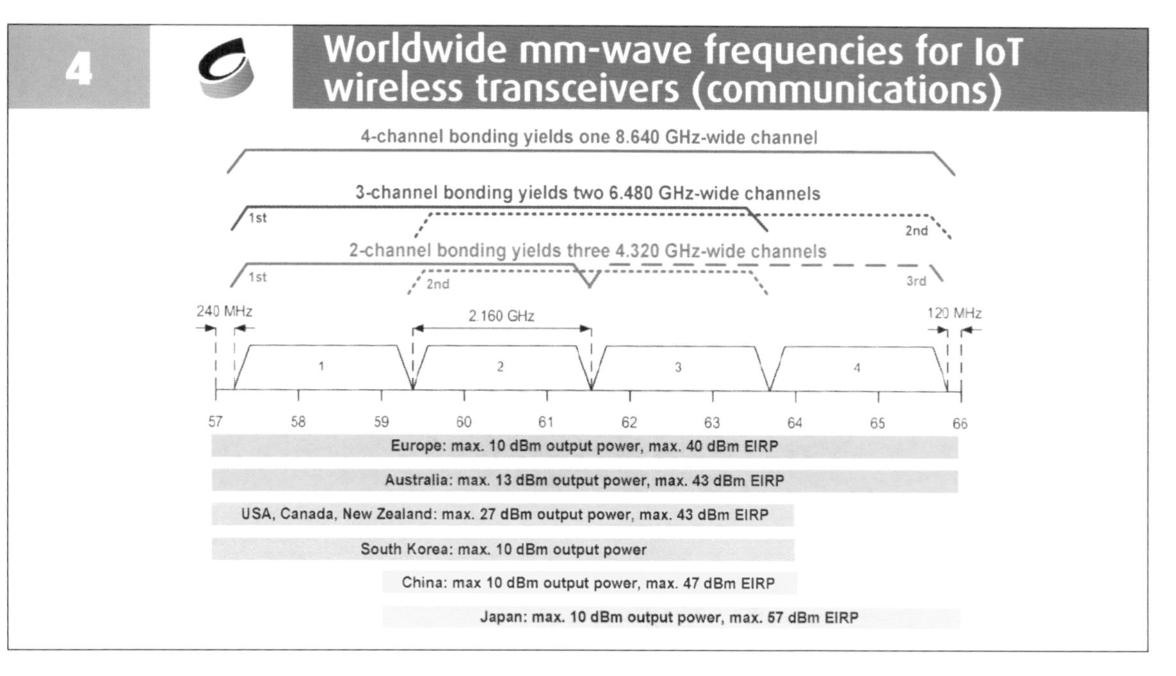

New high-band applications are emerging at mm-waves, exploiting for short-range (due to peak absorption of oxygen) communications the multi-GHz spectrum available worldwide for free at 60 GHz.

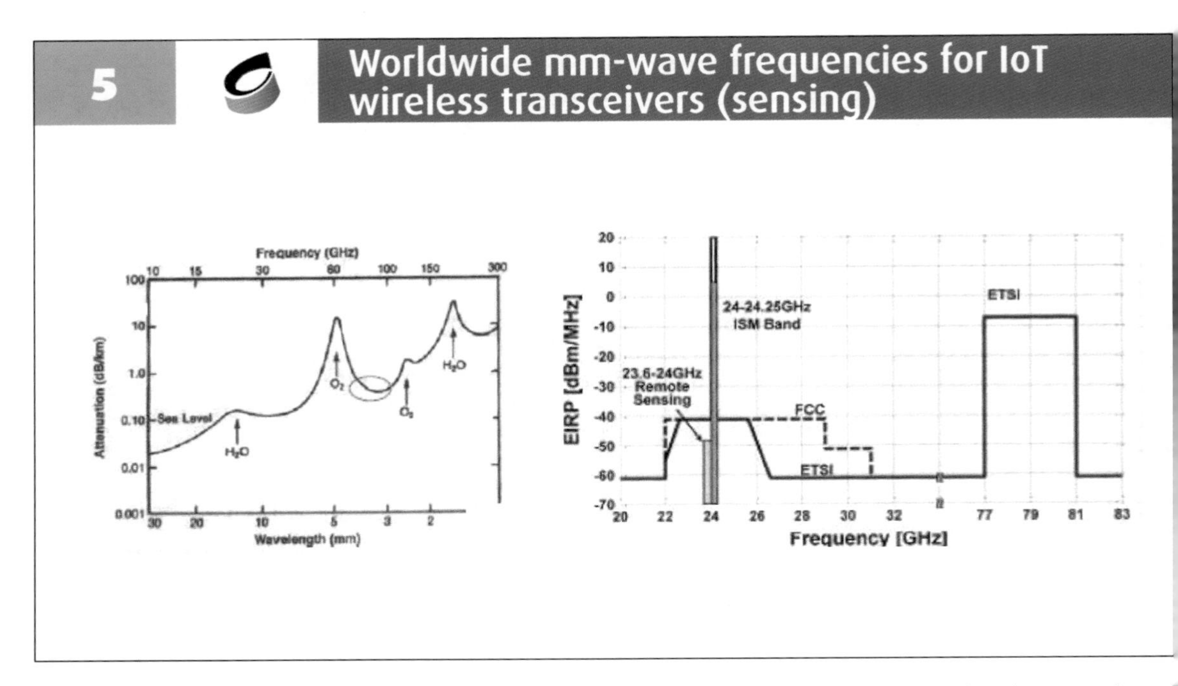

For remote sensing applications at mm-waves, the 24 GHz and the 77 GHz spectrum bands are preferred due to lower attenuation than 60 GHz.

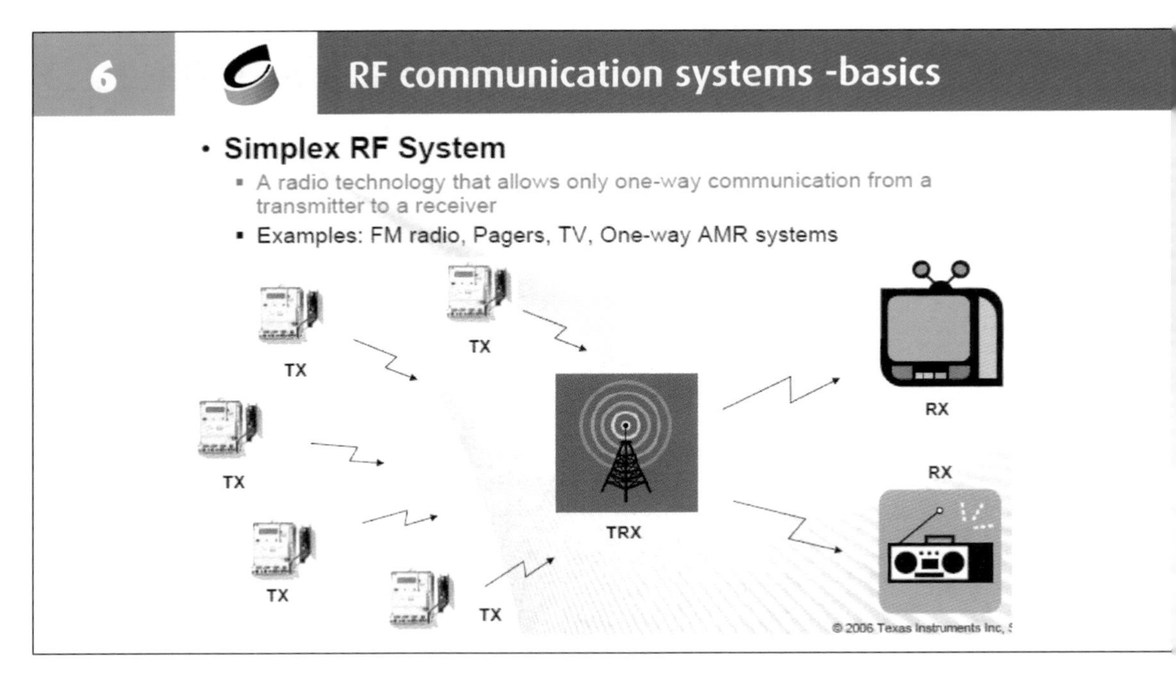

The RF communication system basics include the duplexing mode (simplex, half or full duplex), the latter being preferred for peer-to-peer communication systems, such as cellular phones.

7 

RF communication systems -basics

- **Half-duplex RF Systems**
  - Operation mode of a radio communication system in which each end can transmit and receive, but not simultaneously.
  - Note: The communication is bidirectional over the same frequency, but unidirectional for the duration of a message. The devices need to be transceivers. Applies to most TDD and TDMA systems.
  - Examples: Walkie-talkie, wireless keyboard mouse
- **Full-duplex RF Systems**
  - Radio systems in which each end can transmit and receive simultaneously
  - Typically two frequencies are used to set up the communication channel. Each frequency is used solely for either transmitting or receiving. Applies to Frequency Division Duplex (FDD) systems.
  - Example: Cellular phones, satellite communication

For Half-duplex systems, TDD is usually adopted, while for Full-duplex ones, FDD is adopted.

8

RF transmitter and receiver

- Transmitter

digital data 101101001 → digital modulation → analog baseband signal → analog modulation (radio carrier)

- Receiver

analog demodulation (radio carrier) → analog baseband signal → synchronization decision → digital data 101101001

Rx Tx (a)

Rx Tx Vc (b)

FDD (Frequency Division Duplex) Filters for TX/RX selection (a)
TDD (Time Division Duplex) TX/RX antenna switch (b)

The RF transceiver architecture is based on the modulation of a digital baseband signal with a high-frequency sinusoidal carrier, which is then demodulated at the receiver side when synchronization and maximum likelihood decision is taken to reconstruct the received digital data. Proper antenna switch or filters are used for transmitter/receiver mode selection in the integrated terminal.

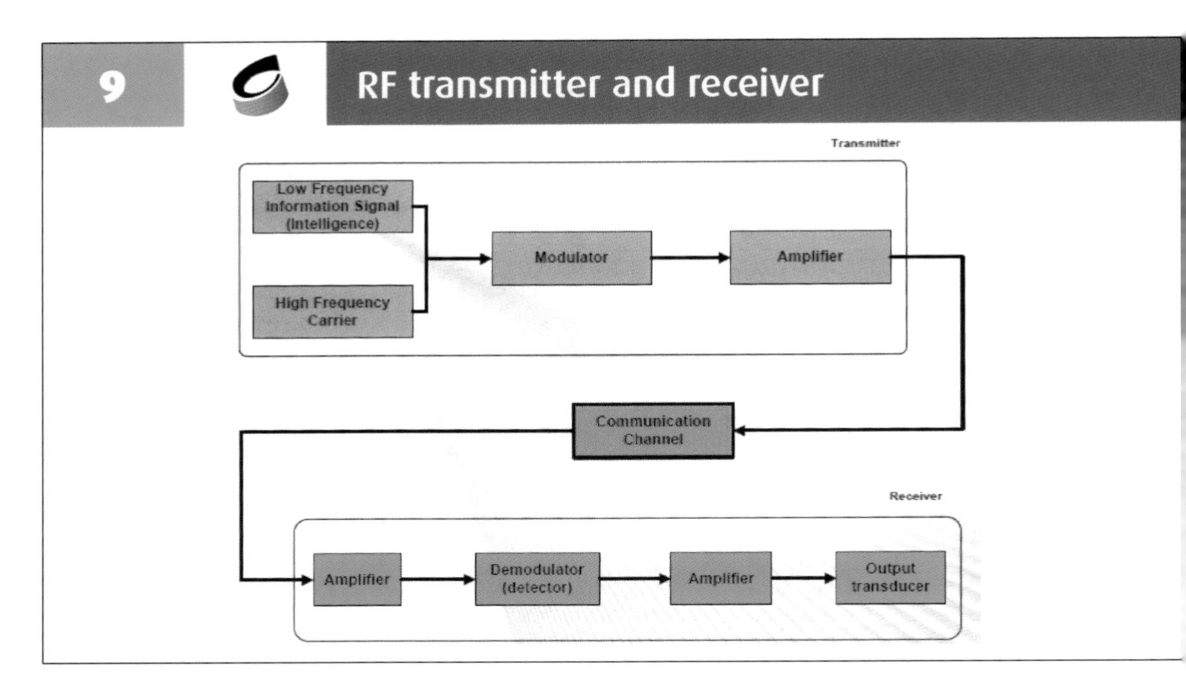

End-to-end simulation/model must take into account the effect on signals of the communication channel.

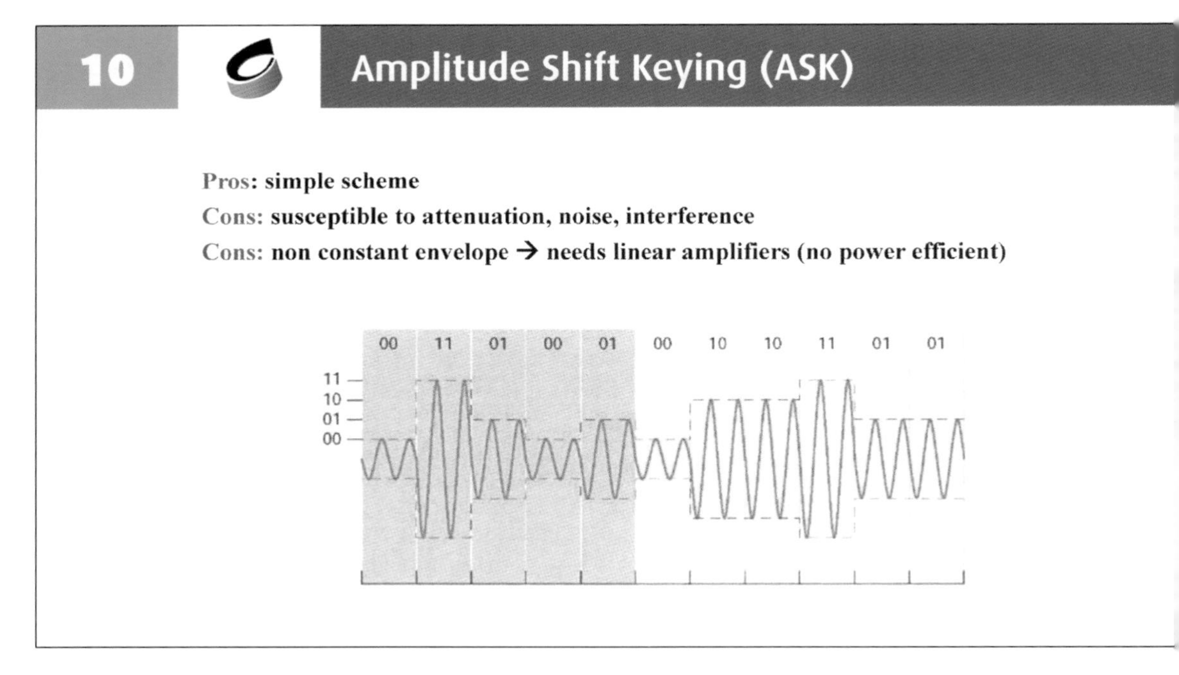

Modulation is the process of superimposing a low-frequency signal (information) onto a high-frequency signal (RF carrier). Modulation of the carrier through digital signals is known as shift keying. In ASK, the Amplitude of the carrier is varying.

11  On-Off Keying (OOK)

Pros: simple scheme, low power consumption

Cons: lack of synchronization

0 → transmitter amplifier OFF

1 -> transmitter amplifier ON

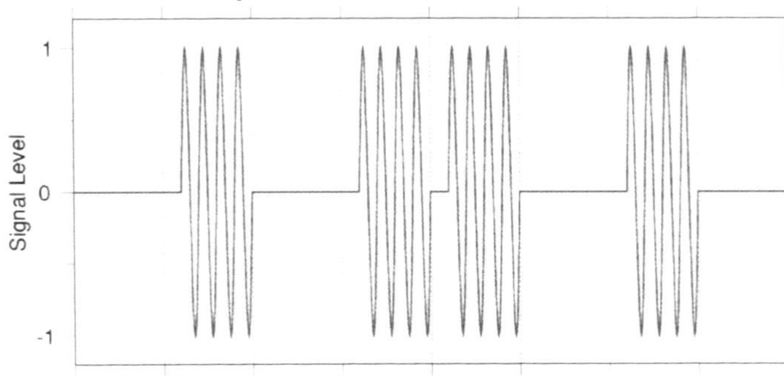

A special case of ASK to reduce power consumption is OOK (On-Off Keying), although missing synchronization.

12  Frequency Shift Keying (FSK)

Pros: less susceptible to noise

Constant envelope → relaxed linearity requirements for the TX amplifier

Cons: theoretically requires larger bandwidth than ASK

Binary FSK (BFSK) as basic

Gaussian FSK (GFSK) used in Bluetooth thanks to better spectral density than BFSK

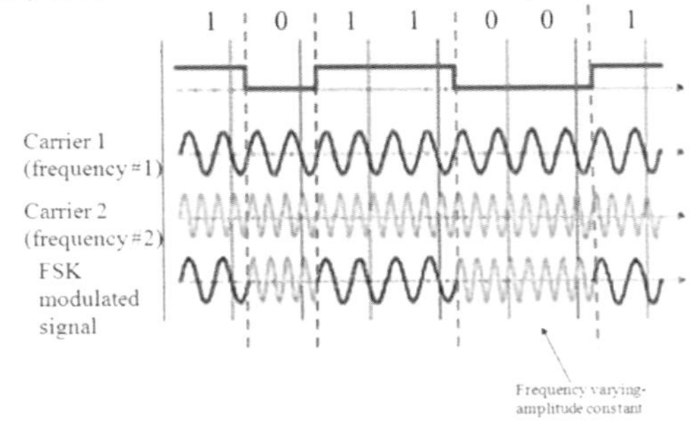

In angular modulations, frequency of the carrier (FSK) or phase of the carrier (PSK) is varying.

13

Phase Shift Keying (PSK)

Pros: **less susceptible to noise, bandwidth efficient**

Cons: **requires synchronization in frequency and phase, thus complicating transmitter and receivers**

Constant envelope → relaxed linearity requirements for the transmitter amplifier

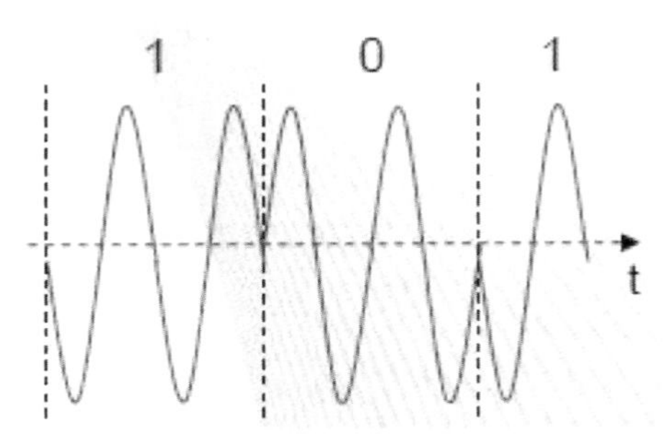

PSK is the modulation scheme used in IEEE 802.15.4 and Zigbee.

14

Quadrature Amplitude Modulation (QAM)

Modulation in amplitude and phase

(e.g. 8 QAM: 0, 90^0, 180^0, 270^0 in phase; VL or VH in amplitude)

Pros: **bandwidth efficient**

Cons: **non-constant envelope (needs linear amplifier), complex scheme**

Baud: transmitted symbol, 1 Baud may include multiple bits

(e.g. bit rate is 3x baud rate in 8 QAM)

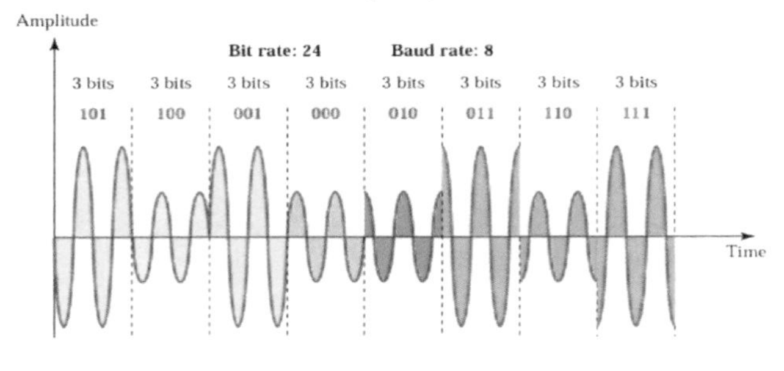

To increase the spectrum efficiency, i.e. to increase the transmitted data-rate (bits/s) for a given spectrum, phase and amplitude modulations have to be used (large QAM formats).

15 Spectral efficiency of modulations

Modulation format	Theoretical bandwidth efficiency limits
MSK	1 bit/second/Hz
BPSK	1 bit/second/Hz
QPSK	2 bits/second/Hz
8PSK	3 bits/second/Hz
16 QAM	4 bits/second/Hz
32 QAM	5 bits/second/Hz
64 QAM	6 bits/second/Hz
256 QAM	8 bits/second/Hz

Large modulation formats entail also reduced robustness vs. interferences and channel impairments.

16 Spread spectrum

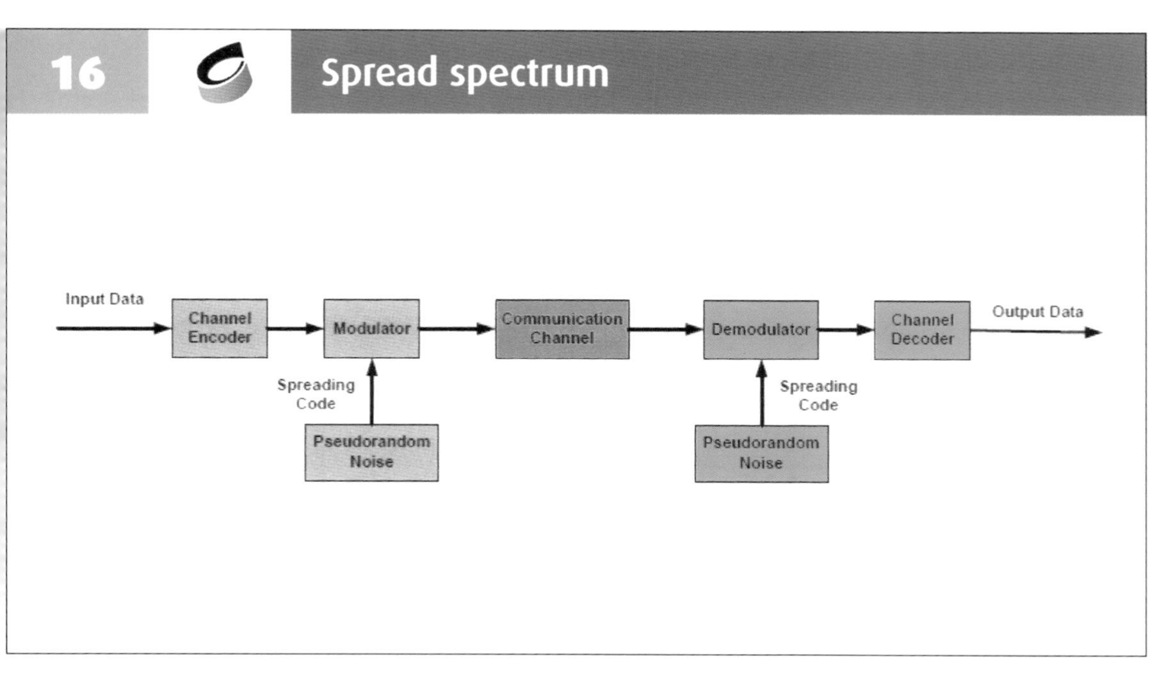

Data sent using Spread Spectrum are intentionally spread over a wide frequency range. Since it appears as noise, the signal is difficult to detect and jam. Thanks to spread spectrum, the communication is resistant to noise and interference, thus increasing the probability that the signal will be received correctly. Moreover, it is unlikely to have interference with other signals.

- Each bit represented by multiple bits using spreading code
- Spreading code spreads signal across wider frequency band
- Good resistance against interferers

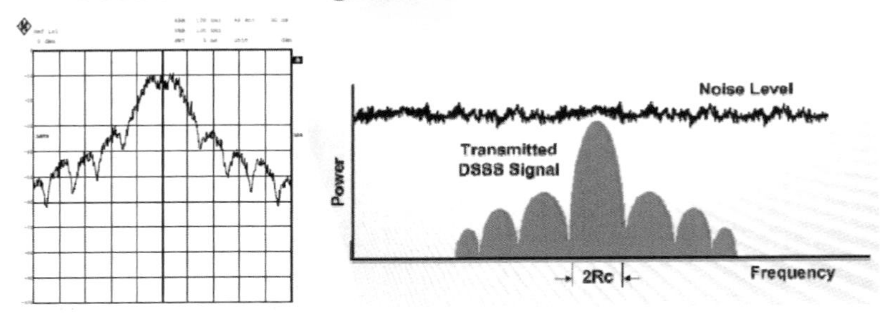

2 types of Spread Spectrum techniques are common in ISM bands: direct sequence spread spectrum (DSSS) and frequency hopping spread spectrum (FHSS).

18  Frequency Hopping Spread Spectrum (FHSS)

- Signal broadcast over a seemingly random series of frequencies
- Receiver hops between frequencies in sync with transmitter
- Jamming on one frequency affects only a few bits

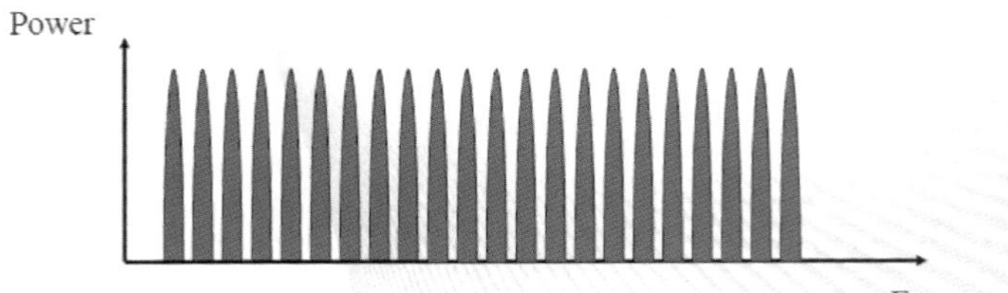

In FHSS, the signal is broadcasted over a seemingly random series of frequencies. Receiver hops between frequencies in sync with the transmitter. Jamming on one frequency affects only a few bits.

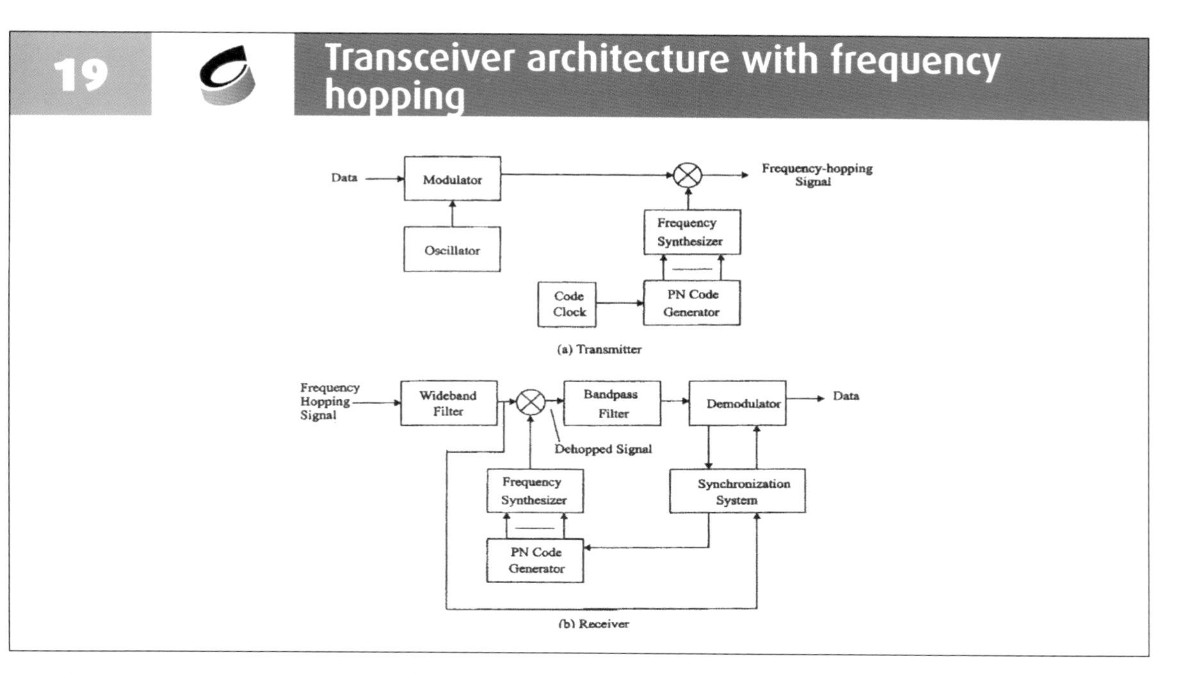

Implementing Spread Spectrum entails more HW complexity for the transceiver, particularly for the frequency generation at transmitter side and for the receiver synchronization.

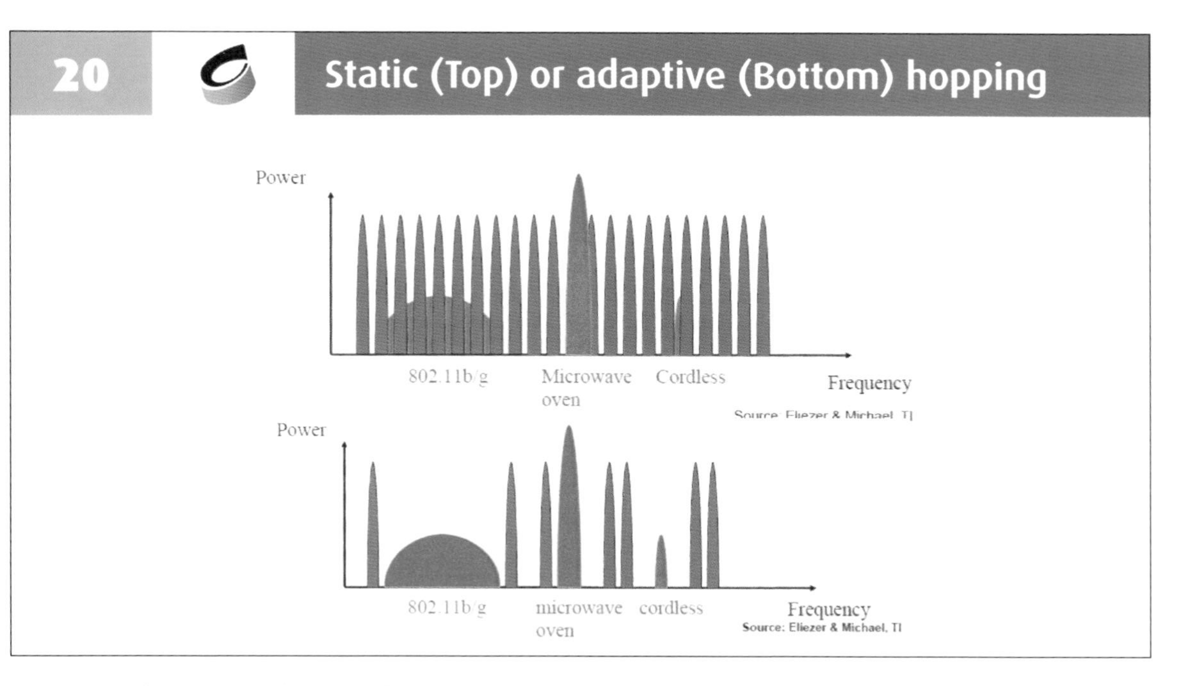

Frequency hopping can be static (utilize a pre-determined set of frequencies with either a repeating hop pattern or a pseudorandom hop pattern, e.g. Bluetooth v 1.0 or 1.1) or adaptive (scan the entire band at start-up and restrict the usage to frequencies with the lowest energy content, as in Bluetooth v1.2 or v2.0).

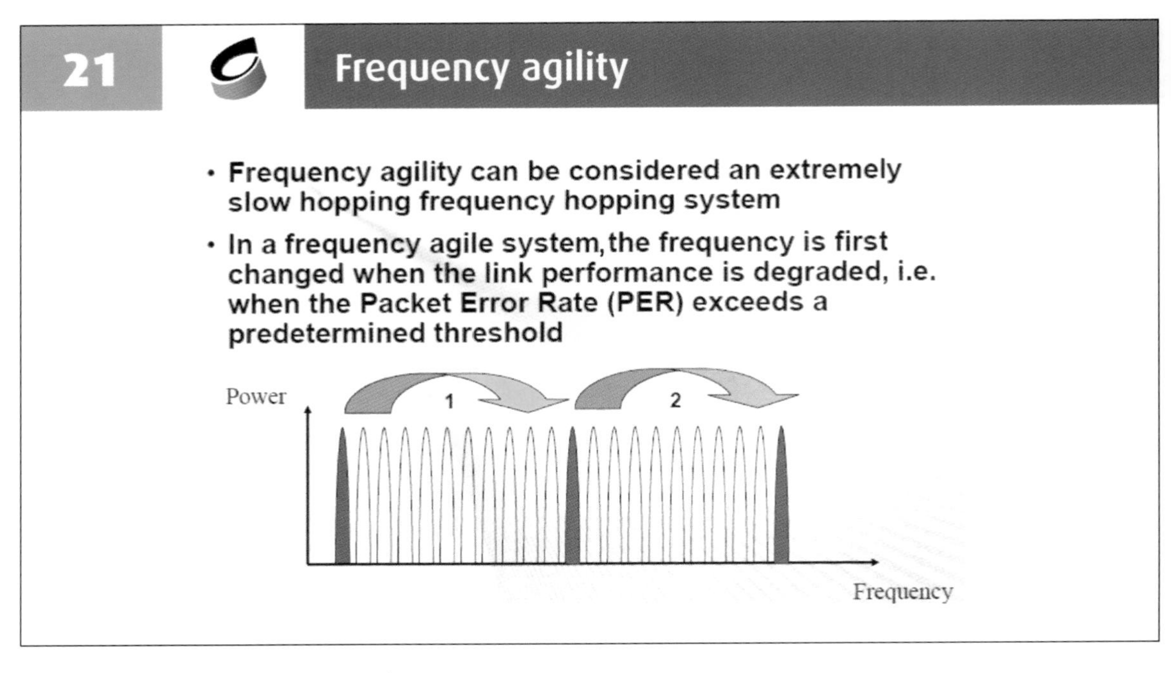

Evolution is toward frequency agile systems.

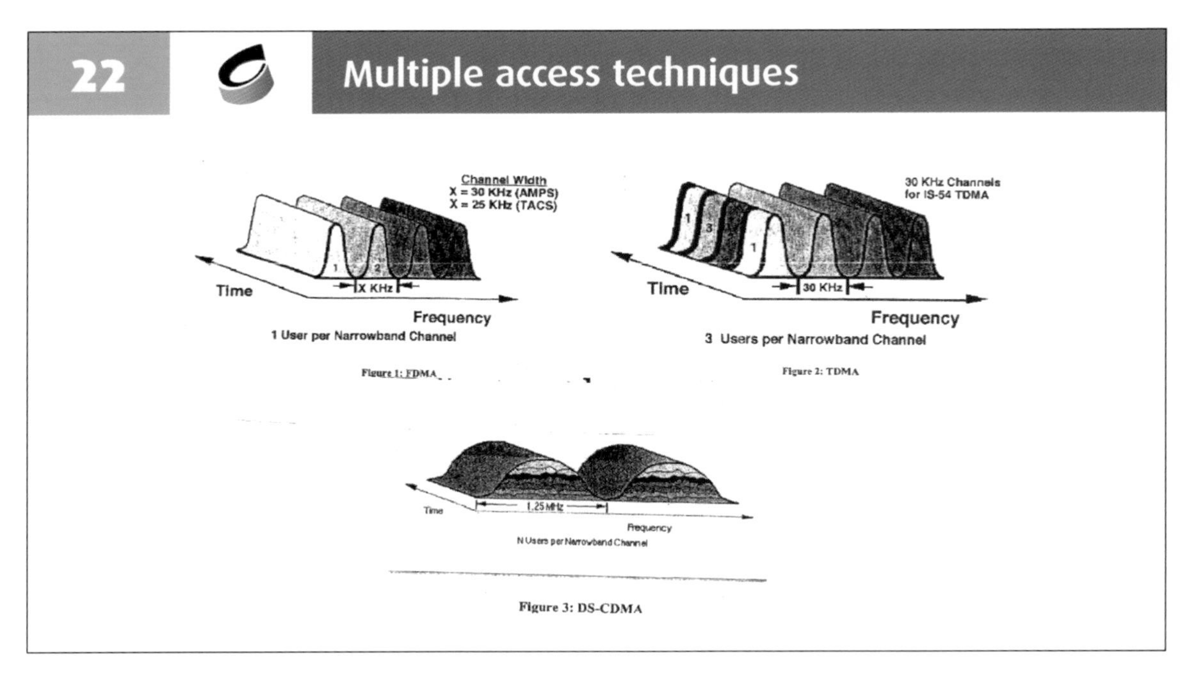

Multiplexing techniques based on Time Division (TDMA), Frequency Division (FDMA), Code Division (CDMA) or Space Division (SDMA) Multiple Accesses.

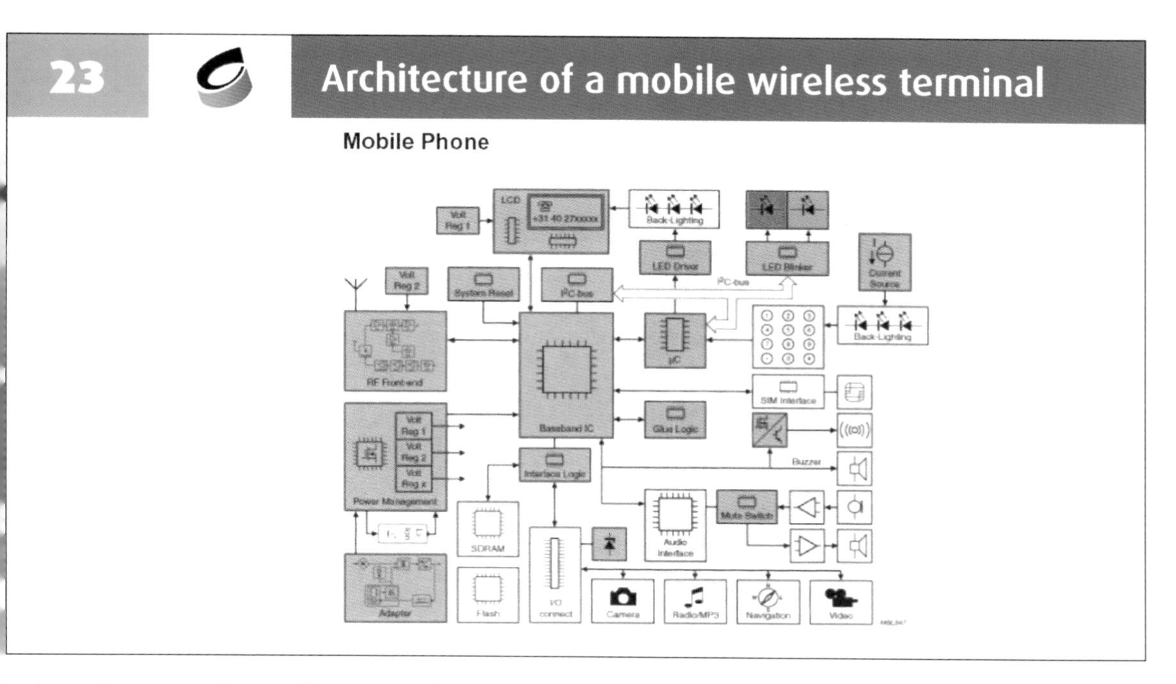

The architecture of a wireless terminal includes the RF front-end, the power management circuitry, the baseband computing and memory units, the audio-video user interface, and dedicated peripherals for GPS, video cameras and HI-FI audio reproduction. The focus of this chapter is on RF front-end.

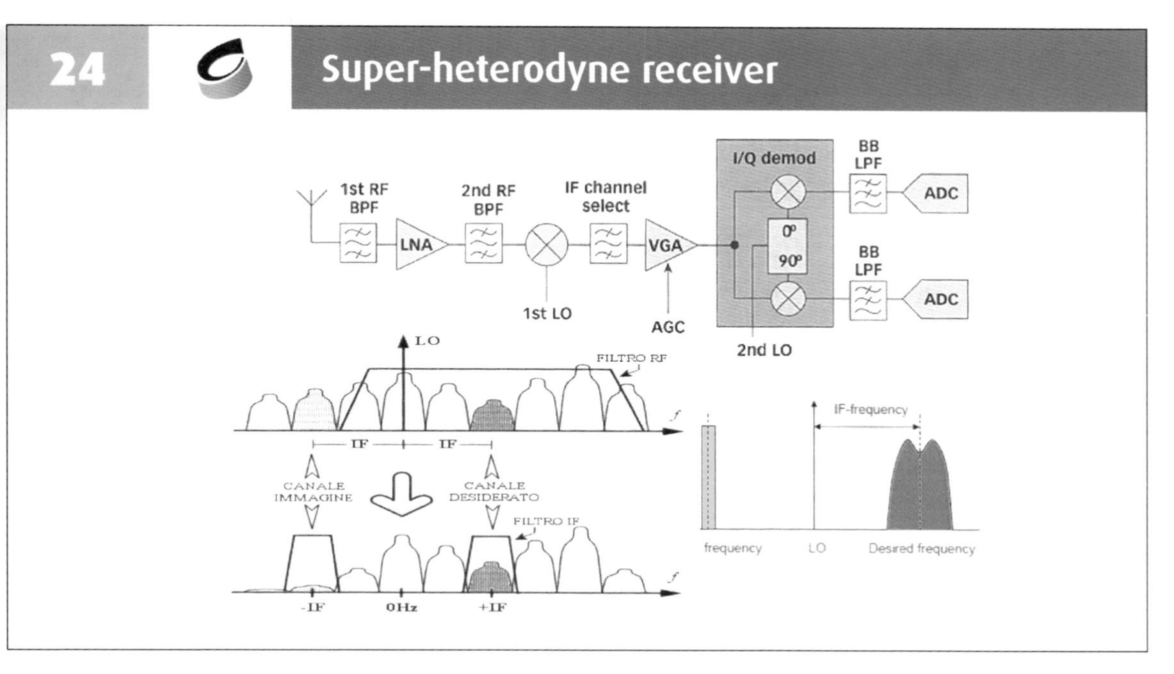

Pros/Cons of super-heterodyne transceiver.

Cons: Needed an external filter for image rejection; Stages driving external components need high current. High-Q LC or SAW filters need off-chip devices.

Pros: High channel selectivity and receiver sensitivity.

25 Double-conversion super-heterodyne receiver

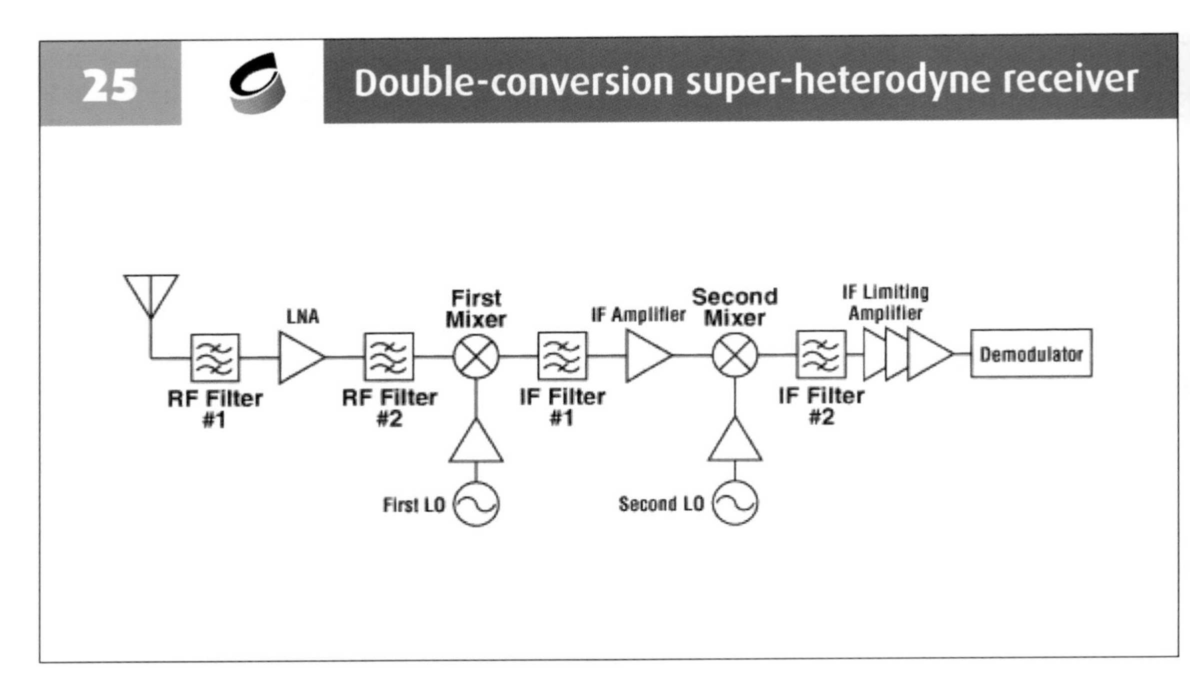

Double-conversion heterodyne receiver can be used to ease the rejection of the "image" channel, but the transceiver complexity and the needs of off-chip component increase.

26 Transceiver with super heterodyne receiver

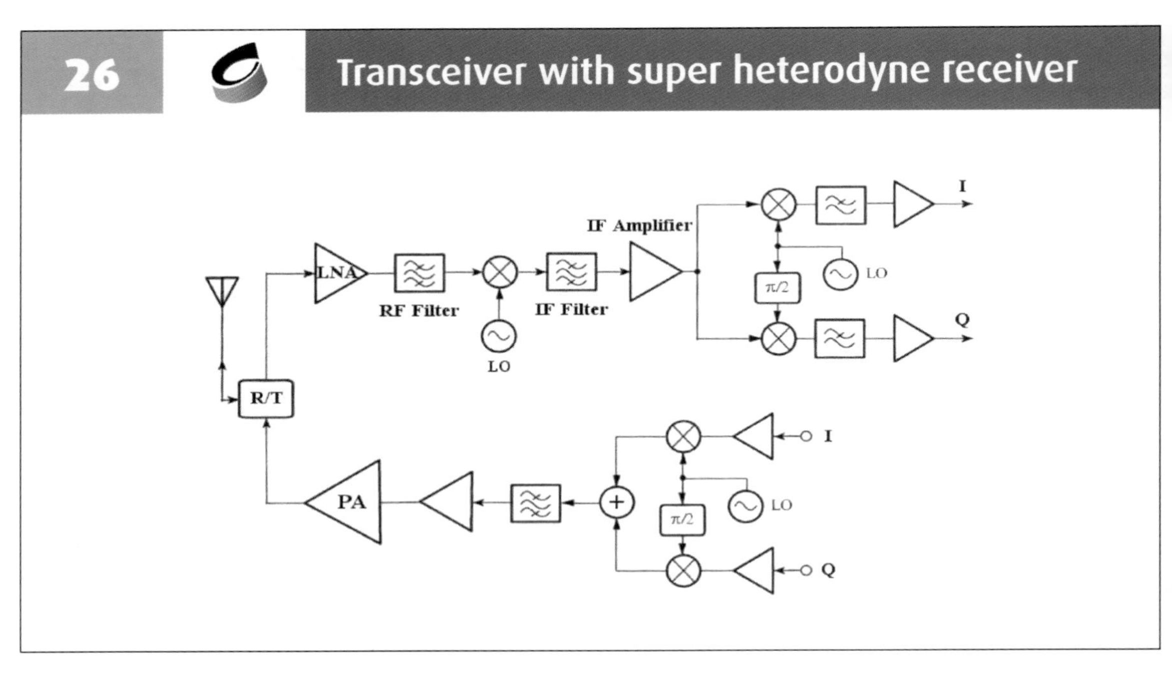

Example of a complete integrated transceiver with a TX/RX antenna switch, homodyne transmitter, and heterodyne receiver.

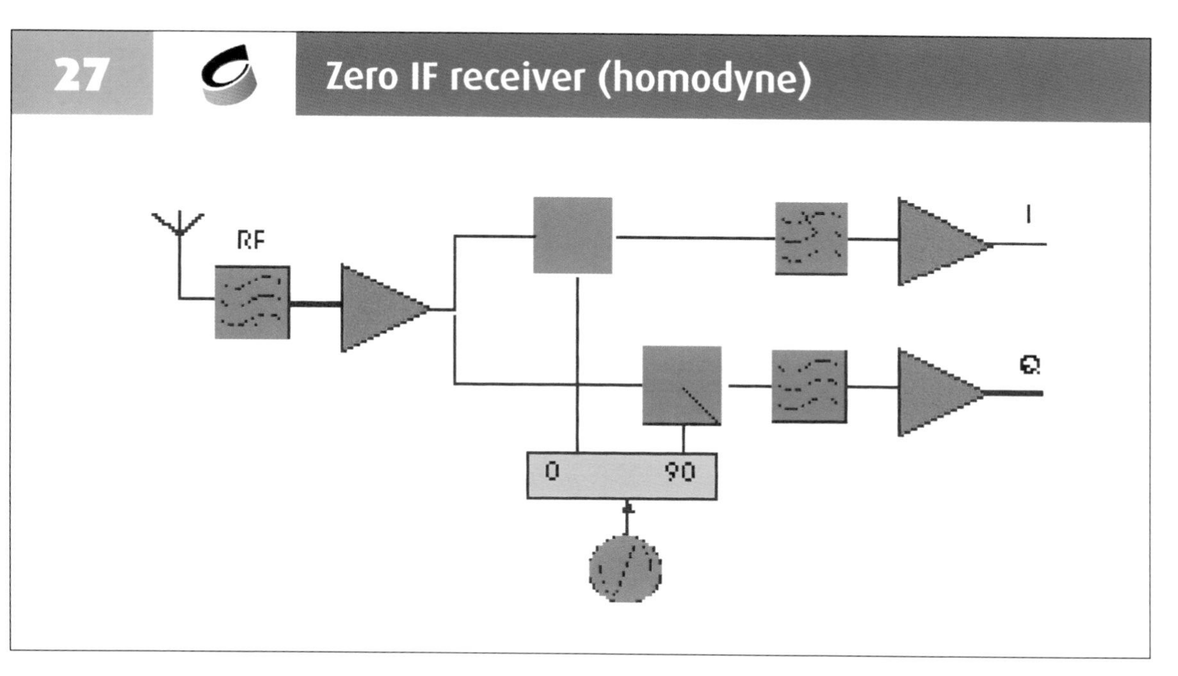

Pros/Cons of Homodyne transceivers.

Pros: Low-power; Compact size; No image frequency.

Cons: DC Offset.

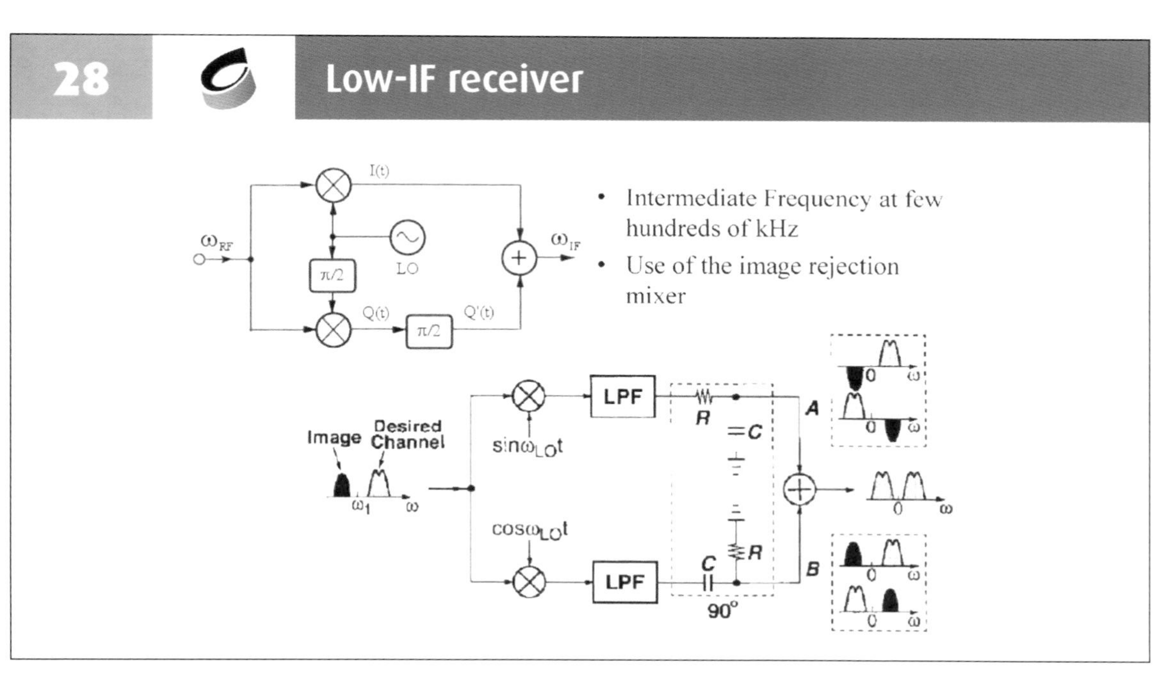

Image rejection mixer enables a new receiver architecture with reduced needs of off-chip components.

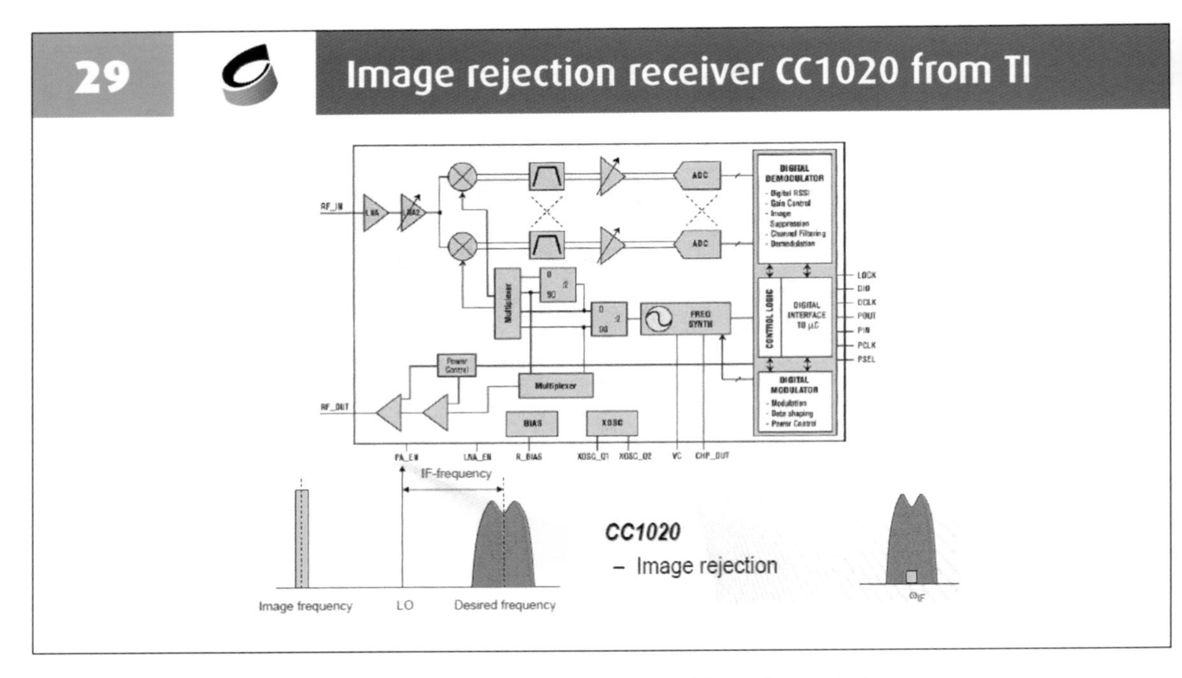

Integrated RF transceivers with image rejection receiver available in the market.

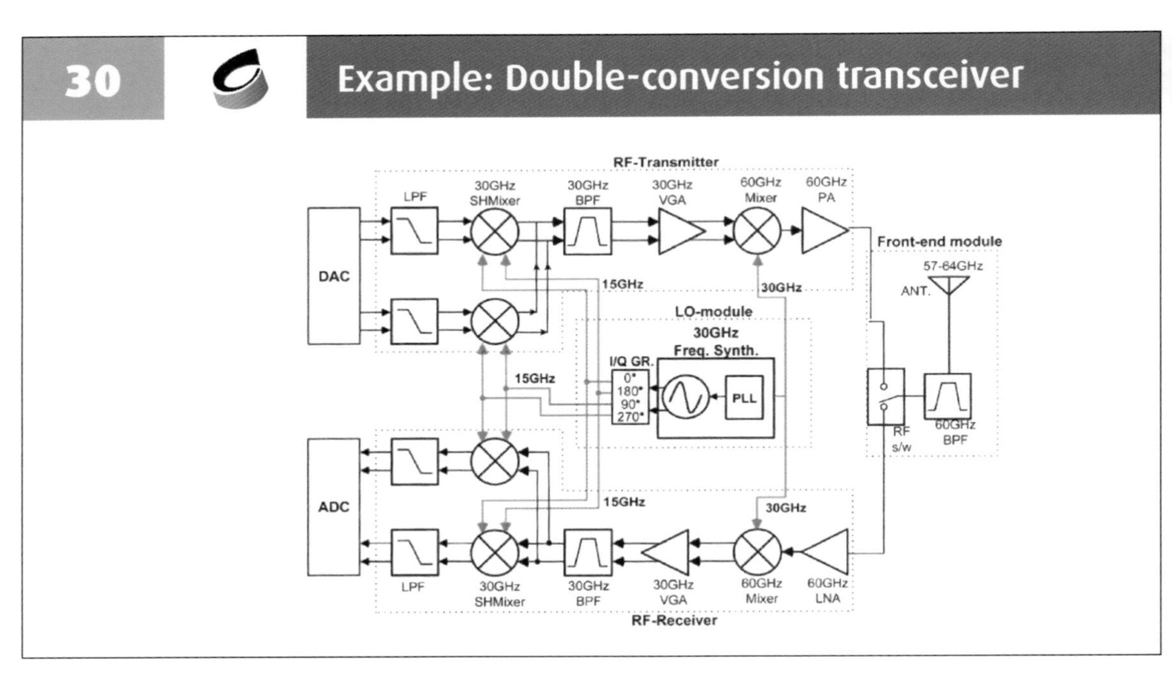

Example of double-conversion transceiver, Lee et al., IEEE Trans. MTT 2008.

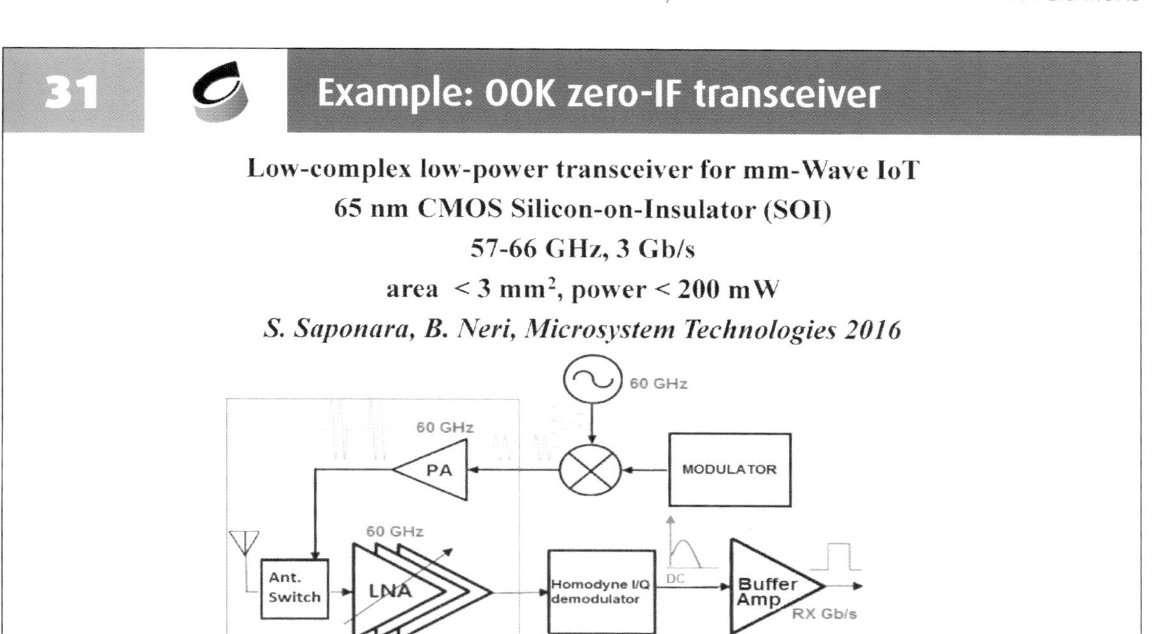

Example of OOK zero-IF transceiver, S. Saponara, B. Neri, Microsystem Technologies 2016.

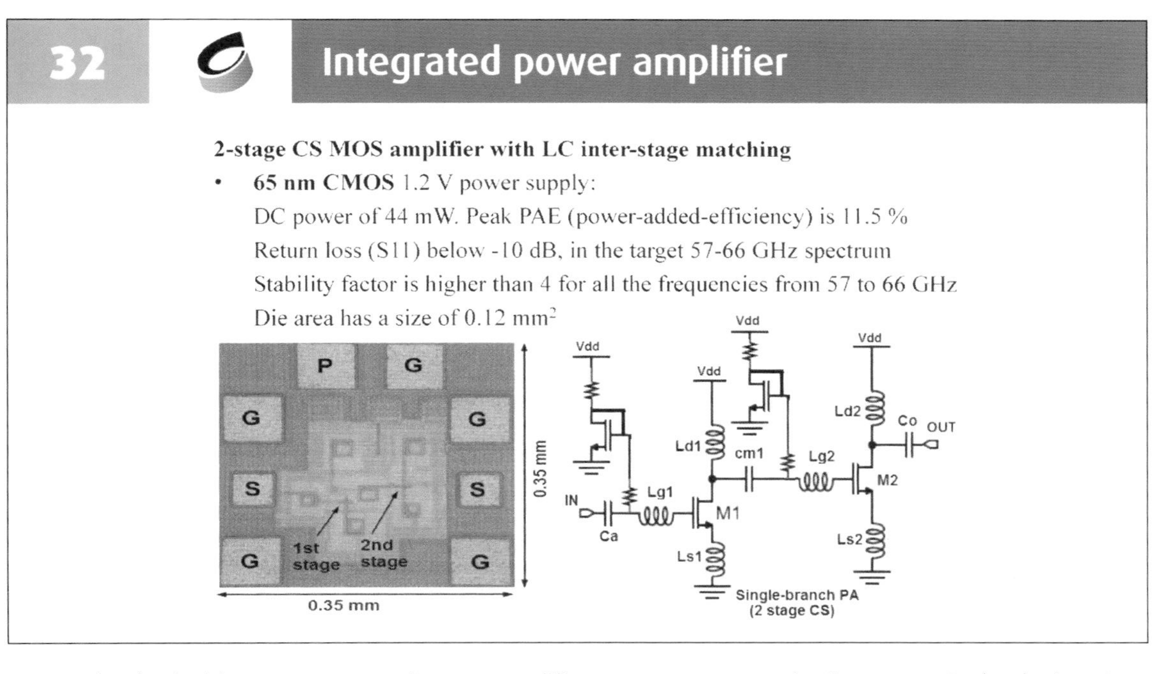

Example of a double-stage integrated power amplifier, S. Saponara, B. Neri, Microsystem Technologies 2016.

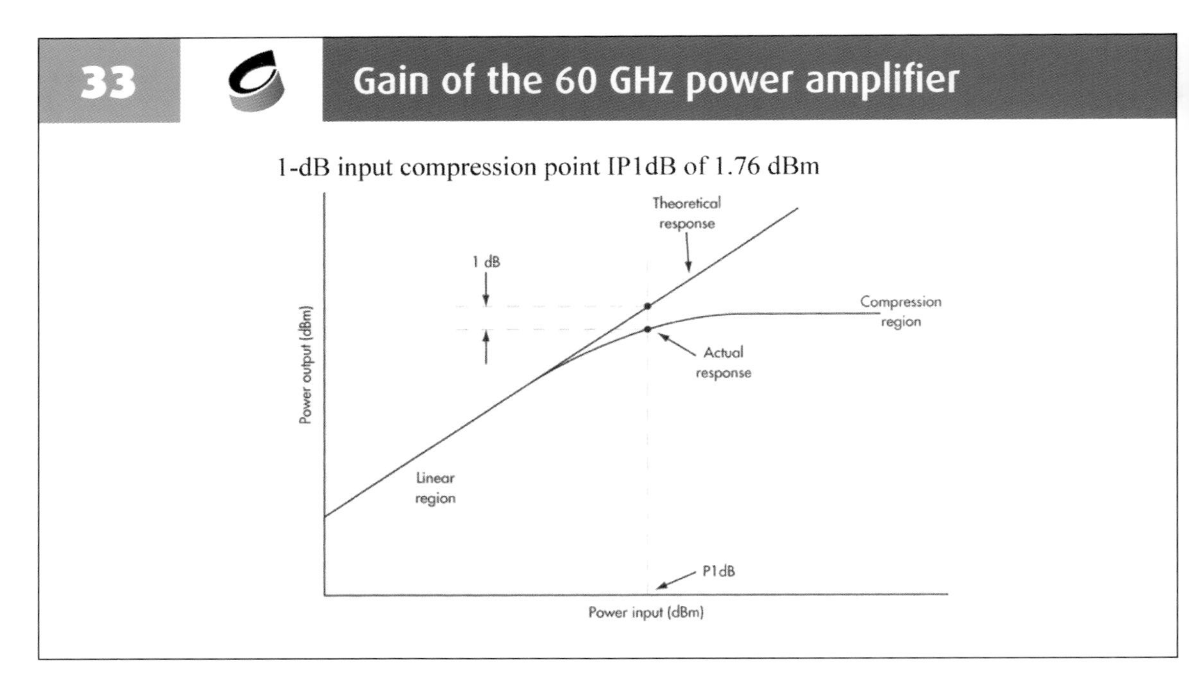

Example of a double-stage integrated power amplifier, S. Saponara, B. Neri, Microsystem Technologies 2016.

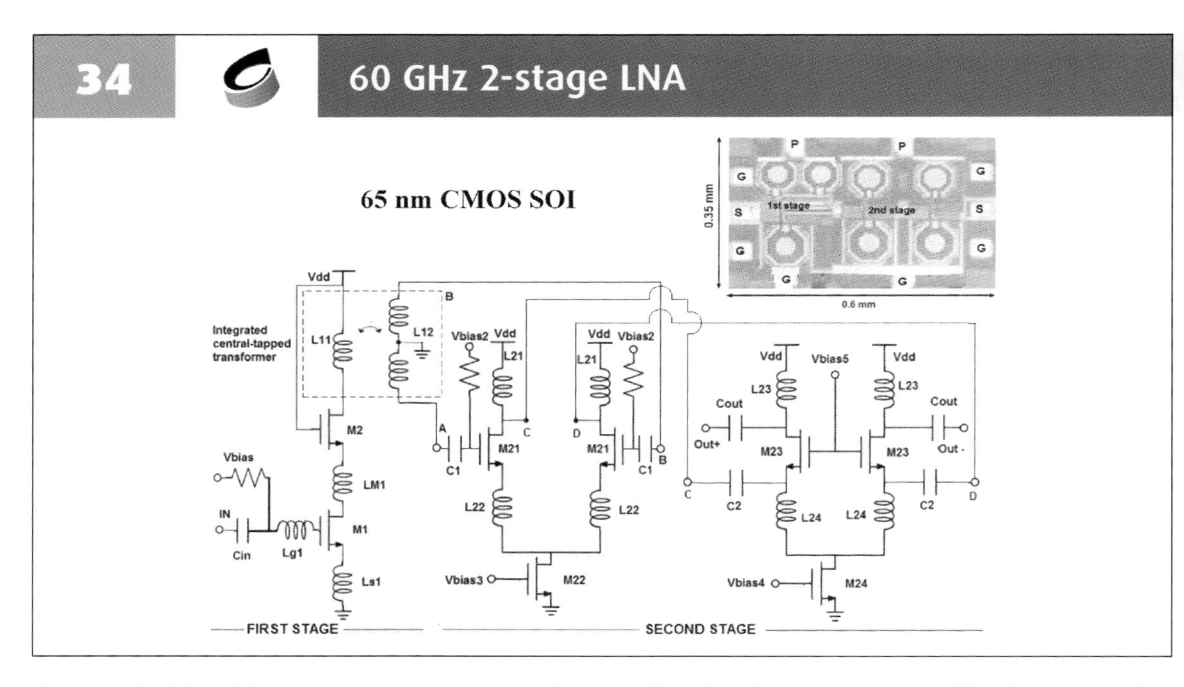

RF key building blocks are Integrated Low-Noise Amplifier (LNA) and Power Amplifier (PA), Cascading amplifiers with boosting performance of RF IC with off-chip PA and LNA; Balun &Matching network, Antenna. The slide reports an example of a 2-stage LNA circuit at mm-wave in CMOS technology.

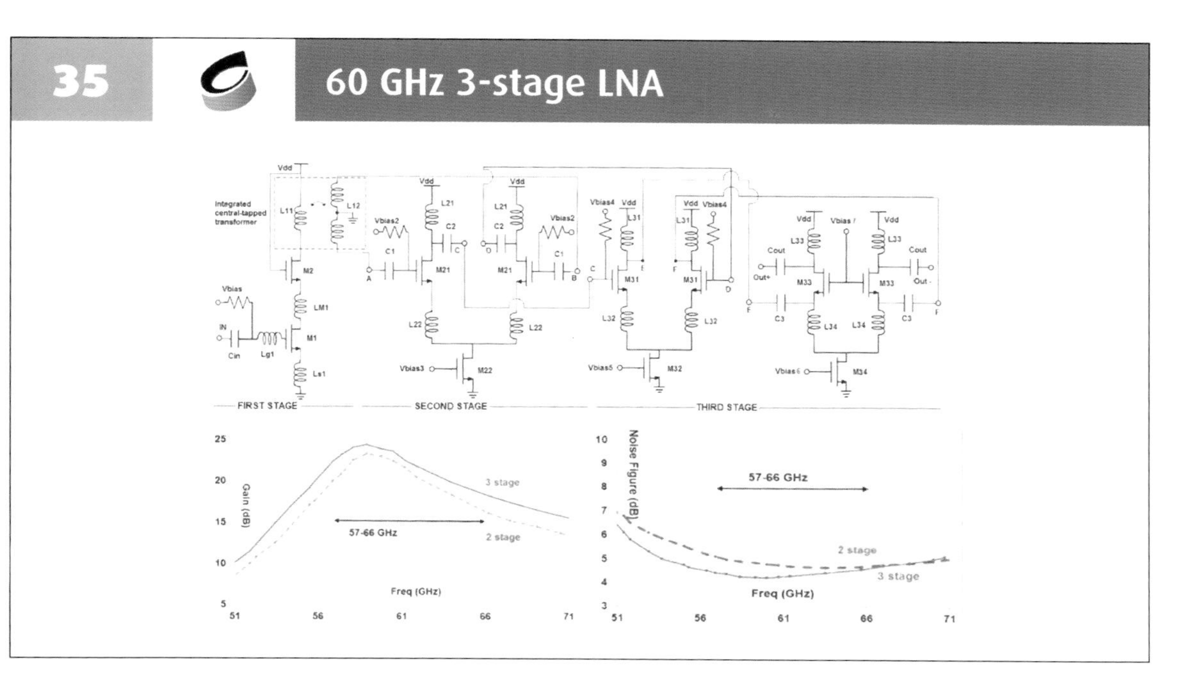

The slide reports an example of a 3-stage LNA circuit at mm-wave in CMOS technology.

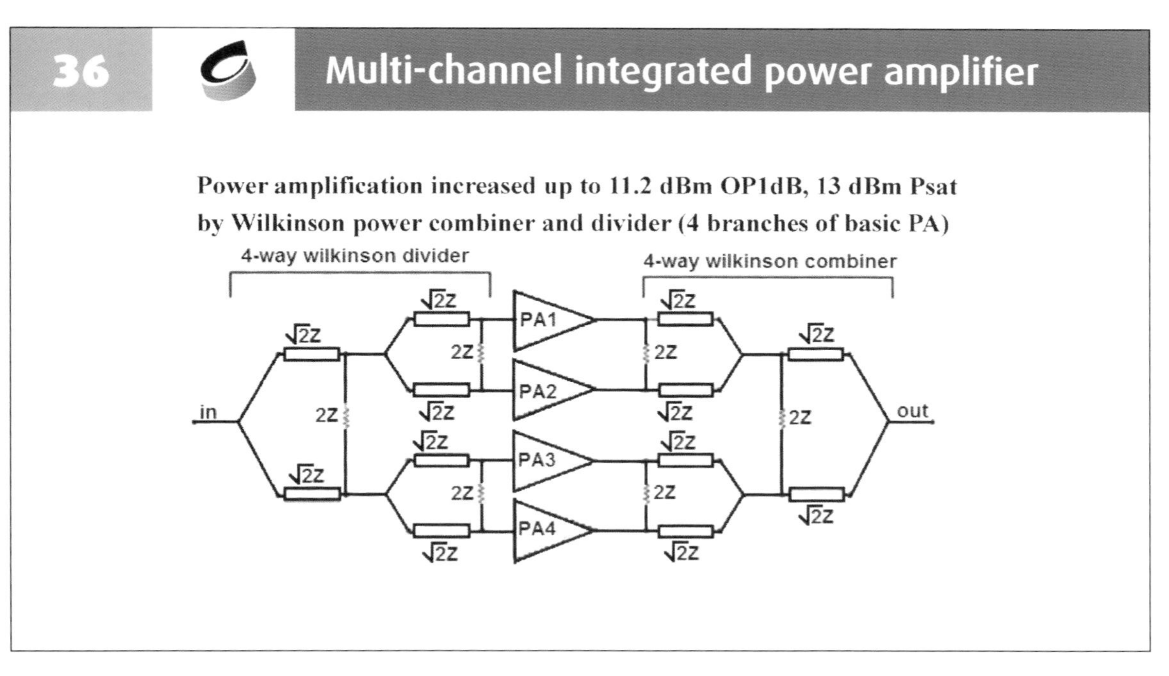

Multiple amplifier stages can be combined at TX and RX sides to improve amplifier performances.

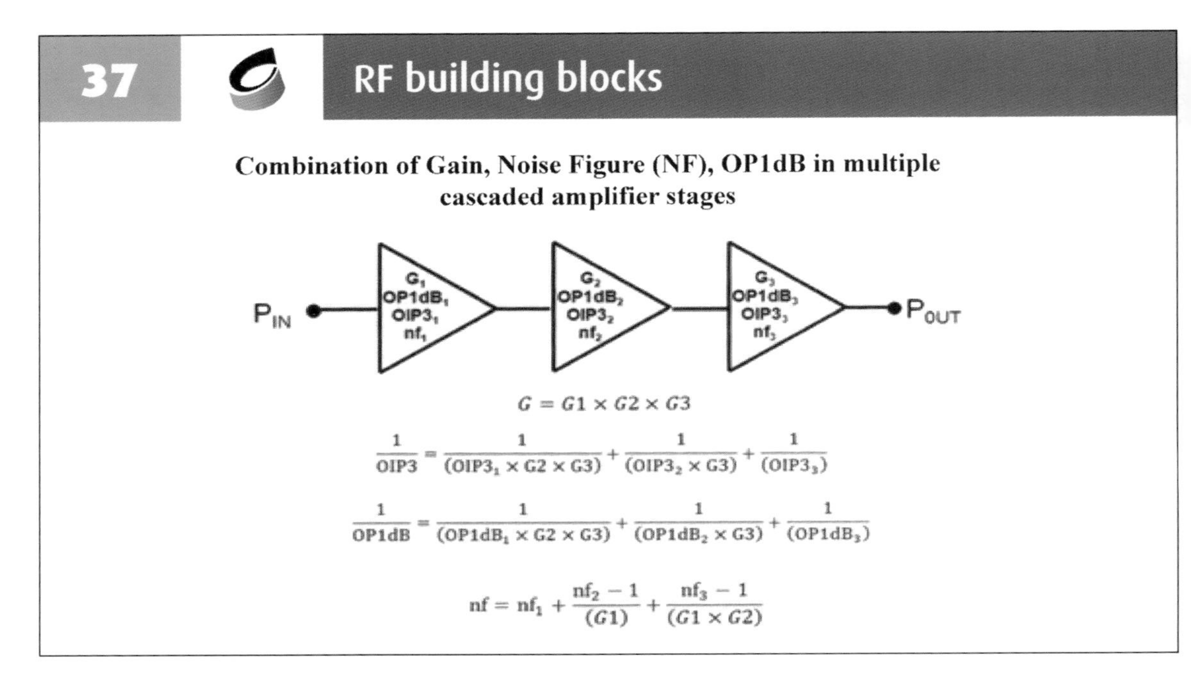

Increase maximum output power or achieve better noise figure (NF) performance.

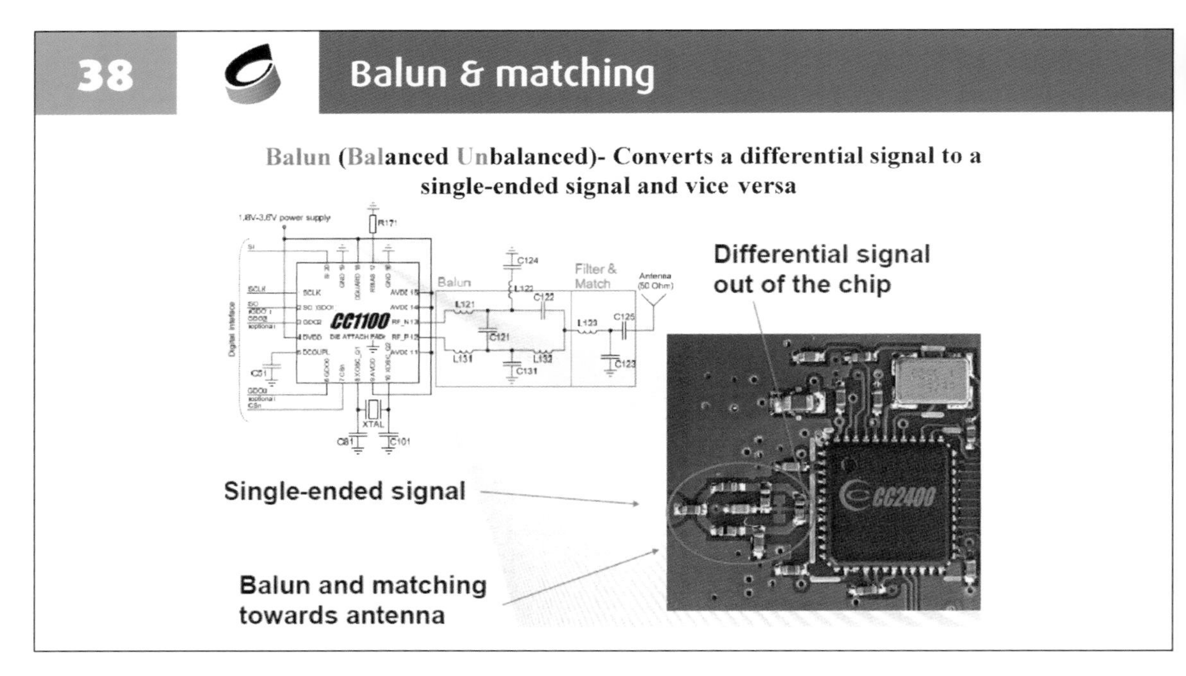

The balun, realized through microwave circuits on board, adapt differential to single-ended stages.

PCB antennas

Commonly used antennas:

- **PCB antennas**
  - Little extra cost (PCB)
  - Size demanding at low frequencies
  - Good performance possible
  - Complicated to make good designs

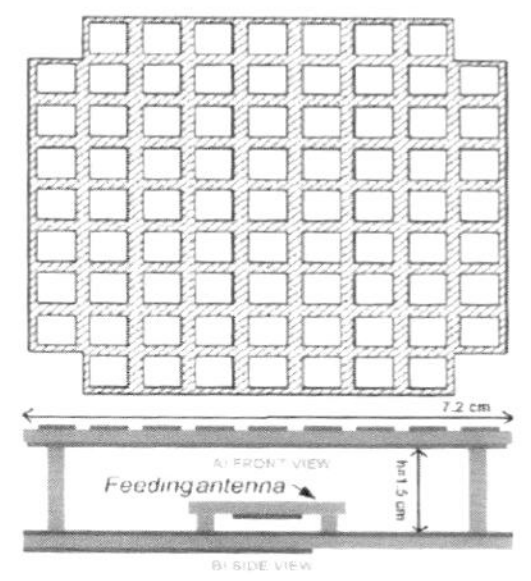

Example of antenna using printed circuit board technology.

40

Boosting performance of RF ICs

1. **Increase the Output power**
   - Add an external Power Amplifier (PA)
2. **Increase the sensitivity**
   - Add an external Low Noise Amplifier (LNA)
3. **Increase both output power and sensitivity**
   - Add PA and LNA
4. **Use high gain antennas**
   - Regulatory requirements need to be followed

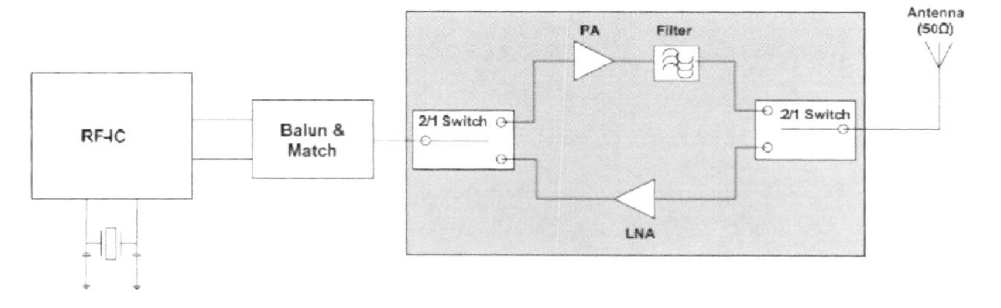

Performance of ICs can be boosted at board level by using off-chip LNA to improve sensitivity and NF of the receive or off-chip PA to improve the maximum transmitted power.

41  RF ICs with off-chip PA

CC2420EM PA DESIGN

- **Signal from TXRX_Switch pin level shifted and buffered**
  - Level in TX: 1.8 V, level for RX and all other modes: 0V
- **CMOS and GaAs FET switches assures low RX current c**
- **Simpler control without external LNA**
  - No extra signal is needed from MCU to turn off LNA in low powe

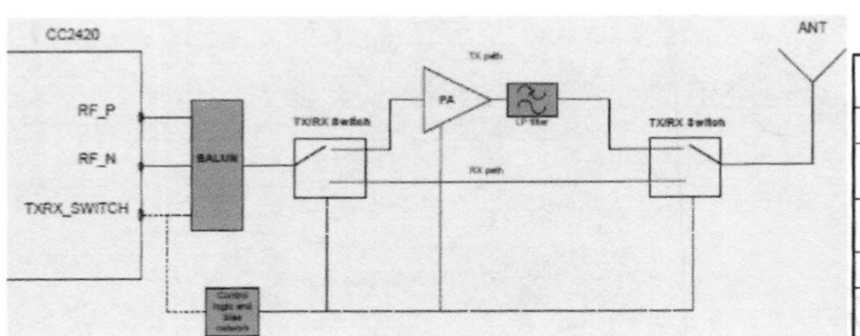

	CC2420EM	CC2420EM w/PA
TX current	17.4 mA	30.8 mA
RX current	19.7 mA	19.7 mA
Output power	0 dBm	9.5 dBm
Sensitivity	-94 dBm	-93.1 dBm
Line of Sight Range	230 meter	580 meter

Example from Texas Instruments.

42  RF link budget analysis

$$P_r = P_t + G_t + G_r + 20\log\left(\frac{\lambda}{4\pi}\right) - 20\log d \quad \text{or} \quad P_r = \frac{P_t G_t G_r \lambda^2}{(4\pi)^2 d^2}$$

- P_t is the transmitted power, P_r is the received power
- G_t is the transmitter, G_r is the receiver antenna gain
- Lambda is the wavelength
- D is the distance between transmitter and receiver, or the range

$$SNR = \frac{P_r}{K \cdot T \cdot NF \cdot BW}$$

BER for a given SNR or Eb/No ratio can be reduced through channel coding techniques to be implemented in the baseband digital domain.

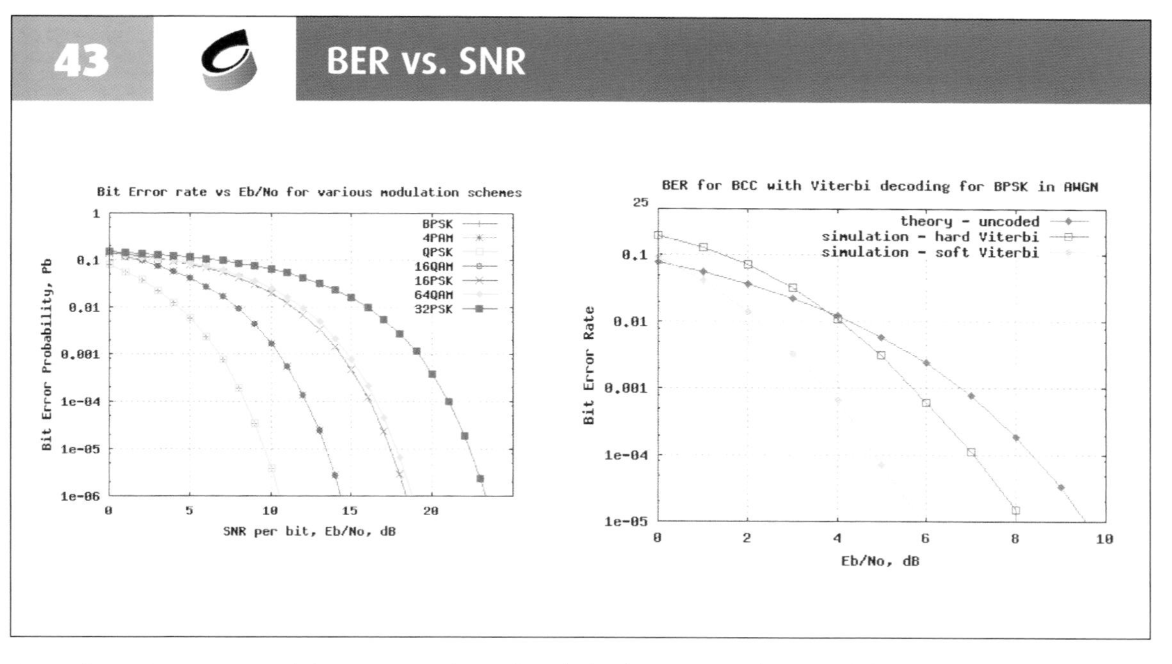

BER for a given SNR or Eb/No ratio can be reduced also by proper selection of the modulation scheme.

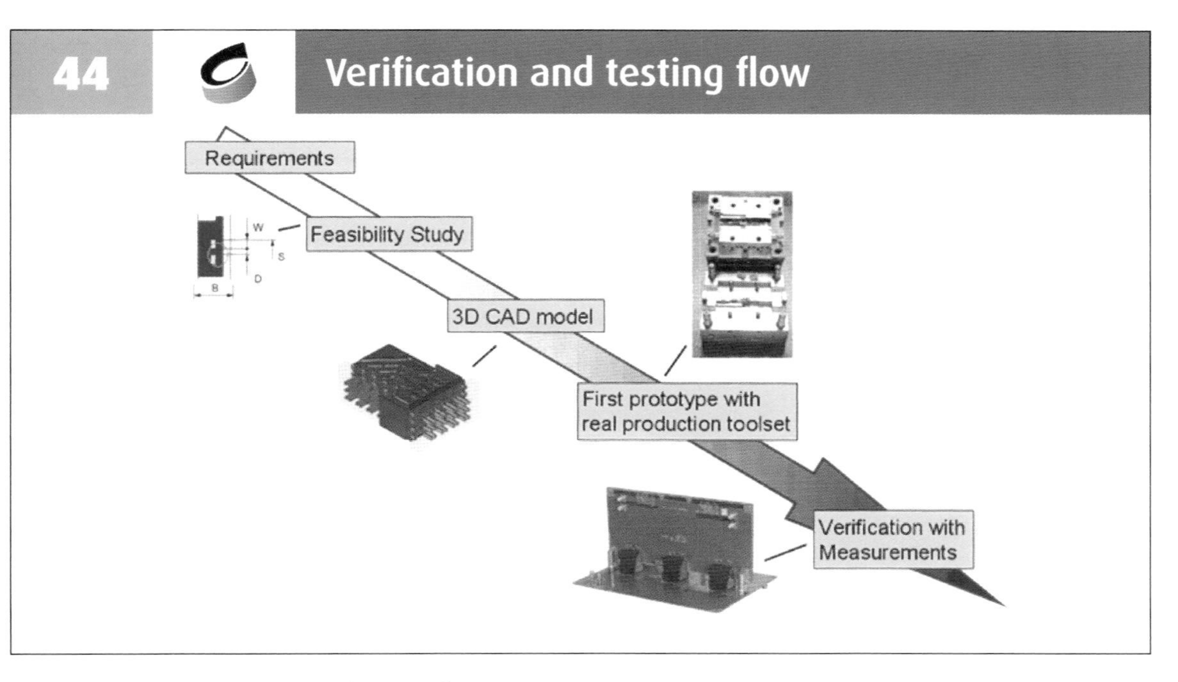

Example of the verification and testing flow.

45 

Equipments for RF measurements

- **Vector Network Analyzers**
  - Component Characterisation – insertion loss
  - S-parameters - matching
- **Spectrum Analyzers**
  - Output Power, harmonics, spurious emission
  - Phase Noise
  - ACP
  - OBW
  - Modulation - deviation
- **Signal Generators**
  - Sensitivity (BER option needed)
  - Selectivity/blocking
  - Two-tone measurements – IP3
- **Power Meters**
  - Output Power – calibration
- **Oscilloscopes**
  - Digital signal analysis
- **Function and Arbitrary Waveform Generators**

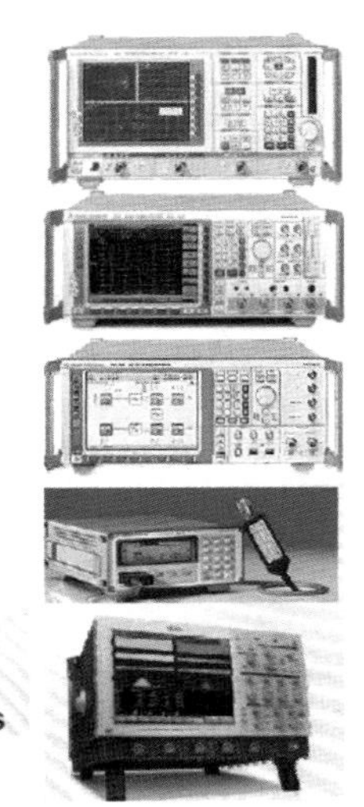

Instrumentation for RF measurements.

46 

References

1. Dag Grini, "RF basics", *MPS430 advanced technical conference* 2006
2. S. Saponara, B. Neri, "mm-wave integrated wireless transceiver: enabling technology for high bandwidth short-range networking in cyber physical systems", *Microsystem Technologies*, 22 (7), 2006
3. B. Neri, S. Saponara, "Advances in technologies, architectures, and applications of highly-integrated low-power radars", *IEEE Aerospace and Electronic Systems*, 27 (1), 2012
4. Ya Jol Lee et al., "A 28.5–32-GHz Fast Settling Multichannel PLL Synthesizer for 60-GHz WPAN Radio", *IEEE Tran. on Microwave Theory and Techniques* 2008
5. S. Saponara, F. Giannetti, B. Neri, G. Anastasi, "Exploiting mm-Wave Communications to Boost the Performance of Industrial Wireless Networks", *IEEE Trans. on Ind. Informatics*, 13, (3), 2017
6. S. Saponara, B. Neri, "Radar Sensor Signal Acquisition and Multi-dimensional FFT Processing for Surveillance Applications in Transport Systems", *IEEE Transactions on Instrumentation and Measurement*, 66 (4), 2017
7. S. Saponara, F. Giannetti, B. Neri, "Design Exploration of mm-Wave Integrated Transceivers for Short-Range Mobile Communications Towards 5G", *Journal of Circuits, Systems and Computers*, 26 (4), 2017

8. S. Saponara, B. Neri, "Design of Compact and Low-power X-band Radar for Mobility Surveillance Applications", *Computers and Electrical Engineering*, 56, 2016
9. D. Davalle, R.. Cassettari, S. Saponara, L. Fanucci, L. Cucchi, F. Bigongiari, W.Errico, "Design, Implementation and Testing of a Flexible Fully-digital Transponder for Low-earth Orbit Satellite Communications", *Journal of Circuits, Systems and Computers*, 23 (10), 2014
10. S. Saponara, G. Ciarpi, B. Neri, "System-level modelling/analysis and LNA design in low-cost automotive technology of a V2X wireless transceiver", *IEEE RTSI* 2017
11. S. Saponara, F. Giannetti, "Radio link design framework for W SN deployment and performance prediction", *Proceedings of SPIE Microtechnologies*, vol. 10246, Barcelona, 8-10 May, 2017
12. S. Saponara, F. Giannetti, B. Neri, "Design exploration for millimeter-wave short-range industrial wireless communications", *IEEE IECON 2016*, pp. 6038-6043
13. S. Saponara, B. Neri, "Radar sensor signal acquisition and 3D FFT processing for smart mobility surveillance systems", *IEEE Sensors Applications Symposium (SAS)*, pp. 417-422, 2016
14. S. Saponara, B. Neri, "mm-Wave Integrated Wireless Transceivers: Enabling Technology for High Bandwidth Connections in IoT", *IEEE WF-IOT conference 2015*, pp.149 – 153
15. S. Saponara, B. Neri, "Gbps wireless transceivers for high band width interconnections in distributed cyber physical systems", *SPIE Microtech., Smart sensors, actuators and MEMS VII and Cyber Physical System*, vol. 9517, 95172L, pp. 1-7, 2015
16. S. Saponara, L. Mattii, B. Neri, F. Baronti, F., L. Fanucci, "Design of a 2 Gb/s transceiver at 60 GHz with integrated antenna in bulk CMOS technology", *IEEE EuMIC* 2014,
17. A. Fonte, S. Saponara, G. Pinto, L. Fanucci, B. Neri, "Design of a Low Noise Amplifier with Integrated Antenna for 60 GHz Wireless Communications", *2011 IEEE MTT-S Int. Microwave Workshop Series on Millimeter Wave Integration Technologies, IMWS 2011*. pp. 160-163
18. A. Fonte, S. Saponara, G. Pinto, B. Neri, "Feasibility study and on-chip antenna for fully integrated μRFID tag at 60 GHz in 65 nm CMOS SOI", *IEEE Int. Conference on RFID-Technologies and Applications (RFID-TA)* 2011. pp. 457-462,
19. A. Fonte, S. Saponara, G. Pinto, L. Fanucci, B. Neri, "60-GHz single-chip integrated antenna and Low Noise Amplifier in 65-nm CMOS SOI technology for short-range wireless Gbits/s applications", *IEEE Conference on Applied Electronics* 2011, pp. 127-132,

CHAPTER 04

ICs and VLSI Architectures for mm-wave/RF Wireless Transceivers in IoT Applications (Remote Sensing)

Sergio Saponara

University of Pisa, Italy

1. Agenda 92
2. Ubiquitous Radar applications 92
3. Ubiquitous Radar design needs 93
4. Ubiquitous Radar design needs 93
5. Radar vs. LIDAR 94
6. Radar vs. LIDAR 94
7. FMCW Radar acquisition & processing@ UniPisa 95
8. FMCW Radar acquisition & processing@ UniPisa 95
9. Specification for the Radar 96
10. X-band Radar transceiver 96
11. Linear-FMCW waveform: moving target 97
12. Frequency analysis for 2D range–doppler 97
13. Received SNR vs. Pcw 98
14. Fabry-Perot resonating antenna 98
15. Microwave board Radar prototype 99
16. X-band Radar transceiver 99
17. 2D FFT processing plus a 3rd FFT along the 4 RX channels for azimuth estimation and peak estimation 100
18. Fast Fourier Transform digital architecture 100
19. CA-CFAR circuit for peak detection 101
20. Baseband processing implementation 101
21. Range-Doppler map 102
22. Example of a network of Radar for crossing monitoring 102
23. Real-time level crossing monitoring 103
24. Car parking real-time monitoring 103
25. V-band automotive radar 104
26. V-band automotive radar 104
27. UWB pulsed Radar 105
28. Biomedical UWB pulsed Radar 105
29. Correlation receiver 106
30. Transmitter 106
31. References 107

This chapter is focused on the analysis of integrated circuits and systems for wireless transceivers operating at several frequencies (mainly at microwaves and mm-waves) targeting remote sensing applications.

Basic concepts of Radar (radio detection and ranging) design for low-power ubiquitous applications are reviewed. A comparison of Radar solutions for remote sensing vs. Lidar and Video Camera solutions is also addressed.

Real examples of RF transceivers and baseband signal processing circuits for Radars from scientific literature and from commercial products are also discussed.

Applications to remote sensing for maritime, automotive, railway, and road-safety fields are also shown.

1 Agenda

- Ubiquitous Radar Applications
- Radar vs. Lidar
- FMCW Radar examples
- Multi-channel X-band Radar transceiver with configurable output power and Fabry-Perot antenna
- FFT-based FMCW Radar range-Doppler processing
- Experimental trails
- V-band automotive Radar
- UWB pulsed Radar for biomedical applications
- Conclusions

Chapter 4 focuses on integrated circuits (ICs) and systems for wireless transceivers operating at RF or mm-waves in remote sensing IoT applications.

2 Ubiquitous Radar applications

Pushed by military applications in II world war with high-power, large-size and long-distance systems, today RADAR can be ubiquitous adopted for:

- Safer transport systems in automotive, railway, ships ...
- Bio-signal detection for health care and elderly/infant monitoring
- Info-mobility in urban, airport or port scenarios
- Civil engineering, (structural health monitoring, landslide monitoring ground penetration for detecting pipes, electric lines,....)
- Distributed surveillance systems (smart cities, airports, banks, schools)
- mm-wave body scanner for security
- Environmental monitoring and civil protection
- Contactless industrial measurements and in harsh environments
- Through-wall target detection

Radar sensing is suited to address societal needs (safety, security, heath, transport) and can be ubiquitous adopted for large-volume applications.

3 

Ubiquitous Radar design needs

W.r.t. conventional RADARs with large transmitted power x antenna aperture product, the realization of highly integrated RADARs with low power consumption, size, weight and cost (using standard technologies) is needed to enable its ubiquitous adoption in large–volume markets

- Transmitted Power < 10-15 dBm
- Short wavelength for miniaturization (3.9 mm@77 GHz)
- Range from < 1m to < 100-200 m
- Detection also with low SNR of 10-20 dB
- Cross section from tens of cm² to m²
- DSP techniques to improve performance and solve range-speed ambiguities
- Receiver sensitivity down to -100 dBm
- Multiple channels may be used for channel diversity gain

$$P_r = \frac{P_t G_t G_r \lambda^2 \sigma}{(4\pi)^3 R^4}$$

At state of art, there is the need of integrated and low-power Radar implementations.

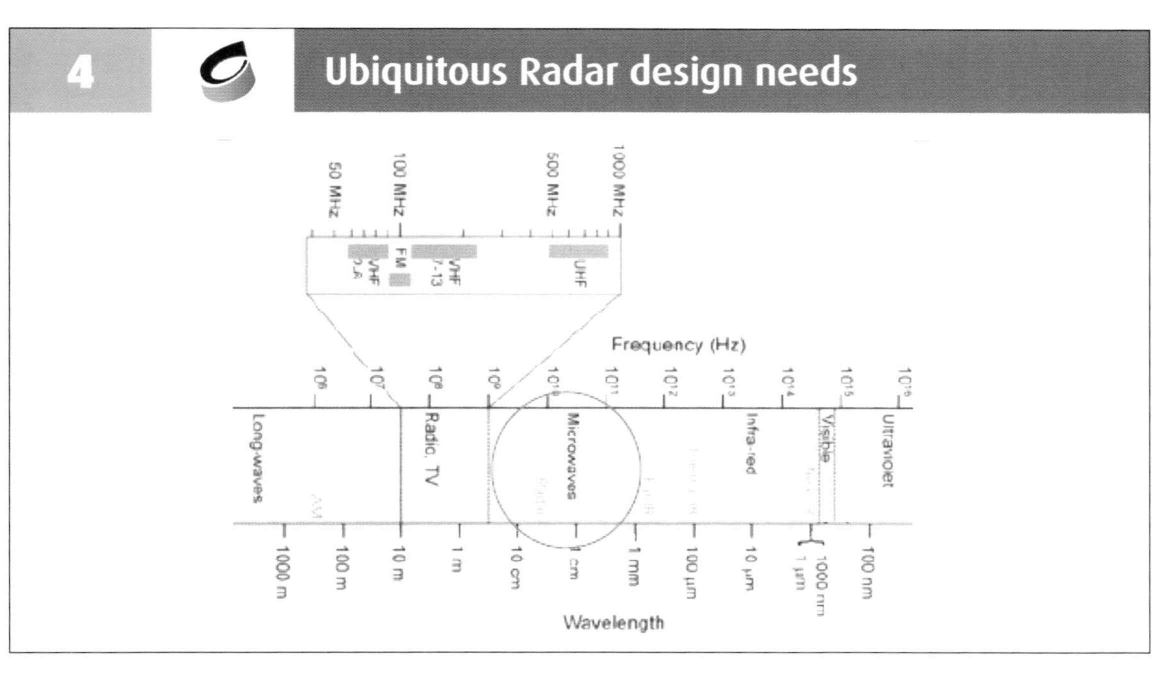

Radar sensing suited to address societal needs (safety, security, health, transport) → ubiquitous adopted for large-volume applications?

5 Radar vs. LIDAR

	Max Distance	Resolution	Power	Cost
HDL-32 [1]	100 m	2 cm	12 W	10000 USD
VLP-16 [2]	100 m	3 cm	8 W	<8000 USD
This work (harbour)	1.5 Km	37.5 cm	12 W	< 1000 USD
This work (railroad&urban road crossing, parking)	300 m	37.5 cm	< 8 W (5 Ch) < 3 W (2 Ch)	<500 USD

Radar vs. Lidar or Video (CMOS or CCD) sensors is more robust for bad weather and bad light conditions

Radar vs. Lidar allows for long ranges at lower cost

Radar vs. Lidar or Video (CMOS, CCD) sensors is more robust for bad weather and bad light conditions, Radar vs. Lidar allows for long ranges at lower cost.

6 Radar vs. LIDAR

Radar	Freq, GHz	Type	Power cost	Range	Output power
IEEE TIM2017	10.3-10.8	FMCW	11.86 W	1.5 km	2 W
			2.56 W	300 m	5 mW
IEEE TBSC2011	3.1-10.6	PulsedUWB	73 mW	<1 m	7 pJ/pulse
ACMMobicom 2015	60	FMCW	N/A	<3.5 m	N/A
MOTL2013	22-26	PulsedUWB	N/A	N/A	2 mW
TERMA2015	12-18	Pulsed	130 W	4 km	8 W
TERMA2015	9.375	Pulsed	N/A	45 km	32 kW
AWC2015	2.48-2.56	FMCW	N/A	100m	100 mW
AMS2013	9.4	FMCW	650 W	50 km	100 W

Performance of different state-of-art surveillance Radars at microwave and mm-wave.

7 

FMCW Radar acquisition & processing@ UniPisa

- X-band Radar for harbor surveillance information system
- Detection & tracking of ships/yachts ingress/egress
- Long distance up to 1.5 km
- 1 Radar for a small harbor
- Network of Radars for large port areas (increase the covered area)
- 1 Tx + 1 Rx speed and distance estimation
- Multiple-channels for speed, distance, angle estimation
- Custom microwave board for imaging sensor front-end in X-band
- DSP via software on a GPP for off-line analysis
- **Real-time DSP to be implemented on FPGA or GPU, FPGA mandatory if power efficiency and compact size are key issues**

Collaboration with CNIT/RASS (Berizzi, Martorella, Lischi, Massini)

Examples of X-band (around 10 GHz) Radar for transport surveillance in maritime applications.

8 

FMCW Radar acquisition & processing@ UniPisa

X-band Radar for railway crossing safety and parking/road crossing safety

Obstacle detection on a railroad or urban road crossing

Up to 4 Radar nodes for high SIL (Safety Integrity Level) in automated railroad crossing

Max detection distances up to 200-300 m

1 Tx + 1 Rx for speed and distance estimation

1 Tx + 3 Rx for speed, distance, azimuth/elevation angle estimation

Real-time power-efficient and compact Radar image processing on FPGA platforms

Custom microwave board for X-band transceiver

Collaboration with I.D.S. spa

Examples of X-band (around 10 GHz) Radar for transport surveillance in railway and road applications.

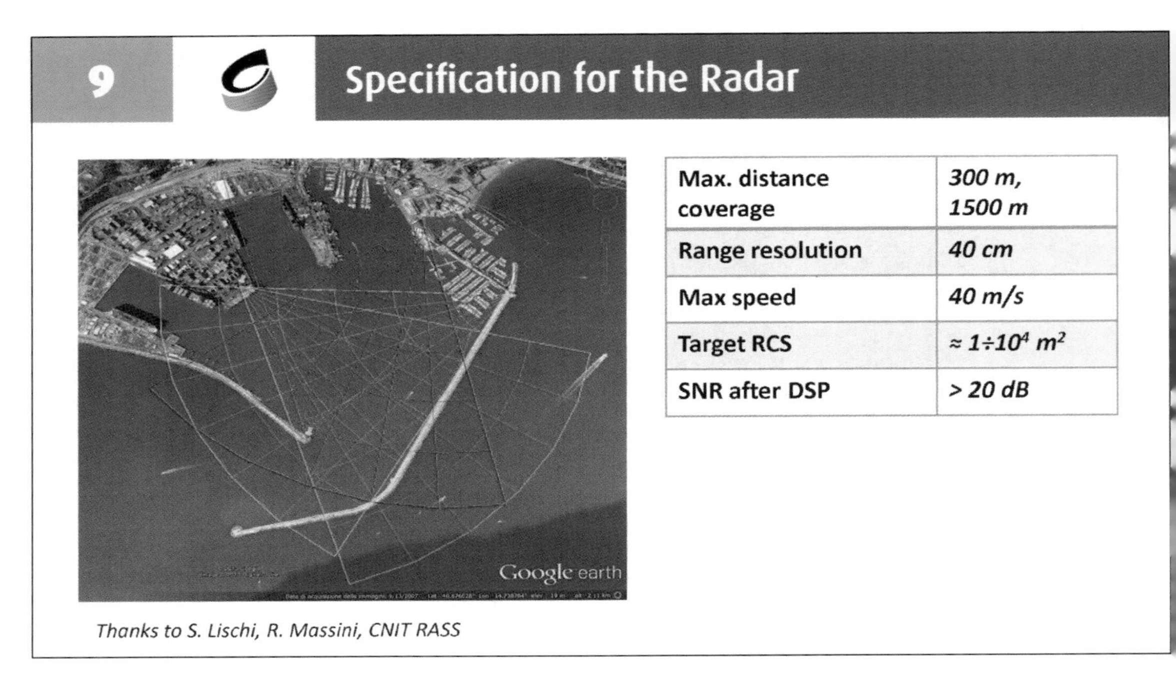

Max. distance coverage	300 m, 1500 m
Range resolution	40 cm
Max speed	40 m/s
Target RCS	≈ $1 \div 10^4$ m^2
SNR after DSP	> 20 dB

X-band (around 10 GHz) FMCW (frequency-modulated continuous wave) Radar for surveillance in mobility systems offers small size, low power consumption, low EMI solutions with detection of target speed and position, and reconfigurability of bandwidth, pulse repetition frequency, and power.

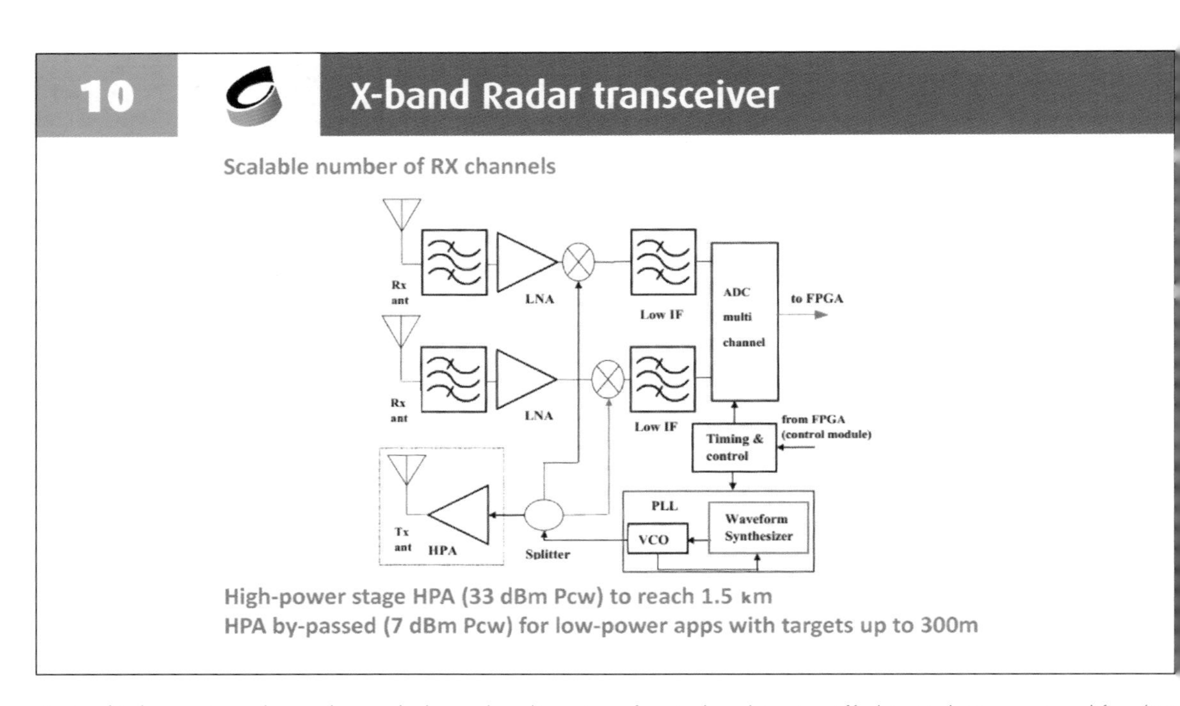

Multiple receiver channels needed to solve direction of arrival ambiguity. Off-chip High Power Amplifier (33 dBm Pcw) needed to reach 1.5 km distance. Max 300 m with a 7 dBm integrated amplifier.

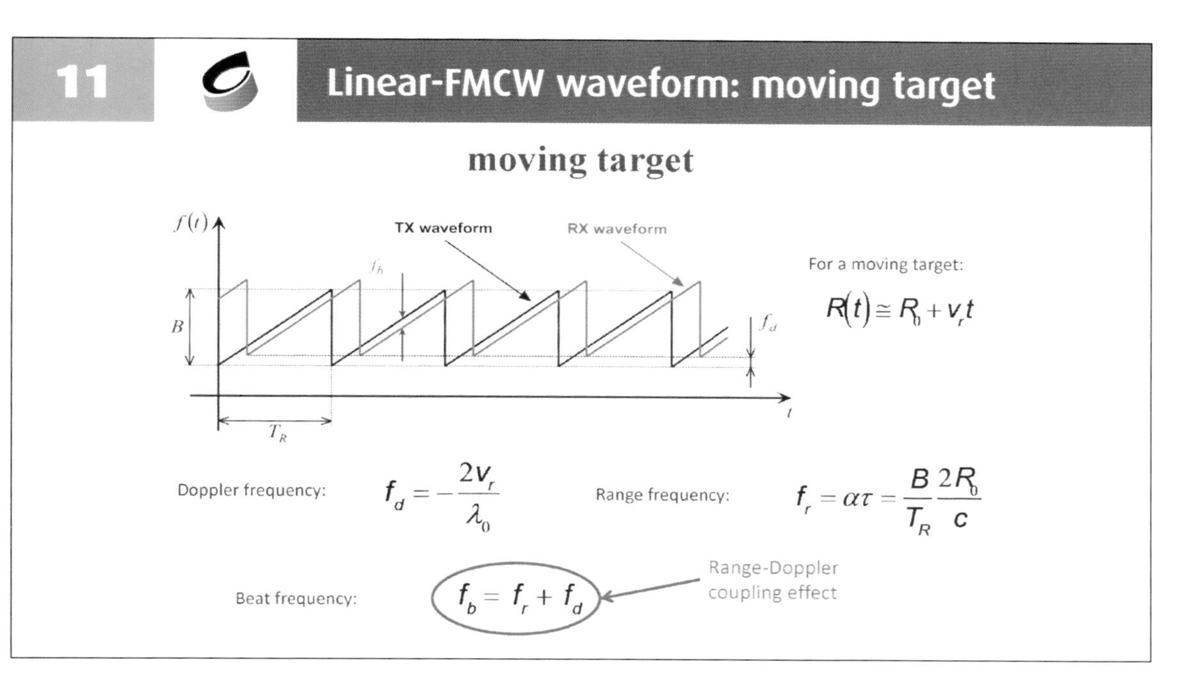

Linear frequency modulated continuous wave: difference between TX and RX waveforms, available at mixer output, contains both time of flight (distance) and Doppler frequency shift (speed) information.

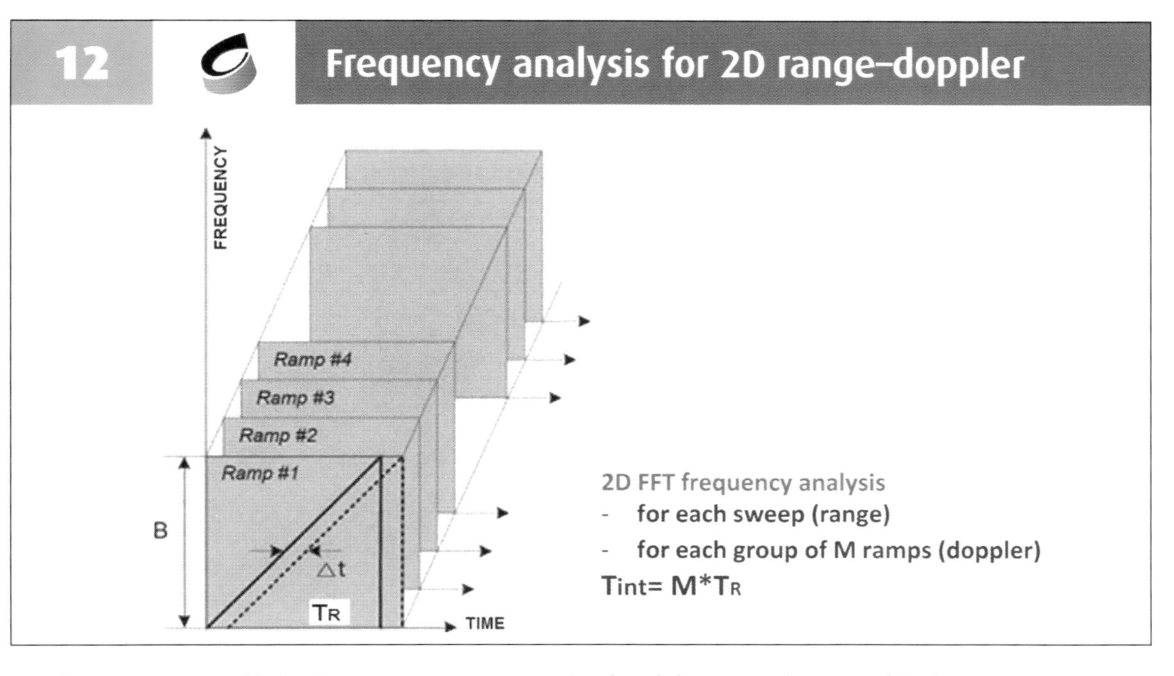

To improve SNR, multiple LFMCW ramps are transmitted and the output integrated in time.

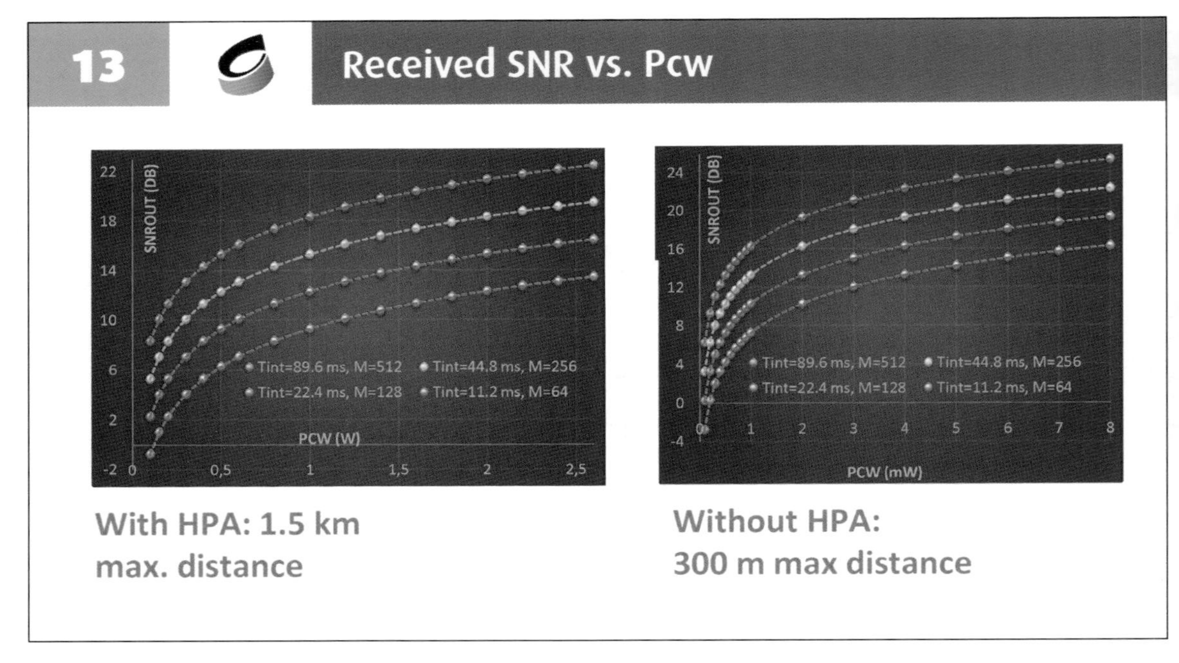

SNR at receiver side is a function of target distance, transmitted power, and integration time.

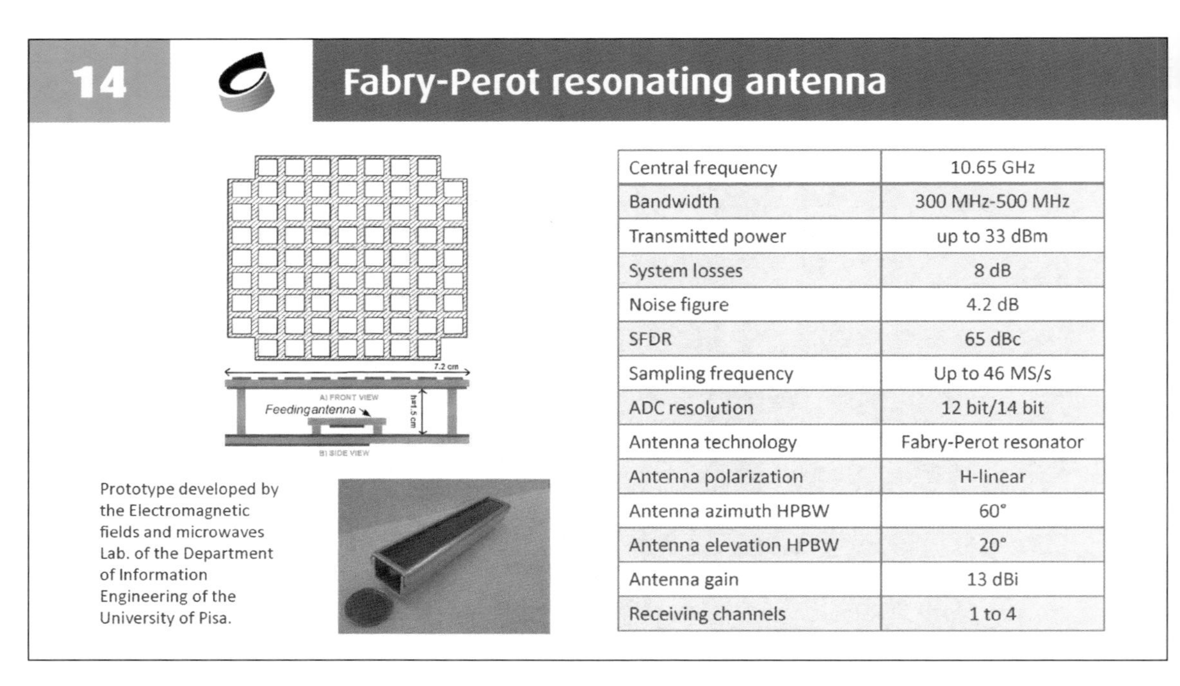

Parameter	Value
Central frequency	10.65 GHz
Bandwidth	300 MHz-500 MHz
Transmitted power	up to 33 dBm
System losses	8 dB
Noise figure	4.2 dB
SFDR	65 dBc
Sampling frequency	Up to 46 MS/s
ADC resolution	12 bit/14 bit
Antenna technology	Fabry-Perot resonator
Antenna polarization	H-linear
Antenna azimuth HPBW	60°
Antenna elevation HPBW	20°
Antenna gain	13 dBi
Receiving channels	1 to 4

Example of a Fabry-Perot resonating antenna (A. Monorchio et al.).

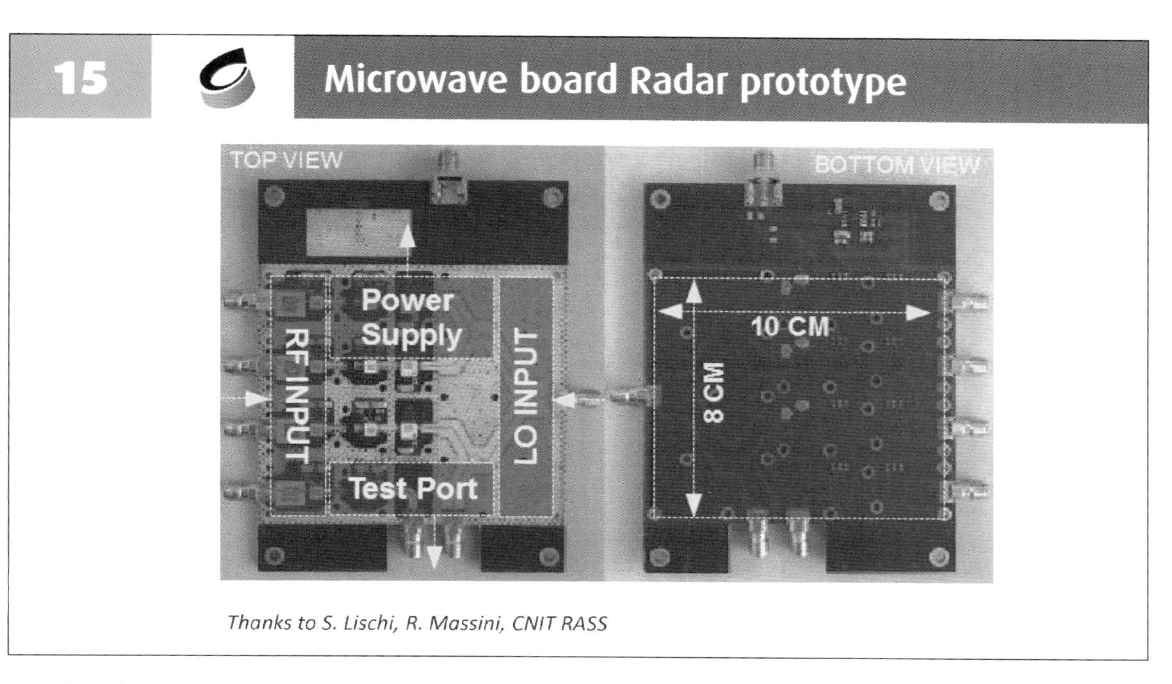

Radar X-band transceiver realized with COTS devices and microwave boards.

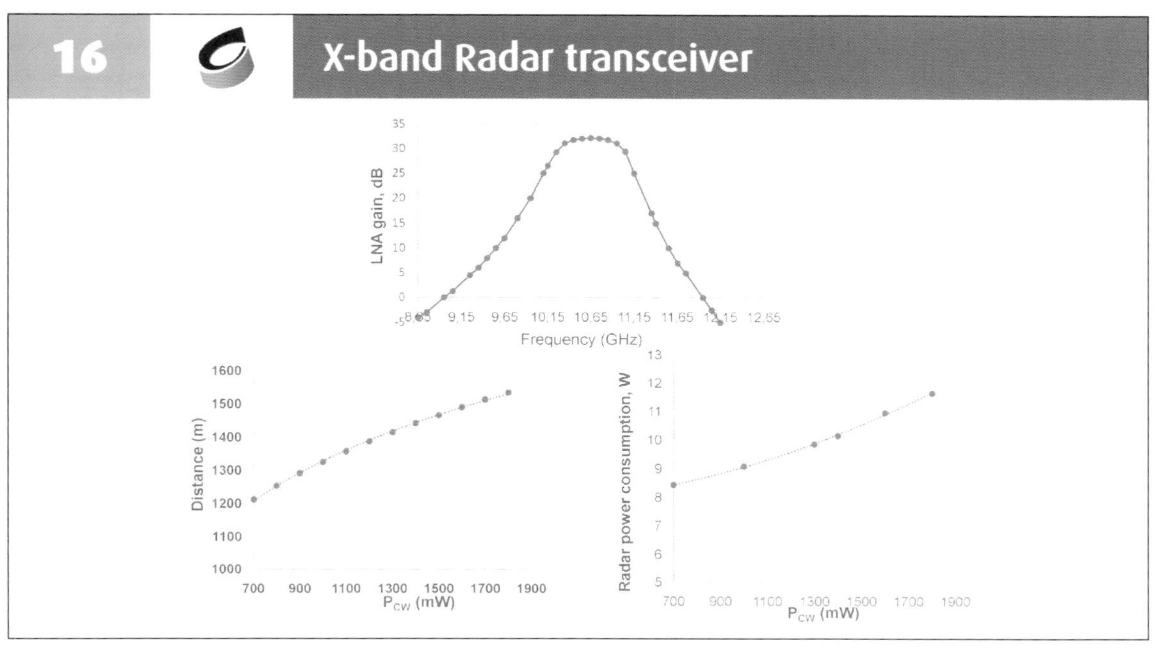

Performance of Radar X-band transceiver realized with COTS devices and microwave boards.

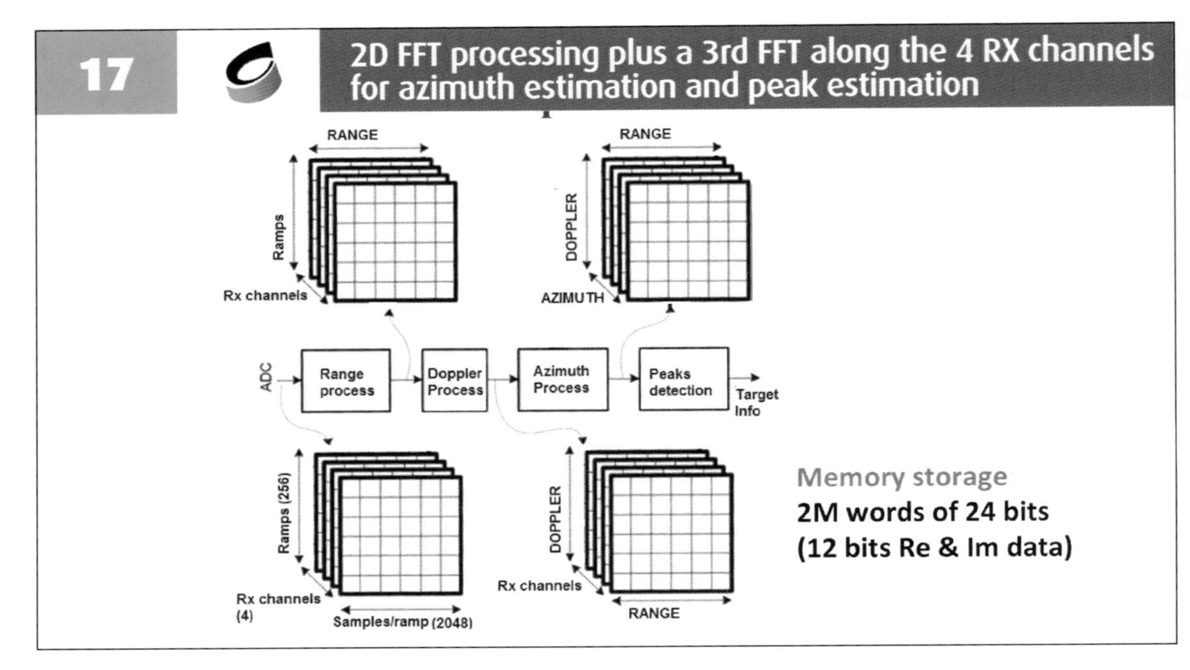

2D FFT processing plus a 3rd FFT along the 4 RX channels for azimuth estimation and peak estimation (Memory storage: 2M words of 24 bits). 3D-FFT by cascading 3 1D-FFT.

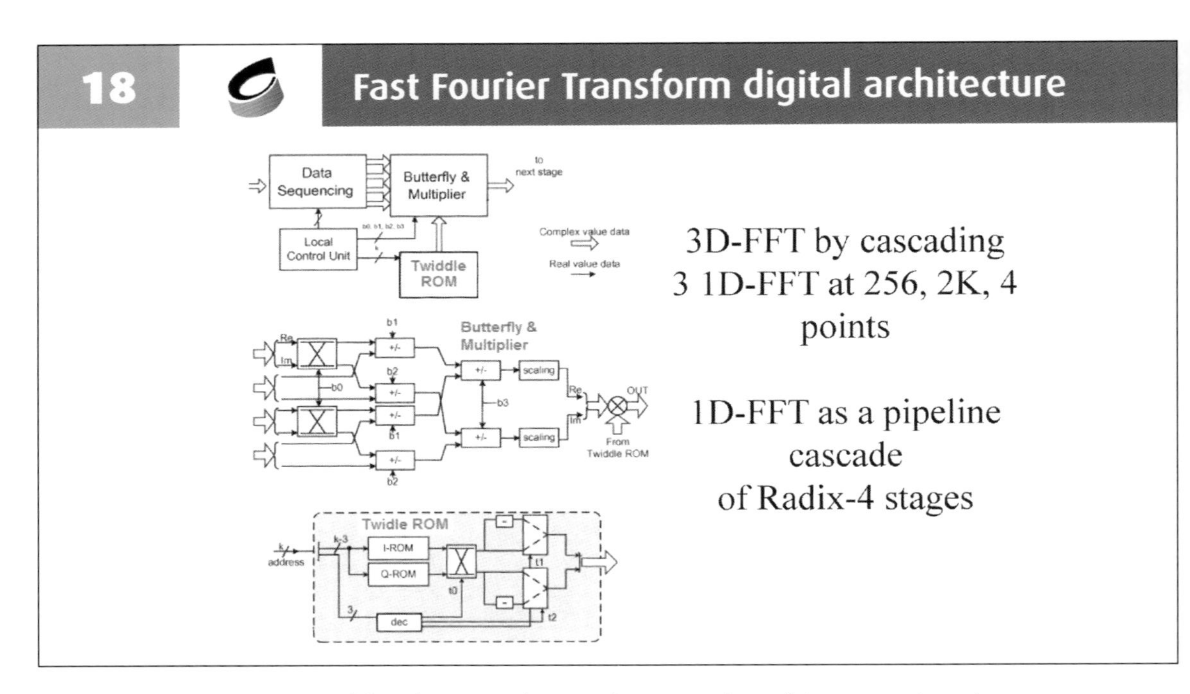

The cell averaging constant false alarm rate (CA-CFAR) circuit reduces false target detections.

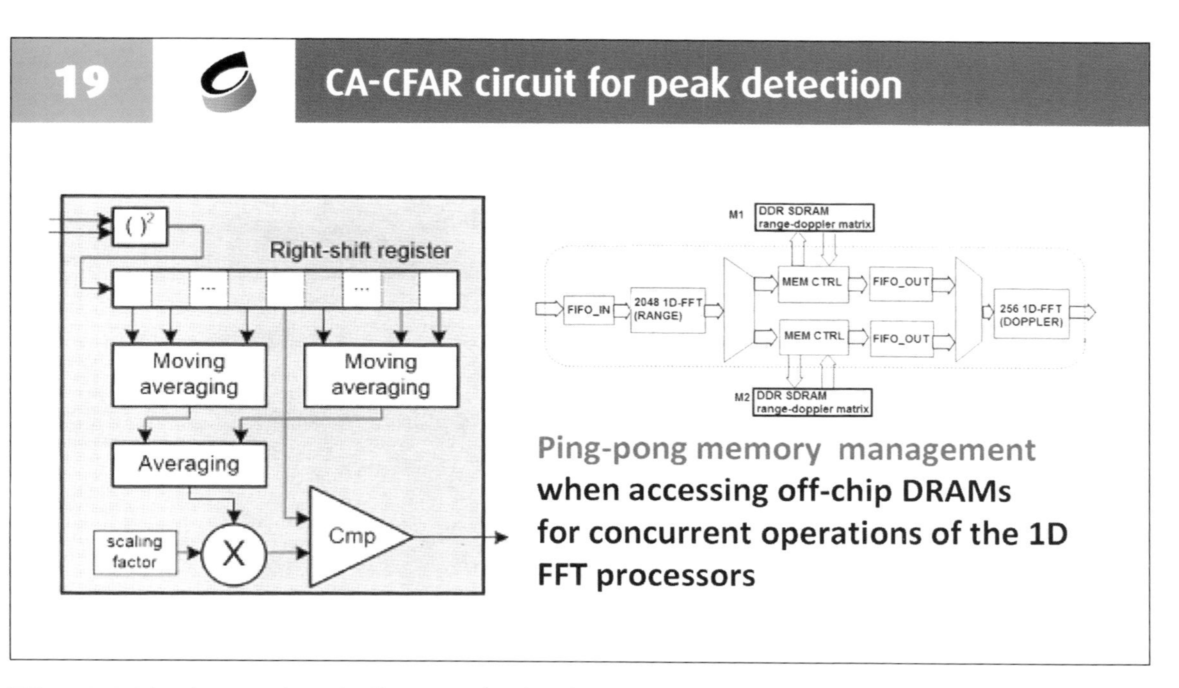

The 1D-FFT implemented as pipeline cascade of Radix-4 stages.

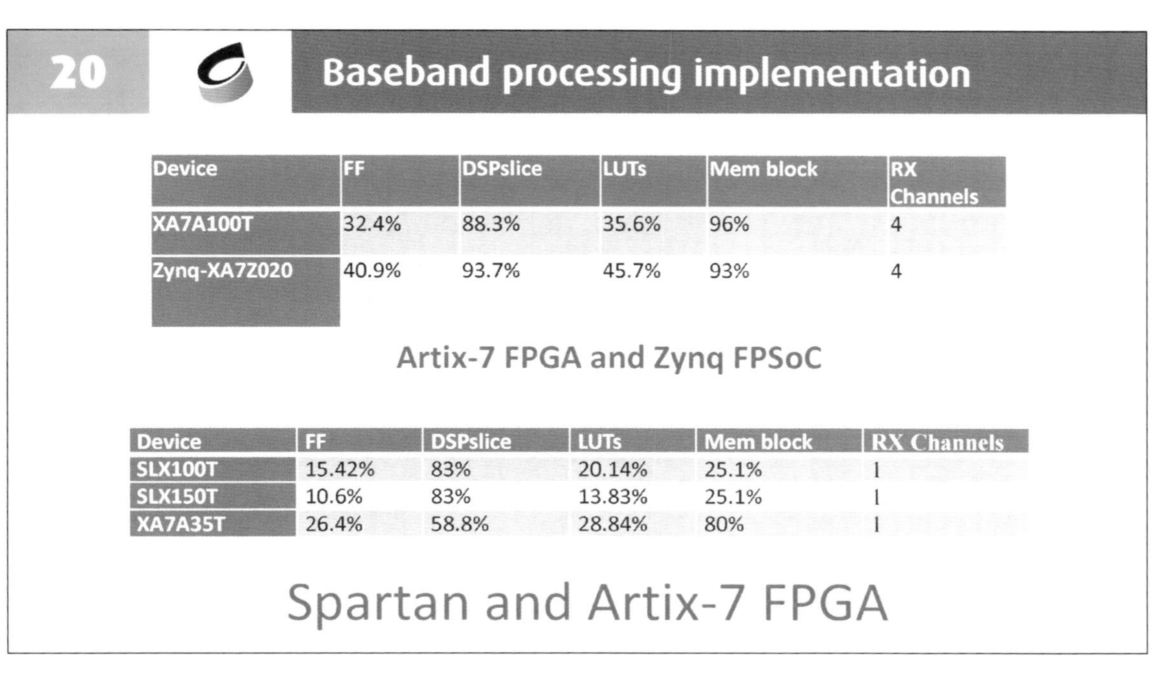

Device	FF	DSPslice	LUTs	Mem block	RX Channels
XA7A100T	32.4%	88.3%	35.6%	96%	4
Zynq-XA7Z020	40.9%	93.7%	45.7%	93%	4

Device	FF	DSPslice	LUTs	Mem block	RX Channels
SLX100T	15.42%	83%	20.14%	25.1%	1
SLX150T	10.6%	83%	13.83%	25.1%	1
XA7A35T	26.4%	58.8%	28.84%	80%	1

Low-complex FPGA used to implement in real time the complete baseband Radar processing.

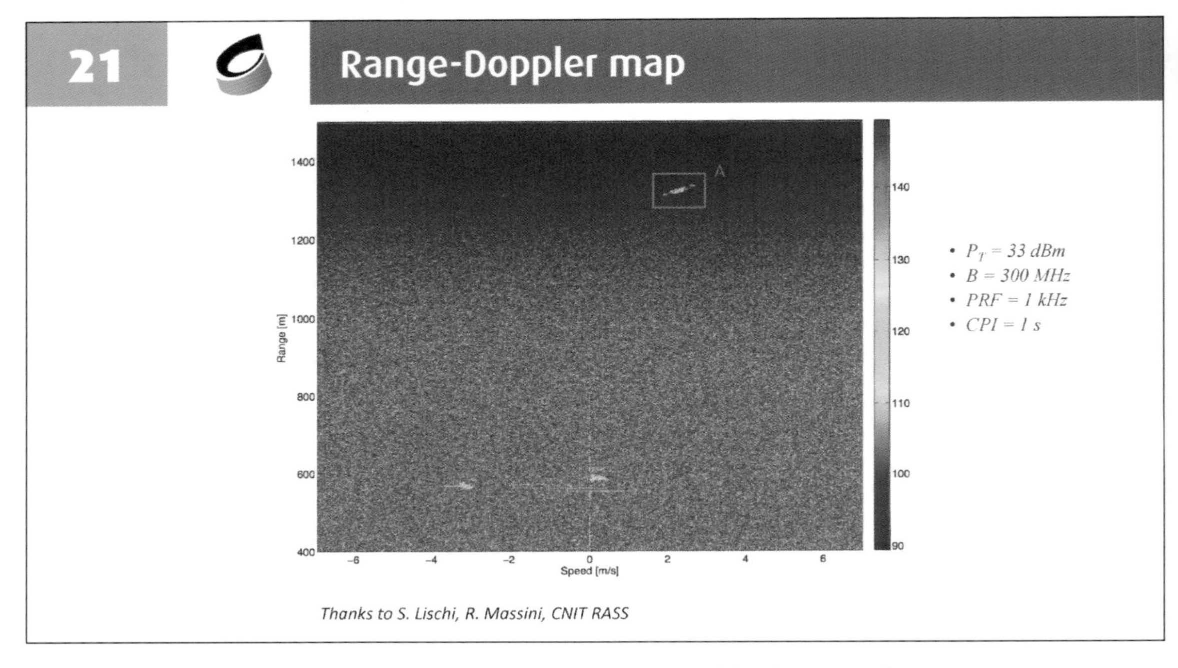

Example of Radar image (range-speed) with 3 different targets (ships) in surveillance maritime app.

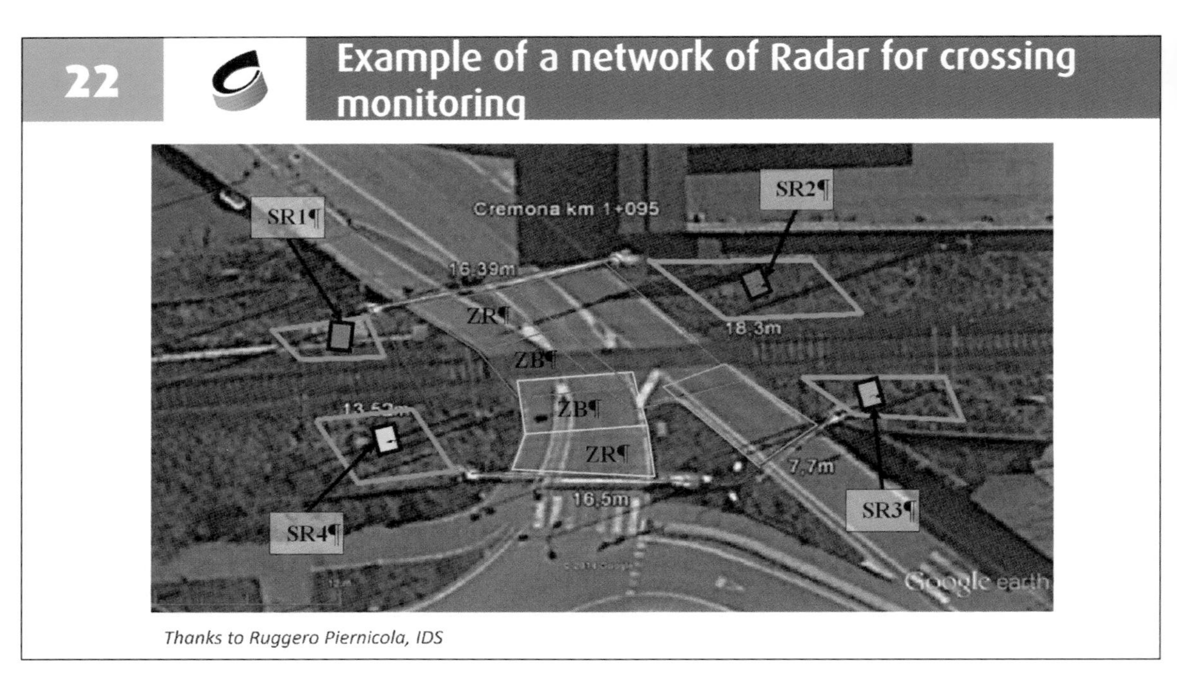

Example of a road crossing surveillance system using 4 Radar sensing nodes.

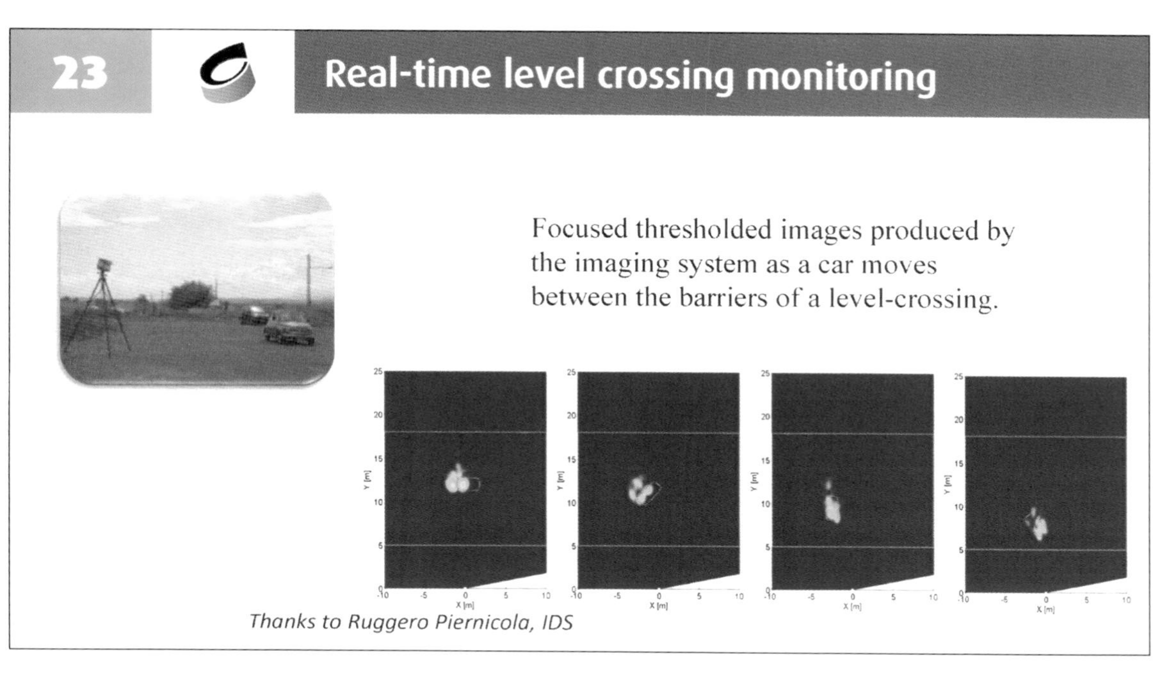

Example of a railroad crossing surveillance system using 1 Radar sensing node.

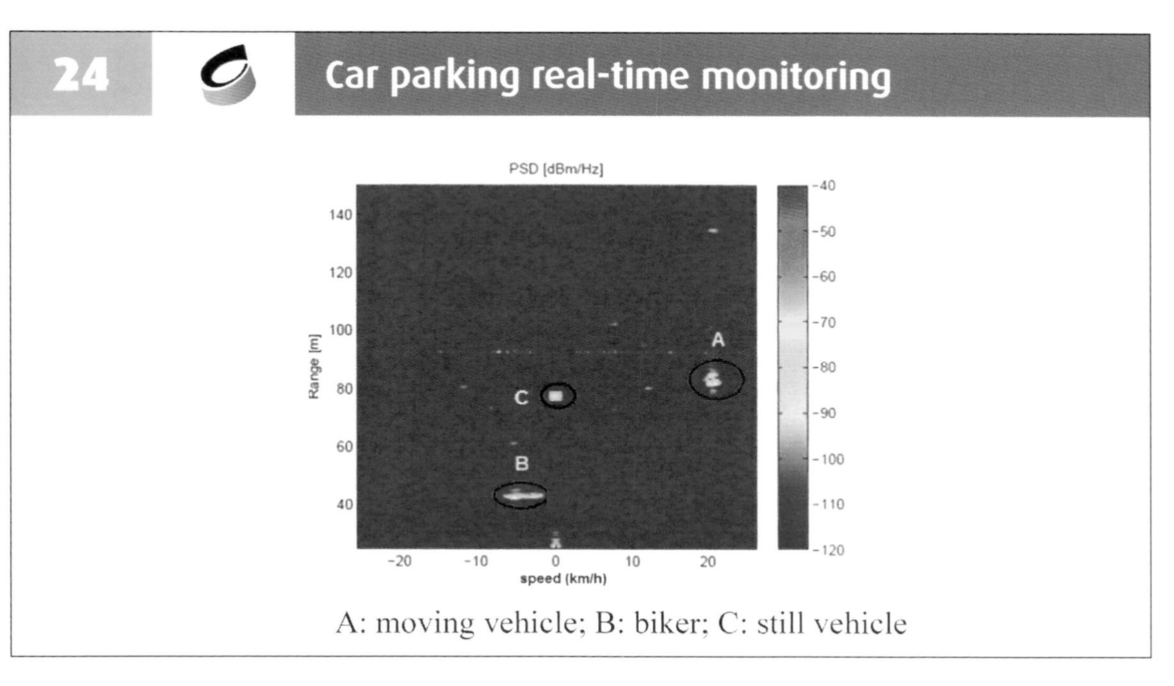

Example of a Radar range-speed image during a car parking monitoring using 1 Radar sensing node.

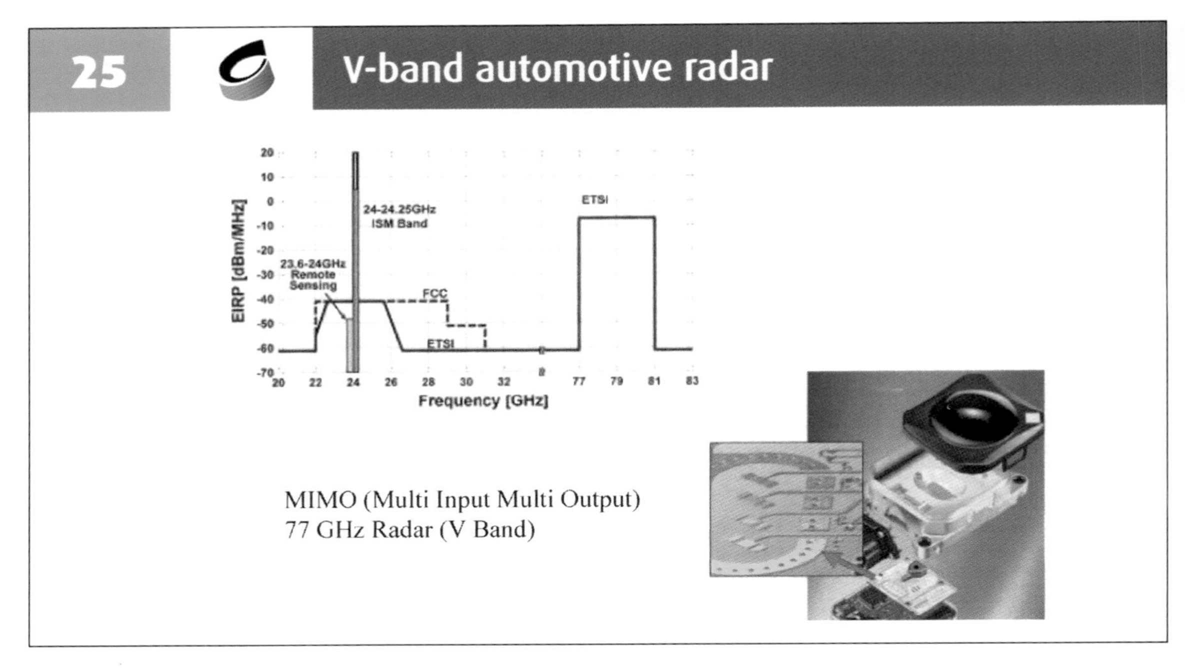

For on-board automotive solutions, fully integrated within the car's chassis, a Radar of 5 cm per side is needed → a Radar with an integrated transceiver at 77-81 GHz is preferred (SiGe technology used although CMOS circuits are also available).

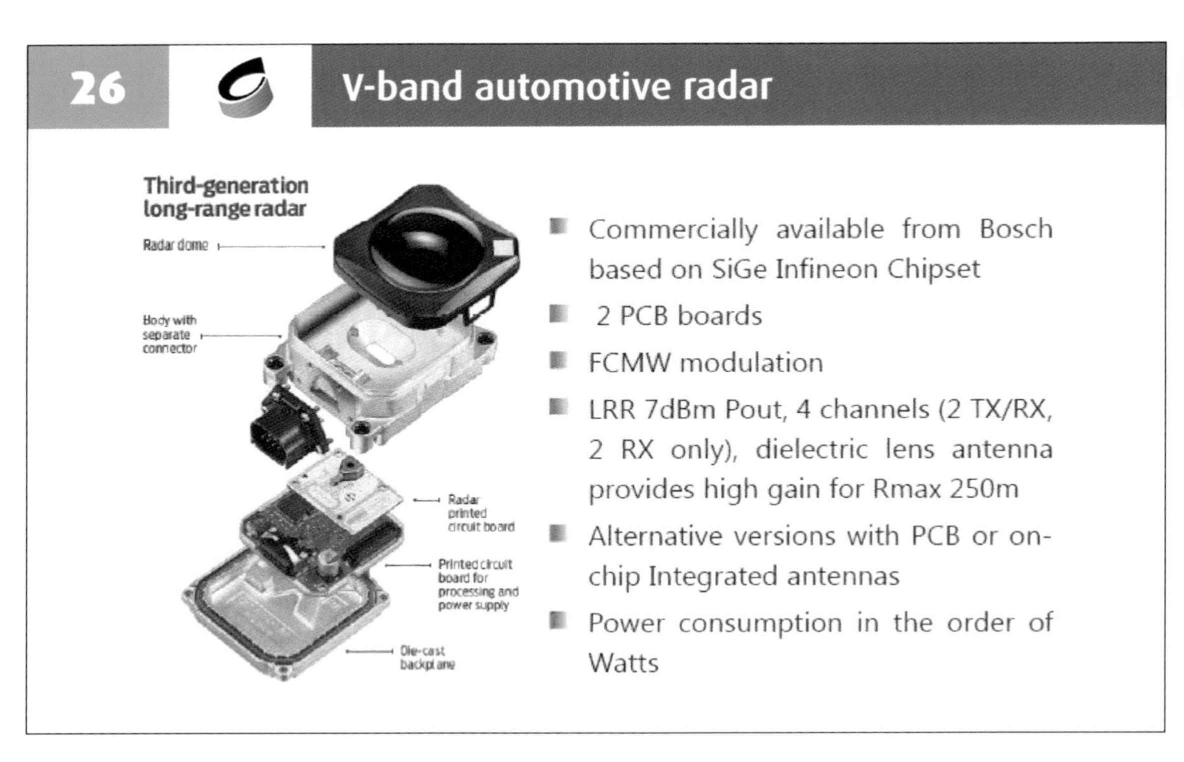

Example of the 3rd generation Long Range Radar from Bosch.

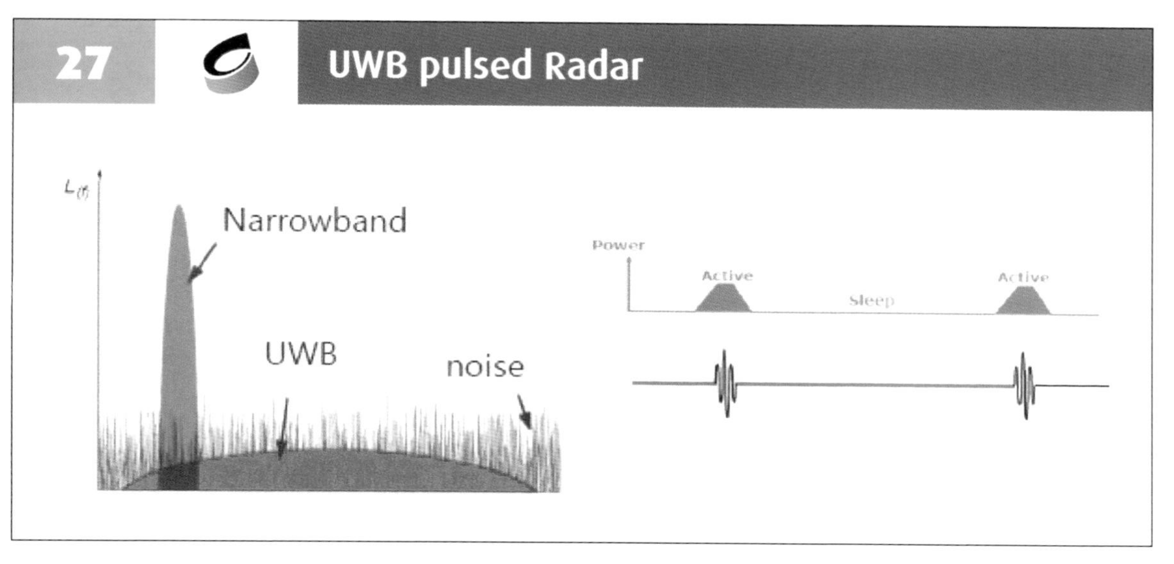

f_{PR} in the range 1-10 MHz, pulses of 300-400 ps, and 7-8 pJ energy.

BB digital processing can be realized with a simple MCU: low-speed ADC required (12b in ISSCC'11), low data rate serial connection, mainly control tasks to be implemented.

Whole chip by Zito et al. in 90 nm CMOS has <2 mm^2 area, < 80 mW power consumption, 40dB SNR integrator improvement, < 1m range.

RADAR packaged in QFN32 and mounted on test-board including antennas (TX nd RX) with 2.3 dBi gain at 3.5 GHz, band 2.8 to 5.4 GHz covering the range of interest from 3 to S GHz.

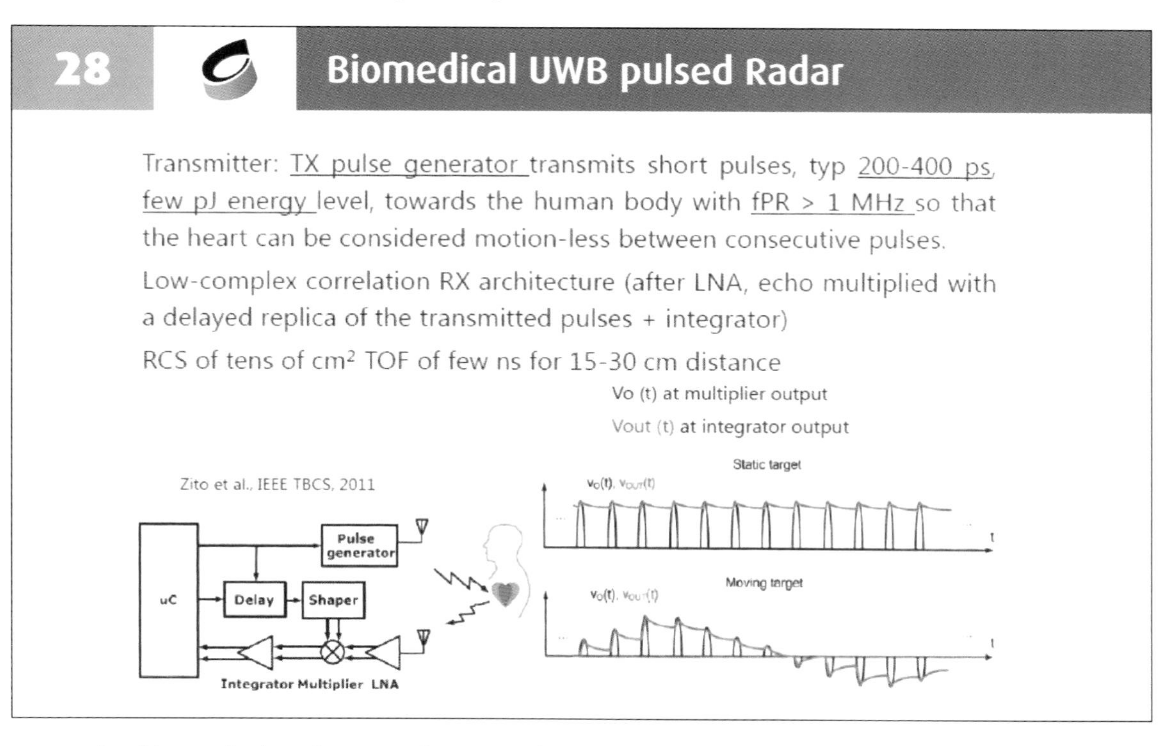

Example of biomedical applications of a UWB radar.

29  Correlation receiver

Averaging several pulses allows increasing SNR (40 dB, 10^4 pulses)

$$SNR_{imp} = 10 \cdot \log(\frac{f_{PR}}{B_{int}})$$

At the low frequency (DC-100 Hz) of the baseband bio-signal the MOS transistors suffer 1/f flicker noise, higher than thermal noise (KTB term)

$$NF_{tot} = NF_1 + (NF_2 - 1)/G_1 + (NF_3 - 1)/G_2G_1$$

To have $NF_{tot} \sim NF_{LNA}$ 20 dB gain required for the LNA if $NF_2 < 15$ dB → 90 nm CMOS LNA: 22.7 dB gain, 6 dB NF, -19 dBm ICP1dB, <35 mW, <0.7 mm²

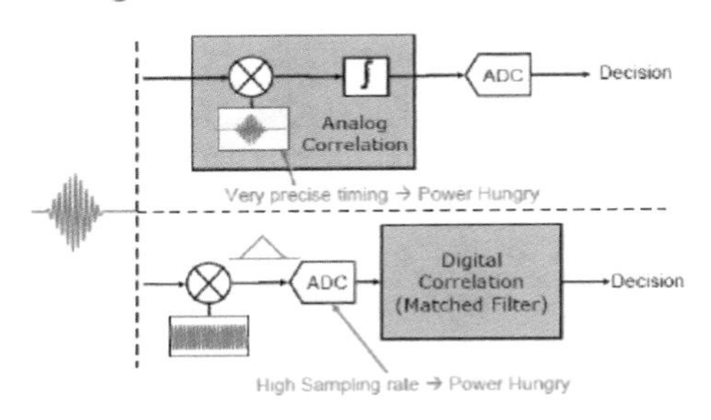

Example of a correlation receiver.

30  Transmitter

Pulse generator based on triangular pulse generation (TPG) and shaping network (SN): two triangular pulsed (delayed by a pulse period) generated and shaped by a CMOS differential pair

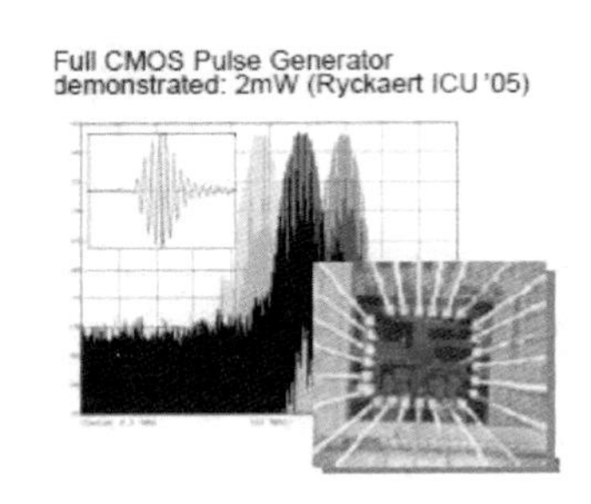

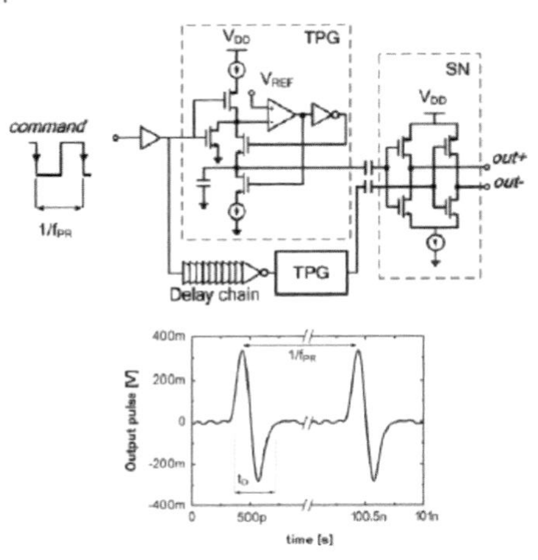

Example of a UWB radar transmitter.

31  References

1. F. Pieri, C. Zambelli, A. Nannini, P. Olivo, S. Saponara, "Consumer electronics is redesigning our cars?", *IEEE Consumer Electronics Magazine* 2018
2. S. Saponara, "Hardware Accelerator IP Cores for Real-time Radar and Camera-based ADAS", *Journal of Real-Time Image Processing*, 2017, pp. 1-18,
3. S. Saponara, B. Neri, "mm-wave integrated wireless transceiver: enabling technology for high bandwidth short-range networking in cyber physical systems", *Microsystem Technologies*, 22 (7), 2006
4. B. Neri, S. Saponara, "Advances in technologies, architectures, and applications of highly-integrated low-power radars", *IEEE Aerospace and Electronic Systems*, 27 (1), 2012
5. S. Saponara, F. Giannetti, B. Neri, G. Anastasi, "Exploiting mm-Wave Communications to Boost the Performance of Industrial Wireless Networks", *IEEE Trans. on Ind. Informatics*, 13, (3), 2017
6. S. Saponara, B. Neri, "Radar Sensor Signal Acquisition and Multi-dimensional FFT Processing for Surveillance Applications in Transport Systems", *IEEE Transactions on Instrumentation and Measurement*, 66 (4), 2017
7. S. Saponara, F. Giannetti, B. Neri, "Design Exploration of mm-Wave Integrated Transceivers for Short-Range Mobile Communications Towards 5G", *Journal of Circuits, Systems and Computers*, 26 (4), 2017
8. S. Saponara, B. Neri, "Design of Compact and Low-power X-band Radar for Mobility Surveillance Applications", *Computers and Electrical Engineering*, 56, 2016
9. D. Davalle, R.. Cassettari, S. Saponara, L. Fanucci, L. Cucchi, F. Bigongiari, W.Errico, "Design, Implementation and Testing of a Flexible Fully-digital Transponder for Low-earth Orbit Satellite Communications", *Journal of Circuits, Systems and Computers*, 23 (10), 2014
10. S. Saponara, F. Giannetti, "Radio link design framework for W SN deployment and performance prediction", *Proceedings of SPIE Microtechnologies*, vol. 10246, Barcelona, 8-10 May, 2017
11. S. Saponara, B. Neri, "Radar sensor signal acquisition and 3D FFT processing for smart mobility surveillance systems", *IEEE Sensors Applications Symposium (SAS)*, pp. 417-422, 2016
12. S. Saponara, B. Neri, "mm-Wave Integrated Wireless Transceivers: Enabling Technology for High Bandwidth Connections in IoT", *IEEE WF-IOT conference 2015*, pp.149 – 153
13. S. Saponara, B. Neri, "Gbps wireless transceivers for high band width interconnections in distributed cyber physical systems", *SPIE Microtech., Smart sensors, actuators and MEMS VII and Cyber Physical System*, vol. 9517, 95172L, pp. 1-7, 2015
14. Saponara, S., Mattii, L., Neri, B., Baronti, F., Fanucci, L., "Design of a 2 Gb/s transceiver at 60 GHz with integrated antenna in bulk CMOS technology", *IEEE EuMIC* 2014,
15. A. Fonte, S. Saponara, G. Pinto, B. Neri, "Feasibility study and on-chip antenna for fully integrated μRFID tag at 60 GHz in 65 nm CMOS SOI", *IEEE Int. Conference on RFID-Technologies and Applications (RFID-TA)* 2011. pp. 457-462
16. S. Saponara, S. Lischi, R. Massini, L. Musetti, D. Staglianò, F. Berizzi, B. Neri, "Low Cost FMCW Radar Design and Implementation for Harbour Surveillance Applications", *Lecture Notes in Electrical Engineering*, vol. 351, pp. 139-144, 2016, Springer
17. S. Saponara, B. Neri, "Fully integrated 60 GHz transceiver for wireless HD/Wigig short-range multi-Gbit connections", *Lecture Notes in Electrical Engineering*, vol. 351, pp. 131-137, 2016
18. F. Pieri, C. Zambelli, A. Nannini, P. Olivo, S. Saponara, "Limits of sensing and storage electronic components for high-reliable and safety-critical automotive applications", *IEEE/AEIT 2017 Int. Conf. of Electrical and Electronic Techn. for Automotive* (Automotive 2017),

CHAPTER 05

Chipless RFID for Identification and Sensing

David Girbau
Antonio Lazaro

University Rovira i Virgili, Tarragona, Spain

Simone Genovesi
Filippo Costa

University of Pisa, Italy

1. Introduction 112
2. Item identification: RFID or Barcode? 112
3. RFID Technologies 113
4. ChiplessRFID Technology 113
5. RFID: Chip vsChiplesstags 114
6. Link Budget: Standard RFID 114
7. Link Budget: Chipless RFID 115
8. Link Budget: comparison (I) 115
9. Link Budget: comparison (II) 116
10. Multipath: Two-ray model (I) 116
11. Multipath: Two-ray model (II) 117
12. Measurement: Example 117
13. Classification of chipless RFID tags 118
14. Chipless RFID tags (I) 118
15. Chipless RFID tags (II) 119
16. Surface acoustic wave (SAW) (I) 119
17. Surface acoustic wave (SAW) (II) 120
18. Surface acoustic wave (SAW) (III) 120
19. Surface acoustic wave (SAW) (IV) 121
20. Left-Handed (LH) transmission lines (I) 122
21. Left-Handed (LH) transmission lines (II) 122
22. Time-coded chipless UWB RFID (I) 123
23. Time-coded chipless UWB RFID (II) 123
24. Time-coded chipless UWB RFID (III) 124
25. Time-coded chipless sensors (I) 124
26. Time-coded chipless sensors (II) 125
27. Time-coded chipless sensors (III) 125
28. Time-coded chipless sensors (IV) 126
29. Applications Semi-passive technology 126
30. Backscattering communication 127
31. Read Range: UHF passive vs BAP tags 127
32. Commercial BAP tags 128
33. NFC tags with Energy Harvesting 128
34. Semi-passive UWB RFID (I) 129
35. Semi-passive UWB RFID (II) 129
36. Semi-passive UWB RFID sensors (I) 130
37. Semi-passive UWB RFID sensors (II) 130
38. Semi-passive UWB RFID sensors (III) 131
39. Semi-passive UWB RFID sensors (IV) 131
40. Semi-passive UWB RFID sensors (V) 132
41. Semi-passive UWB RFID sensors (VI) 133
42. Semi-passive UWB RFID sensors (VII) 133
43. Semi-passive UWB RFID sensors (VIII) 134
44. Tags based on Modulated FSS 134
45. FSS impedance 135
46. Frequency Selective Surfaces 135
47. Dielectric effects 136
48. Tags based on Modulated FSS (I) 136
49. Tags based on Modulated FSS (II) 137
50. Tags based on Modulated FSS (III) 137
51. Tags based on Modulated FSS (IV) 138
52. Distance measurement with a backscattered FSS using FMCW (I) 138
53. Distance measurement with a backscattered FSS using FMCW (II) 139
54. Distance measurement with a backscattered FSS using FMCW (III) 139
55. Distance measurement with a backscattered FSS using FMCW (IV) 140
56. Wearable Sensors Based On FSS 140
57. Wearable Sensors Based On FSS (I) 141
58. Wearable Sensors Based On FSS (II) 142
59. Wearable Sensors Based On FSS (III) 142
60. Wearable Sensors Based On FSS (IV) 143
61. FD chipless RFID tags 143
62. Frequency domain chipless tags 144
63. FD chipless RFID tags (I) 144
64. FD chipless RFID tags (II) 145
65. FD chipless RFID tags (III) 145
66. Artificial Impedance Surfaces (AIS) (I) 146
67. Artificial Impedance Surfaces (AIS) (II) 146
68. A scattering-based Chipless RFID (I) 147
69. A scattering-based Chipless RFID (II) 147
70. Exploiting amp/phase/pol (I) 148
71. Exploiting amp/phase/pol (II) 148
72. Exploiting amp/phase/pol (III) 149
73. Exploiting amp/phase/pol (IV) 149
74. Exploiting amp/phase/pol (V) 150
75. Exploiting amp/phase/pol (VI) 151
76. Exploiting amp/phase/pol (VII) 151
77. Cross-pol generation 152
78. Limitations and current challenges (I) 152
79. Limitations and current challenges (II) 153
80. Limitations and current challenges (III) 153
81. Limitations and current challenges (IV) 154
82. Limitations and current challenges VI) 154
83. Chipless RFID sensor based on HIS (I) 155
84. Chipless RFID sensor based on HIS (II) 155
85. Chipless RFID sensor based on HIS (III) 156
86. Chipless RFID sensor based on HIS (IV) 156

This chapter deals with an overview of the chipless RFID technology. The topic is initially introduced by presenting a comparison of this item-tagging technique with existing technologies such as the UHF RFID and the barcode. The advantages and drawbacks are discussed. Afterward, the most popular approaches to synthesize a chipless RFID are addressed both in time domain (TD) and in frequency domain (FD). The former approach is described in detail with several examples related to information encoding schemes and sensing applications. Both passive or semi-passive TD tags are described. The last part of the chapter is dedicated to the description of FD tags, where the information is usually encoded in the presence or absence of a frequency peak at a predetermined frequency. It is also shown that FD tags can be employed as sensors if the resonator is loaded with a material sensible to an environmental variable such as the humidity level.

1 

Introduction

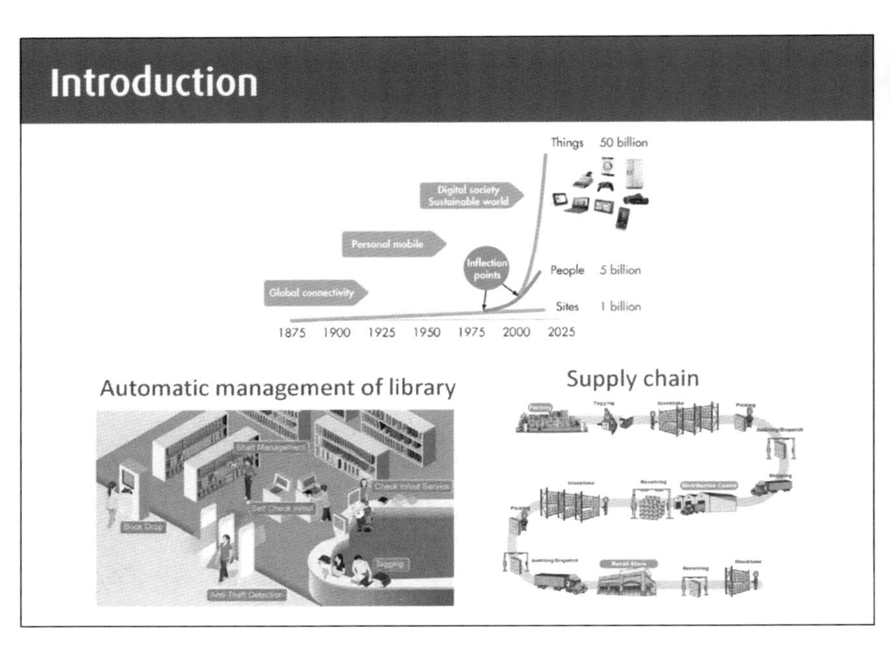

The growth of Internet has bought to a huge number of devices connected to the Web. The number of devices connected to the internet has already overcome the number of people, and this trend is still growing with an exponential behavior. This phenomenon is addressed as Internet of Things (IoT). In order to connect a device to the internet, it is necessary to install on it some hardware. This hardware usually comprises a CPU or a microcontroller and a network interface that need to be fed with a DC power supply. However, in order to further increase the number of objects connected, it is necessary to adopt much lower-cost technologies through less intelligence to link low-value objects to the global infrastructure. RFID is surely a suitable candidate for this purpose. RFID allows the automatization of a supply chain or a library by using a 10 cents label on every object.

2 

Item identification: RFID or Barcode?

	RFID	Barcode
Rewritable	Yes	No
Reading distance	Up to 9m	Up to 90 cm
Line of sigth	Not necessary	Necessary
Reading speed	Up to 1500 tags/sec	2 bar codes/sec
Reliability	Difficult to damage and counterfeit	Easy damage and anticounterfeit
Cost	0.1 – 1 $	0.001 – 0.01 $
Equipment cost	Similar	

Currently available technologies for the identification of objects are RFID and barcode. Barcode is certainly advantageous in terms of cost, but there are some limitations that have brought the industry to invest on radio frequency identification technologies. Indeed, RFID allows a much faster and non-line-of-sight reading. The tag can be detected at a much longer distance with respect to barcode and it is reprogrammable. Moreover, RFID labels are also rewritable. Barcode is also prone to damaging. However, in the end, the lower cost of barcode is a fundamental advantage that makes it often preferable when a huge number of objects need to be identified and tracked.

3

RFID Technologies

Technology	Passive RFID	Battery-Assisted Passive (BAP)	Active RFID
Tag Power Source	Energy harvesting from the reader via RF	Tags use internal power source to power on, and the energy transferred from the reader via RF to backscatter	Internal to tag
Communication principle	Backscattering	Backscattering	Transmission
Tag Battery	NO	YES	YES
Availability of Tag Power	Only within field of reader	Only within field of reader	Continuous
Required Signal Strength from Reader to Tag	Very high (must power the tag)	Moderate (does not need to power tag, but must power backscatter)	Very low
Read Range	Short range (<10 m)	Moderate range (up to 50 m)	Long range (100 m or more)
Sensor Capability	Ability to read and transfer sensor values only when tag is powered by reader	Ability to read and transfer sensor values only when receives RF signal from the reader	Ability to continuously monitor and record sensor input
Cost	0.05-1€	1-10€	>10€

RFID tags can be classified according to their power source in passive, semi-passive, and active.

Passive tags are battery free. Therefore, these tags can be considered the greenest devices. The tags harvest the power from the incoming RF signal and therefore the read range is short. Passive tags do not transmit any signal, and the communication is by backscattering. A special case of passive tags are the chipless tags, which do not contain chip or electronics.

In the opposite side, in active RFID systems, tags have their own transmitter and power source. The read range is typically higher than 100 m since they transmit a large signal.

In between active and passive, there are the semi-passive tags or battery-assisted passive tags. These tags are not active because they do not transmit any signal: the communication is by backscattering as in the passive tags. But they have a battery; therefore, the RF signal is used to wake up the tag from a low-power consumption state in sleep mode. Once the tag is waked up, it uses the internal battery to feed the electronic circuits. As a consequence, the read range is moderate, larger than passive but smaller than active.

4

Chipless RFID Technology

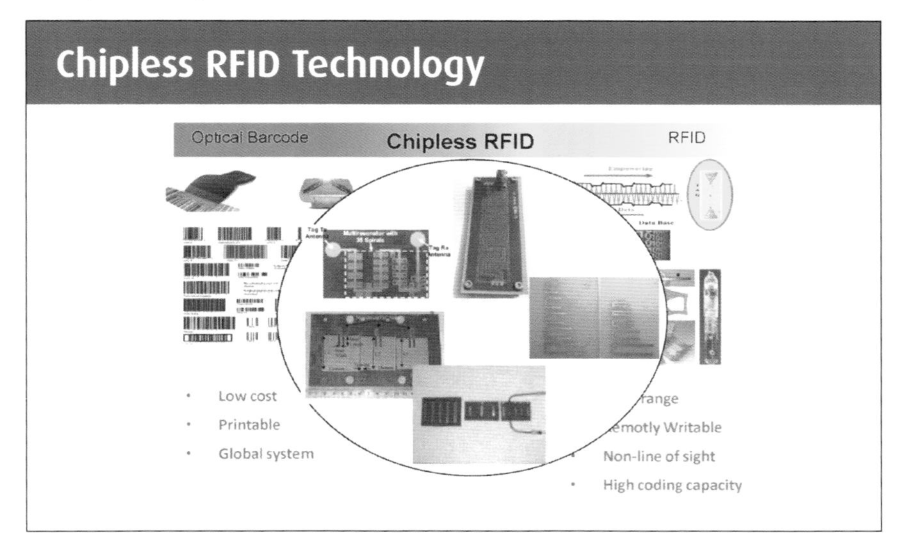

The idea behind the chipless RFID is to introduce a compromise between barcode and radio frequency identification (RFID). The main problems of the barcode and the RIFD as the non-line-of-sight reading and the price can be potentially solved by using an entirely passive structure without any chip. The information is encoded in the electromagnetic footprint of the label. The other potential advantage of chipless RFID is the operation in harsh environments and the possibility to perform sensing. On the other hand, there are a number of drawbacks, e.g. number of bits encoded and reading procedure, which need to be solved before thinking a realistic implementation on the market. Chipless RFID can be potentially employed in some niche applications as an alternative to the other two most popular technologies.

5

RFID: Chip vs Chipless tags

CHIP-EQUIPPED TAGS

- ☺ Reprogrammable
- ☺ Single-frequency operation
- ☹ High cost compared to optical barcode (not below 5 cents);
- ☹ Radiated power
- ☹ Operation in harsh environment;

CHIPLESS TAGS

- ☺ Low-cost
- ☺ Potentially low-power
- ☺ Operation in harsh environment;
- ☹ Not reprogrammable;
- ☹ Usually requires wideband operation;
- ☹ Limited bit storing capacitance;

RFID tag

Radar analogy

This slide presents a comparison between standard RFID and chipless RFID technology. The standard RFID tags offer some important advantages such as reprogrammability and single-frequency operation, but they also have some problems such as relatively high cost and operation in harsh environment, which impede their widespread diffusion. On the other hand, chipless RFID tags can be of potentially low cost since not any chip has to be installed on the tag and, for this reason, they are also suitable for operating in harsh environments. Chipless tags are fixed structures and thus they are not reprogrammable and usually require a large frequency band to operate. The maximum number of encoded bits is also a relevant problem now.

6

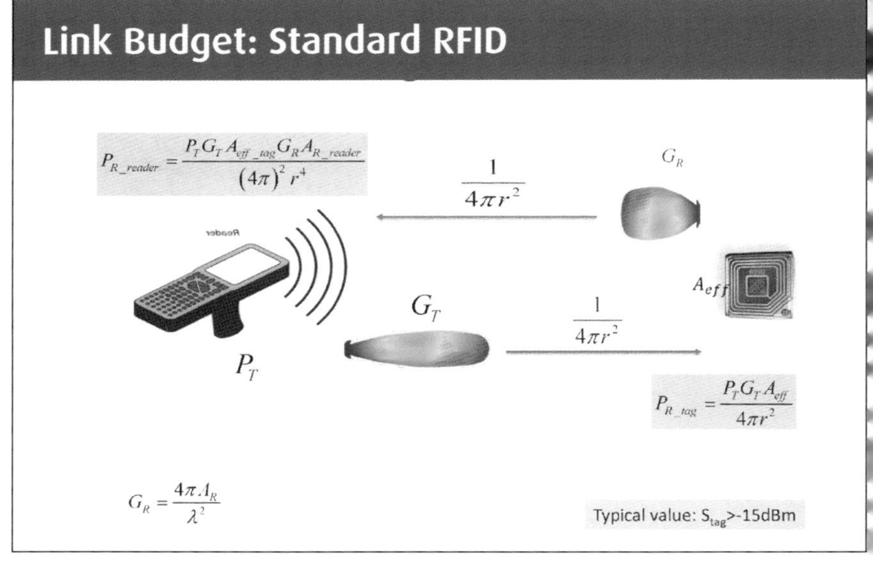

The slide reports the link budget of a standard RFID system. The most important aspect in a conventional RFID system is the power received at the tag input. Indeed, the tag is usually a passive structure without any battery and thus needs to get the DC input from the RF electromagnetic signal. The impinging signal is received and, part of it, is rectified to provide the necessary power supply to the tag. The level of the input power must be higher than a certain threshold, called sensitivity of the tag. If the impinging RF power is below this threshold, the tag is not activated and thus it does not respond to the reader. In conclusion, the reading range of a tag is limited by the tag sensitivity. This parameter has been greatly improved over the last years thanks to the research effect of the industry. Nowadays, a typical value is around -15 dBm.

7

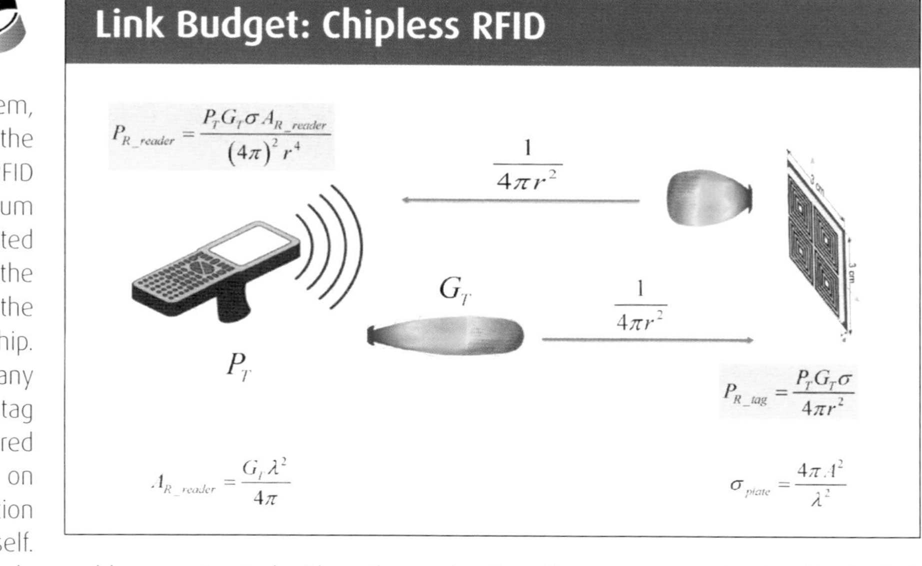

In a chipless system, differently from the conventional RFID system, the maximum reading range is dictated by the sensitivity of the reader and not by the sensitivity of the chip. Indeed, there is not any chip installed on the tag and the backscattered field depends only on the Radar Cross Section (RCS) of the tag itself. The RCS of the tag can be roughly approximated with the RCS of the plate of the same dimension of the tag. Therefore, the larger the tag, the larger will be the power scattered by the tag and thus received by the reader. The other parameters involved in the link budget are clearly the gain of the reader antenna and the distance between the reader and the tag.

8

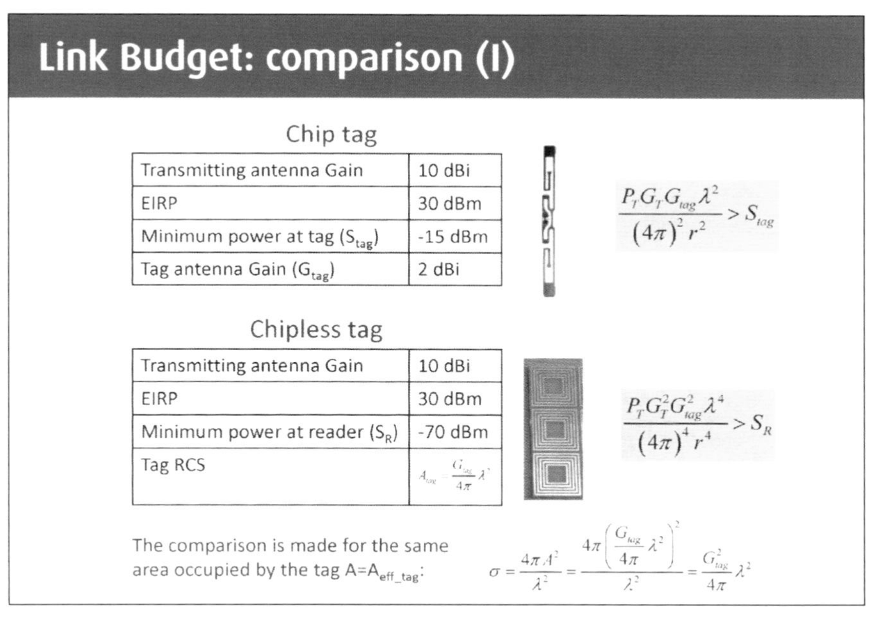

Chip tag

Transmitting antenna Gain	10 dBi
EIRP	30 dBm
Minimum power at tag (S_{tag})	-15 dBm
Tag antenna Gain (G_{tag})	2 dBi

Chipless tag

Transmitting antenna Gain	10 dBi
EIRP	30 dBm
Minimum power at reader (S_R)	-70 dBm
Tag RCS	$A_{tag} = \frac{G_{tag}}{4\pi}\lambda^2$

An interesting study, hypothesizing of having a tag of certain fixed dimension, is to understand what is the amount of power required to detect a conventional RFID tag or a chipless RFID tag. Some typical parameters are set for the reader antenna gain, the tag antenna gain, and the sensitivity of the tag and the reader.

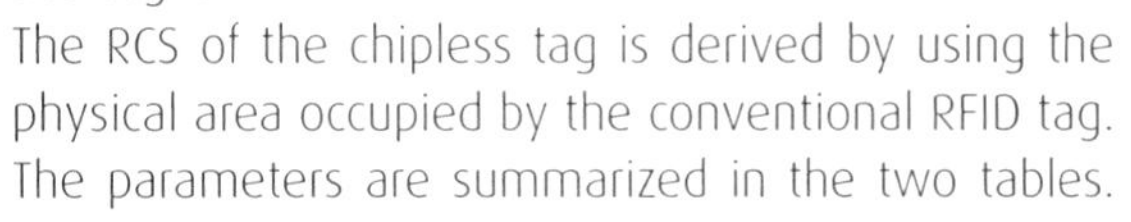

The RCS of the chipless tag is derived by using the physical area occupied by the conventional RFID tag. The parameters are summarized in the two tables. The power needed to detect the conventional RFID tag and the chipless one must satisfy the inequalities reported in the green boxes.

9

Link Budget: comparison (II)

f=3 GHz

f=10 GHz

The result of the comparison is reported in the two figures. The calculation has been performed at two different frequencies. At 3 GHz, the power required by using the chipless tag, with the parameters hypothesized in the previous tables, is much lower than the power required in the conventional RFID system. By increasing the operating frequency, and keeping fixed the other parameters, the chipless tag remains advantageous only at moderated reading distances. However, this is clearly a preliminary qualitative study, since chipless readers are not currently available in the market and the chipless systems usually work by using multiple frequencies.

10

Multipath: Two-ray model (I)

d = Antenna separation; h_t = Transmitter height; h_r = Receiver height

$$x + x' - l = \sqrt{(h_t + h_r)^2 + d^2} - \sqrt{(h_t - h_r)^2 + d^2}$$

One important aspect of radio frequency identification technique is the impact of the multipath. To understand the phenomenon, a simple two-rays model can be initially employed. Let us suppose to have a transmitting antenna located on a metallic ground plane and a receiving antenna at a certain distance l. The electric field received is composed of at least two contributions: the direct ray and the ray reflected by the ground plane. The ray reflected by the ground plane, for the image theorem, can be as high as a direct ray radiated by an imaginary antenna place below the ground plane. While the amplitude of the direct ray and the reflected ray can be considered to be of the same level, the phase of the two contributions is instead very different and depends on the length of the two paths. The total electric field is derived as the summation of two complex numbers after the dot product along the direction of the receiving antenna.

11  Multipath: Two-ray model (II)

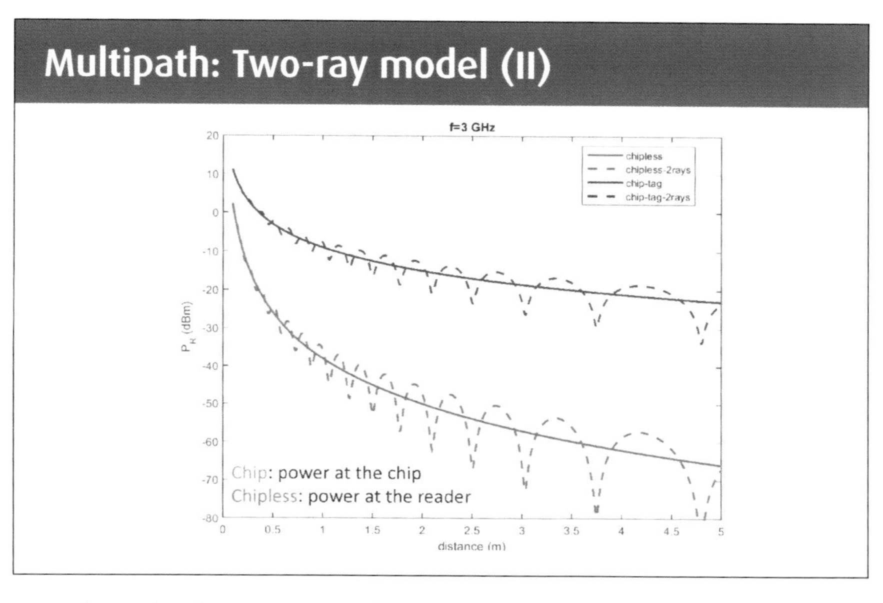

The total power received at the receiver side (for chipless case) and at the tag side (for chip case) is analyzed according to link budget models previously presented. The single direct path and the two-ray model are considered for evaluating the power. It is evident that, by using the single-ray model, the power decreases proportionally to the distance. On the contrary, by using the two-ray model, the received power oscillated around the previous trend. The distance between contiguous minima of the signal is proportional to the frequency. Indeed, as the frequency increases, the wavelength decreases, and a certain fixed distance corresponds to additional multiples of the phase.

12  Measurement: Example

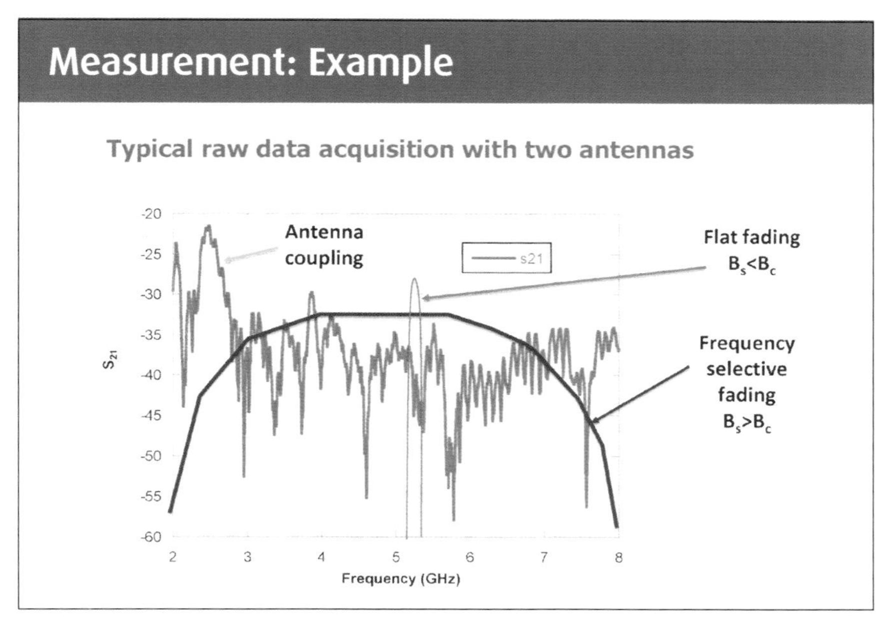

In a practical scenario, the received signal is the summation of several contributions. The summations of contributions arrive with different delays. One important parameter defined in a practical scenario is the delay spread, that is, the difference between the arrival time of the first and the last contributions. The larger is the dalay spread, the larger is the fluctuation of the signal as a function of frequency. In the figure, a typical received signal measured in a laboratory scenario is represented. If the bandwidth of the useful signal is small compared to the variation of the signal, we can speak of flat fading, whereas, if the signal bandwidth is large compared to the signal variation, we have frequency selective fading. The latter case is the case of frequency-coded chipless RFID.

13

Classification of chipless RFID tags

Chipless RFID tags
Time domain (TD)
Frequency domain (FD)
non printable
SAW tags
printable
TFTC
delay lines based tags
chemical
Nanometric materials
planar circuit
SAR based
LC resonant
antennas with multiresonators
resonant particles based

S Preradovic, NC Karmakar, "Chipless RFID: Bar code of the future", Microwave Magazine, IEEE, 2010

This slide reports a classification of the different chipless tag configurations.

The first level of discrimination is the domain of operation: time or frequency domain. In the former case, the information is encoded in the time delay of the received signal. A popular example of this type of tag is the Surface Acoustic Wave (SAW) tag. Time domain tags based on delay units are usually very limited in terms of bit capacitance. In the frequency-coded tags, the information is encoded in the spectrum. A bit is usually associated with the presence or absence of a certain resonant peak at a predetermined frequency. The frequency-coded chipless tags can be realized using different techniques. The most popular one is based on printed circuit boards.

14

Chipless RFID tags (I)

Chipless RFID tags
Time Domain
Frequency Domain
Hybrid SAR Harmonics

Key aspects of passive tags:

- Absence of any source of power supply but powered by the EM wave provided by the reader;
- The collected energy is used for both data "processing" and communication.

We are going to analyze the different approaches employed for the design of chipless RFID tags. Within this framework, we will report the two typical paradigms for information encoding, namely the Time Domain Reflectometry and the Spectral Signature, although other mixed approaches are possible. We will also address the limitations and challenges related with the reading system and calibration procedures of chipless RFID systems. Finally, we will show that a chipless RFID tag can be transformed into a chipless RFID sensor by exploiting a chemical-interactive material (CIM).

As already mentioned, a chipless RFID tag does not require any power supply other than that provided by the reader by means of an EM wave. The collected energy is exploited for processing the data and retransmitting the information.

15 

Chipless RFID tags (II)

Chipless RFID tags

Time Domain

Frequency Domain

Hybrid SAR Harmonics

- For chipless RFID tags exploiting the Time Domain,the reader sends a pulse to the tag and then collects the backward echoes of the pulse coming from the tag. The information is encoded within the reflected pulse train.
- For chipless RFID tags operating in the Frequency Domain,the data can be embedded in the amplitude, phase polarization (or a mix of them).

→ Chipless RFID systems that exploit the Time Domain for transferring the information employ a reader to send a pulse to the tag and then collect the backward echoes of the pulse coming from the tag. On the other hand, chipless RFID systems that operate in the Frequency Domain embed the information in the backscattered electromagnetic field as a function of the signal amplitude or phase, or a mix of them. Other strategies have been recently proposed. Among these, the use of higher-order harmonics can be adopted for designing harmonic tags. The reader sends a signal at frequency f_0 and the tag produces an echo signal at $2f_0$ by means of a non-linear device and retransmits to the reader. The collected energy is used for both data "processing" and communication. Thus, the harmonic is generated and detected by the reader, only in the presence of the tag. In a more recent approach, an image-based RFID tag surface is illuminated by an EM signal and the reflected signal with orthogonal polarization is collected to generate the EM image of the tag. In this technique, the presence or absence of each polarizer in the image represents 1-bit of encoding data.

16 

Surface acoustic wave (SAW) (I)

- SAW tags are the most successful example of chipless RFID which are on the market.
- They require photolithographic process for their fabrication.
- They completely rely on piezoelectricity: when applying a voltage to such a crystal, it will deform mechanically, converting electrical energy into mechanical energy. The opposite occurs when such a crystal is mechanically compressed or expanded. Charges form on opposite faces of the crystalline structure, causing a current to flow in the terminals and/or voltage between the terminals.
- These materials have a high dielectric constant (between 50 and 100).

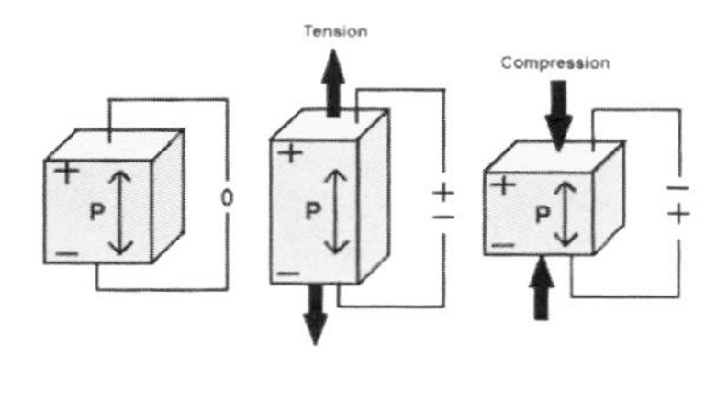

→ One of the most remarkable examples of chipless RFID exploiting the time domain signature is the Surface Acoustic Wave tag. SAW tags are fabricated by using a photolithographic process and rely on materials that, among other properties, exhibit a high dielectric constant (permittivity between 50 and 100). The working principle of these tags is based on piezoelectricity, which is exhibited especially by crystals and ceramics, both natural and synthetic. Due to this phenomenon, when applying a voltage to such a material, it will deform mechanically, converting electrical energy into mechanical energy. The opposite occurs when the material is mechanically compressed or expanded. Charges form on opposite faces of the crystalline structure, causing a current to flow in the terminals and/or voltage between the terminals.

17 

Surface acoustic wave (SAW) (II)

- The SAW tag receives the EM wave emitted by the reader;
- A transduction between the interrogated pulse and the (much slower) acoustic wave is achieved by unidirectional interdigital transducer (IDT);
- The SAW pulse propagates along the surface of the piezoelectric material substrate and it is partially reflected by each of the metal-based reflectors;
- The train of reflected SAW pulses are reconverted into an electrical signal by the IDT and retransmitted by the tag antenna;
- Finally, the backscattered signal is collected by the reader and decoded.

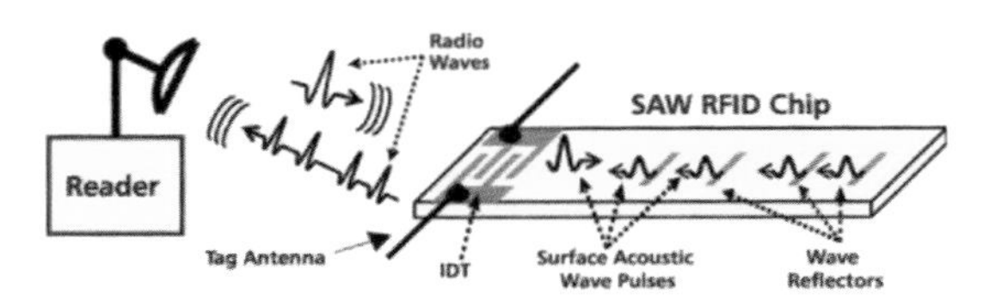

→ The working principle of a chipless tag that exploits the features offered by piezoelectricity to realize a SAW tag can be summarized as reported in the figure. Passive circuit components, such as the interdigital transducer (IDT) as well as reflectors, are printed on a slab of piezoelectric material. An antenna is connected to the IDT to guarantee the harvesting of the power provided by the radio waves. More in detail, the SAW tag is illuminated by an electromagnetic pulse produced by the reader. The tag antenna collects this signal and the unidirectional IDT connected to it realizes the transduction between the interrogated pulse and the (much slower) acoustic wave. The acoustic wave propagates along the surface of the piezoelectric material substrate and the pulse undergoes several partial reflections caused by the metal-based reflectors. Therefore, the single pulse is transformed into a train of reflected SAW pulses. These pulses are then reconverted into an electrical signal by the IDT and retransmitted by the tag antenna. The backscattered train of pulses are finally collected by the reader and thus the information can be recovered.

18 

Surface acoustic wave (SAW) (III)

Encoding techniques

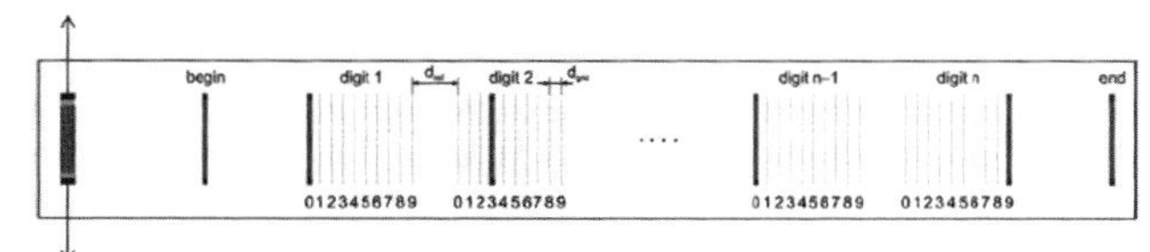

Time positioning

- Tag identification is performed by measuring the time of travel of surface waves to reflectors and back to the antenna, and fitting the computed delays to the known code grid;
- To ensure reliable identification, an additional spacing d_{ref} is used between code blocks so that interference between reflections is prevented;
- Additional reflectors are used at the beginning and end of the tag to remove reflections that produced by the surrounding environment. Relatively large distances before the first and after the last code block assure reliable scaling of the measured delays to the code grid.
- Only a single reflector can be placed in each group of *n* possible positions.
- By using *m* reflectors, the data capacity in bit is equal to *m* log2(*n*)

More details in: A. Stelzer, M. Pichler, S. Scheiblhofer, and S. Schuster, "Identification of SAW ID-tags using an FSCW interrogation unit and model-based evaluation," *IEEE transactions on ultrasonics, ferroelectrics, and frequency control*, vol. 51, no. 11, pp. 1412–1420, 2004.

Different options are available to encode the information. One option is represented by the "time positioning". In this case, the information content is embedded in the time required by the surface waves to travel to reflectors and back to the antenna. It is then possible to compare the computed delays to the known code grid to recover the information. In order to increase the reliability of the reading process, a spacing d_{ref} is used between code blocks to prevent interference (merge) between reflections of two different blocks. Moreover, additional reflectors are used at the beginning and end of the tag to remove reflections produced by the surrounding environment and set a reference in time. Relatively large distances before the first and after the last code block assure a more reliable scaling of the measured delays to the code grid. This scheme requires that only a single reflector can be placed in each group of n possible positions and thus the data capacity in bit is equal to m log2(n) if m reflectors are employed.

19 Surface acoustic wave (SAW) (IV)

Encoding techniques

Phase modulation

- Recovering the phase of the reflected pulses can increase the data capacity.
- This scheme consists of placing the reflectors more precisely within their slots. Phase shifts of 0°,−90°, −180°, and −270° are obtained by placing the reflector at multiples of λ/8.
- When the combined time position and phase encoding is used, each reflector has 4 possible time positions and 4 possible phases. This sums up to 16 different states and corresponds to 4 bits of data. For example, a 10-code-reflector tag can achieve a data capacity of 40 bits.
- All reflector positions for the time-position-encoded tags lie at multiples of λ from each other and correspond to a phase of 0°, although phase information does not play any role in time position encoding.

λ/8 λ/8 λ/8
0°
-90°
-180°
-270°
nλ
pλ
qλ
(n+1/8)λ
(p+1/4)λ
(q+3/8)λ

More details in: S. Harma, W. G. Arthur, C. S. Hartmann, R. G. Maev, and V. P. Plessky, "Inline SAW RFID tag using time position and phase encoding," IEEE Transactions on Ultrasonics, Ferroelectrics and Frequency Control, vol. 55, no. 8, pp. 1840–1846, Aug. 2008.

→ Another interesting approach is related to a "phase modulation" since the knowledge of the reflected pulse phase can increase the data capacity. In this solution, the reflectors are placed within separate slots. Phase shifts of 0°,−90°, −180°, and −270° are then obtained by placing the reflector at multiples of λ/8. Therefore, when the combined time position and phase encoding is used, each reflector has 4 possible time positions and 4 possible phases. This sums up to 16 different states and corresponds to 4 bits of data. For example, a 10-code-reflector tag can achieve a data capacity of 40 bits. SAW tags have been developed at 433 MHz, 868 (previously 856 MHz) and 2.4 GHz ISM. The number of different codes is determined by the *B T* product (*B* is the frequency bandwidth and *T* is the coding time). If a data capacity of 32 bits (or, better, 64 or 128 bits) is requested, a frequency band of 16 MHz (or 32 MHz or 64 MHz) is needed. A recent proposal is to exploit the UWB band, which is much wider than the 2.45-GHz ISM band (but different allowed power levels). A certain value of BT product (tag capacity) can now be obtained with a significantly shorter coding delay, which means a considerable reduction of tag size. A shorter coding time also implies lower losses.

20

Left-Handed (LH) transmission lines (I)

- The dispersion diagram shows that, below a certain frequency, the group velocity of the left-handed (LH) line is smaller than the group velocity of the RH line built up with identical elements.

- Since there is no way of building a TL with a distributed left-handed lumped element model, one has to build a structure that shows the behavior of a left-handed TL, a so-called artificial line. The simplest approach is to arrange unit cells given by the lumped element model in a ladder structure.

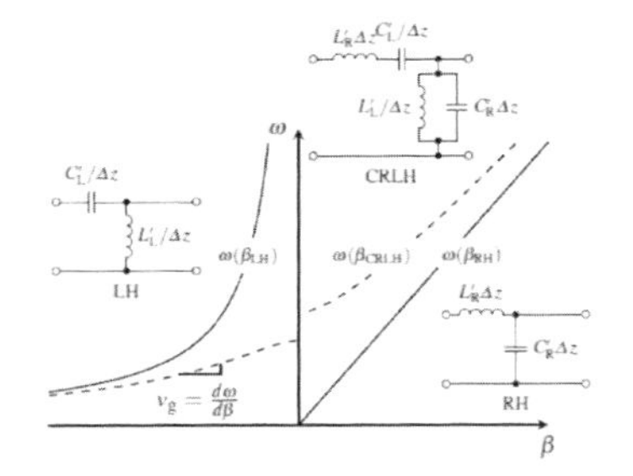

More details in: C. Mandel, M. Schussler, M. Maasch, and R. Jakoby, "A novel passive phase modulator based on LH delay lines for chipless microwave RFID applications," in *Wireless Sensing, Local Positioning, and RFID, 2009. IMWS 2009. IEEE MTT-S International Microwave Workshop on*, 2009, pp. 1-4.

It is apparent that the maximum number of bits that can be stored is limited by the number of pulses that can be stored on the line. Assuming a line length of L and the length L_p of a single pulse, num_bits is limited by their ratio. Since L_p depends on the products of the group velocity v_g and the pulse duration τ, which corresponds to the inverse of the pulse bandwidth Δf , the upper limit for the maximum value of ***num_bits*** is: ***num_bits*** $= L/L_p = L\, Df/v_g$. For real systems generally, the bandwidth is rather limited as well as the physical line length. Therefore, in order to increase ***num_bits***, the group velocity, v_g, has to be minimized. A possible solution to overcome this drawback could be the use of high-permittivity substrates. However, they may suffer high losses, be expensive, and pose design problems.

21

Left-Handed (LH) transmission lines (II)

- Physically realized LH structures always show an additional RH behavior that is caused by parasitic. The CRLH (Composite Right/Left-Handed) exhibits a group velocity lower than that of the LH line, both the right-handed and the left-handed parts contribute to the entire delay.

- Practically, depending on the pulse bandwidth and center frequency, 2-4 times larger values can be obtained with the drawback of dispersion effects on the LH line.

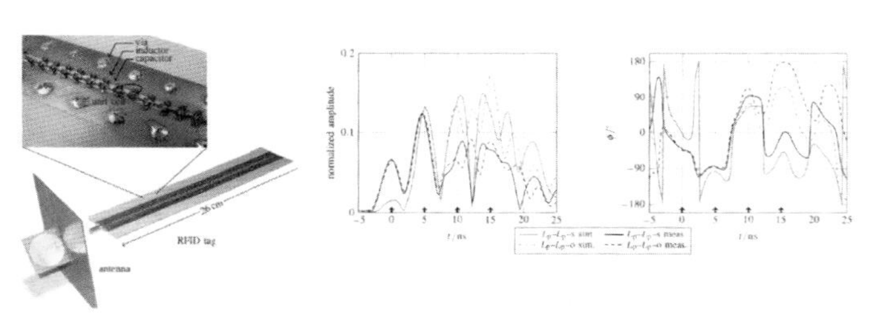

More details in: C. Mandel, M. Schussler, M. Maasch, and R. Jakoby, "A novel passive phase modulator based on LH delay lines for chipless microwave RFID applications," in *Wireless Sensing, Local Positioning, and RFID, 2009. IMWS 2009. IEEE MTT-S International Microwave Workshop on*, 2009, pp. 1-4.

→ Recently, metamaterials have been proposed as an alternative solution. In fact, the smaller group velocity provided by left-handed (LH) medium with respect to a right-handed (RH) one can be helpful. An artificial line with such properties can be obtained by arranging unit cells in a ladder structure. However, when manufactured, these structures do not exhibit a purely LH response but always show an additional RH behavior that is caused by parasitic effects. Therefore, the transmission line is a mix of the two, namely a Composite Right/Left-Handed (CRLH) one. It is important to note that this guiding structure determines a group velocity that is even lower than that of the LH line since both the right-handed and left-handed parts contribute to the overall delay.

22

Time-coded chipless UWB RFID (I)

Tag: UWB antenna + delay line (length L)

- **Structural Mode:** shape, size, material of tag
- **Tag (or antenna) Mode:** radiation properties and load of the antenna (**here: open-ended**)

→ **ID: Time** between Structural and Tag Modes

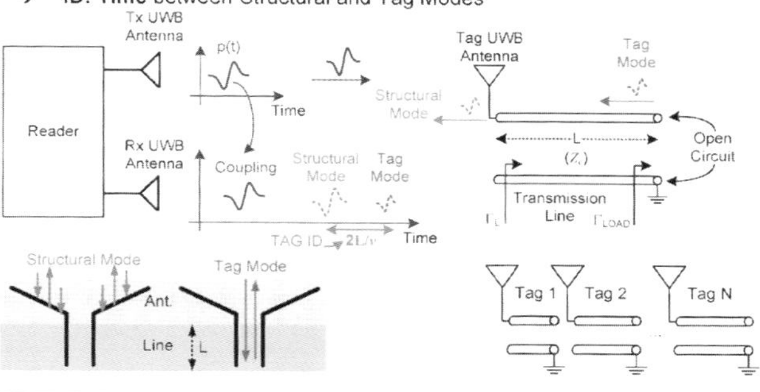

Time-coded chipless UWB RFID addresses a topology of chipless tags which integrate a UWB antenna connected to an open-ended delay line with length L. The information is coded in the length of this delay line. When a wave arrives at the tag from the reader (normally a UWB radar), two modes are generated: **1)** a structural mode, which depends on the shape, size, and material of the tag, which is basically that portion of the energy that is reflected at the physical tag toward the reader and **2)** a tag mode, which contains the tag information. This tag mode is the portion of energy captured by the antenna, which travels along the delay line, that is reflected at the end of the line and is re-radiated by the antenna toward the reader. The time difference between the structural and the tag modes at the reader depends on the length of the delay line: in this time difference, the identification information of the tag is coded.

23

Time-coded chipless UWB RFID (II)

- Tag antenna modeled as a 2-port model:

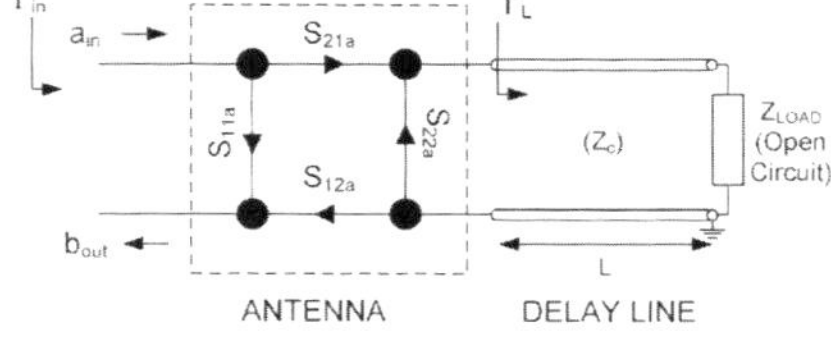

- Signal at reader: **sum** of structural and tag modes

$$s(t) \approx s_{Str.}(t) + s_{Tag}(t) \begin{cases} s_{Str.}(t) = \alpha S_{11}(t) * p(t) * \delta(t - \tau_p) \\ s_{Tag}(t) = \alpha \Gamma_{LOAD}\, g(t) * p(t) * \delta\left(t - \tau_p - \frac{2L}{v}\right) \end{cases}$$

- α is the atennuation due to propagation
- $g(t) = \Im^{-1}[S_{21a} S_{12a}]$
- v is the propagation speed in the line

Time-coded chiples tags can be modeled as 2-port antennas connected to a transmission line. The signal at the reader is the sum of the structural and tag modes. The main differences between the two modes analytical expressions are two: **1)** the tag mode amplitude depends on the reflection coefficient G_{LOAD} at the end of the transmission line and **2)** the tag mode is delayed with respect to the structural mode depending on the delay line length L. In the first case, the parameter G_{LOAD} can be used to insert a sensor, while in the second case, the length of the line can be changed to introduce different ID tag codes, or its electrical length can be changed depending on a physical parameter to act as sensor. The time resolution that can be coded and detected (coding capacity) is the limitation of this topology in front of other chipless topologies.

24

Time-coded chipless UWB RFID (III)

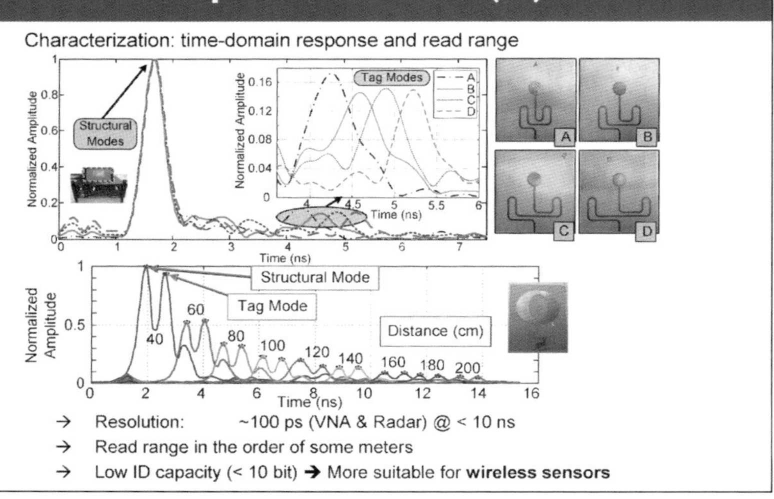

Four time-coded chipless tags with different line lengths are compared in this slide (A,B,C,D). The structural mode can be observed, superimposed the ones to the others, while the four tag modes can be observed at different time positions. Since the tag size is large, a much higher structural modes than tag modes can be observed. The measurement has been obtained by using a commercial UWB radar as reader. The signal at reception has been time gated in order to separate reflections from surrounding elements and the wavelet transform has been applied as matching filter in order to suppress noise. At the bottom of the slide, the read range of a time-coded tag is also presented. It can be observed that distances up to around 2 m can be read. Read range is a strong advantage of this topology in front of other chipless topologies.

25

Time-coded chipless sensors (I)

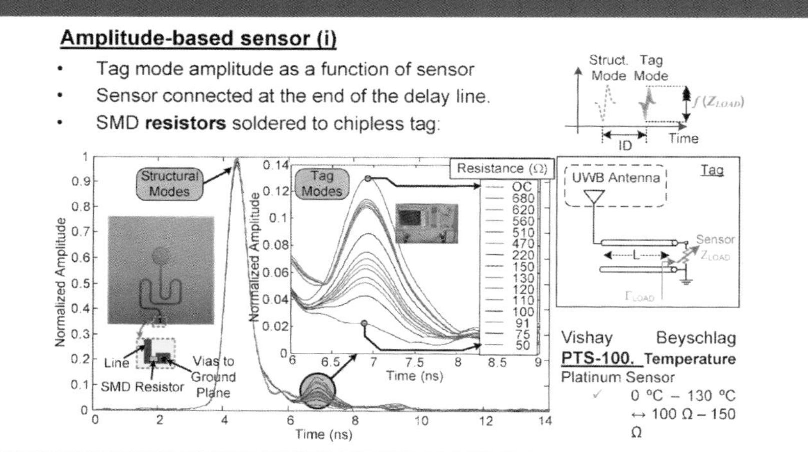

Since this topology offers a long read range but a small coding capacity, the most interesting application for these tags is to use them as sensors. Information can be coded in time (changing the propagation velocity along the transmission line) and in amplitude of the tag mode (changing the reflection coefficient at the end of the delay line). In the amplitude-based sensor shown in this slide, the reflection coefficient is changed by varying the value of a resistor connected at the end of the delay line. A change in the tag mode amplitude can be detected, and this opens the door to replace these lumped resistances by resistive sensors, as, for instance, a Vishay Beyschlag PTS-100 temperature platinum sensor. In consequence, in this sensor topology, the delay between modes is used as ID coding element and the ratio between the amplitudes of the tag and the structural modes is used as sensing parameter, which changes with temperature.

26

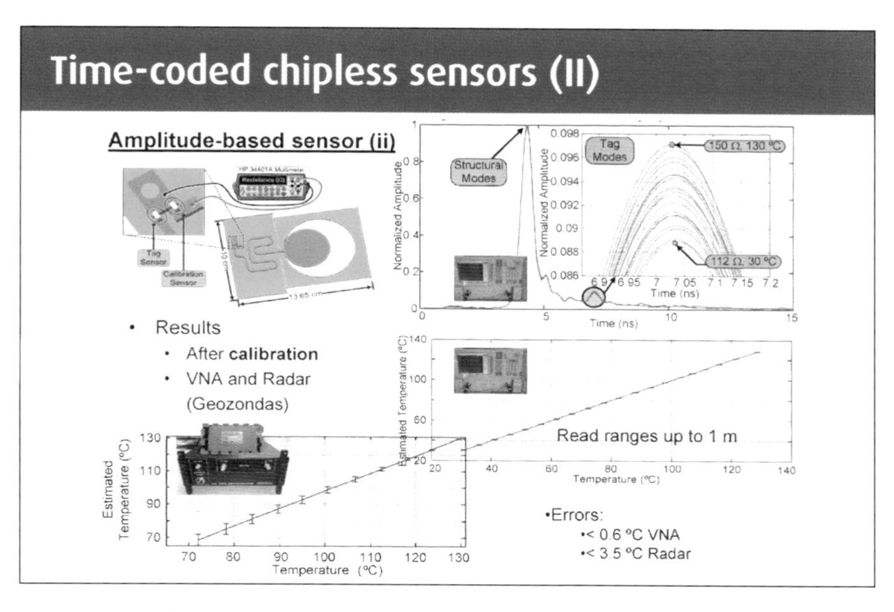

Here is the measurement of the time-coded chipless tag working as temperature sensor. Two PTS-100 are placed side-by-side. One of them is connected at the end of the delay line, and the other is connected to a multimeter for reference and calibration purposes. The sensor is read with a UWB radar and also with a VNA. The time response in the latter is obtained by applying the inverse Fourier Transform. The variation in the tag mode amplitude is observed and the the measured response of the sensor at distances up to 1 m is shown as a function of the real temperature (obtained from the measurement with the multimeter). Errors of 0.6° and 3.5° are obtained when using the VNA and the UWB radar, respectively, for a measurement range between 72° and 130°.

27

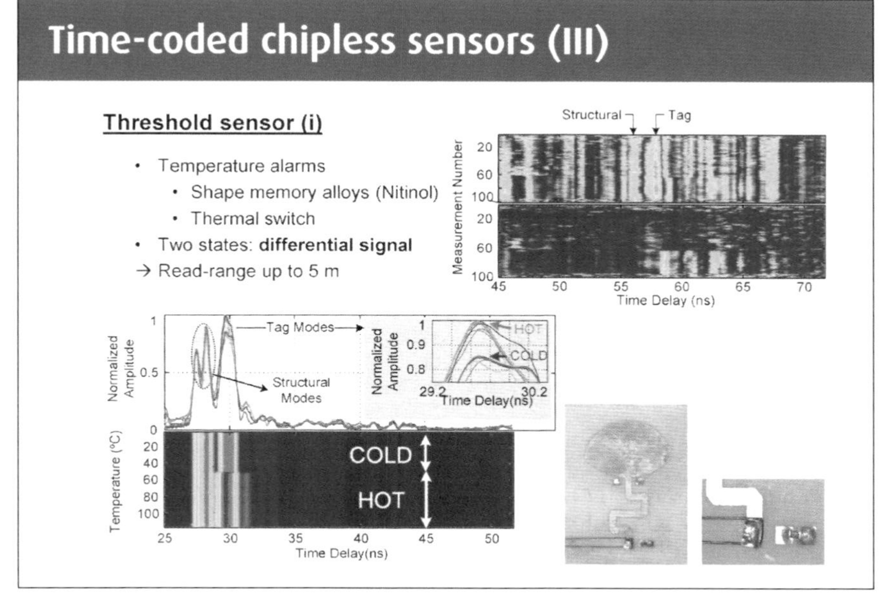

Another possibility is to use a temperature threshold sensor. To this end, a shape memory alloy (in this case nitinol) is used as mechanical element. This alloy can be memorized with a determined shape when a temperature is surpassed. When the threshold temperature is surpassed at some time, the alloy recovers the memorized position, leading to a mechanical variation. In this prototype, this mechanical variation takes place at the end of the delay line: an elastic, adhesive copper strip at the end of the delay line is de-attached, and the reflection coefficient changes. This application does not require high sensitivity, it is just an on/off application. By using differential calibration (detecting changes, not states), a long read range up to 5 m is obtained.

28 

Time-coded chipless sensors (IV)

Delay-based sensor (i)

- Concrete composition sensor
  - Civil engineering: concrete $\varepsilon_r \sim 5$ + dry Sand $\varepsilon_r \sim 2.7$
  - Percentage of each affects structure integrity
  - **Metal plate** sets stable reference peak

Struct. Mode, Tag Mode, Time, $f(\varepsilon_{efr})$

UWB Antenna, Tag, Material (ε_{r2}), L, Tr. Line (ε_{r1}), Medium (ε_{eff})

Metal Plate, Concrete (ε_{r2}), Antenna (on Air), Embedded Delay Line (ε_{r1})

Air ε_0, W, ε_{r2}, h_2, h, ε_{r1}

Metal Plate

Normalized Amplitude; Time (ns); Fs, A, B, C; Norm. Ampl.; Time (ns)

A delay-based sensor is presented here. The operation principle is to change the propagation velocity in the delay line depending on the material that is in contact with the tag. Depending on the permittivity of this material, the velocity changes and so does the time difference between tag and structural modes. As a proof of concept, it has been applied to the measurement of concrete composition for civil engineering. The tag is inserted inside a concrete block. Since the tag must be read in perpendicular, a Vivaldi antenna has been used. A metal plate perpendicular to the tag is placed in order to establish a stable reference plane (which does not depend on the insertion of the tag inside the mixture). Several samples where the mixture between concrete and sand is changed have been tested. It can be observed in the measurements the change in the delay between structural and tag modes depending on the percentage of concrete and sand.

29 

Applications Semi-passive technology

Semi-passive High Frequency (13.56 MHz):
-NFC/ISO14443 data loggers (Temperature, pressure...)
Semi-passive Ultra High Frequency (860-960 MHz) tags
-Track and monitor *temperature sensitive products*
Semi-passive Microwave Frequency (5.8 GHz) tags
- *Highway toll collection (Teletac, via-T)*
- *Short ranges but very fast data transfer rates*

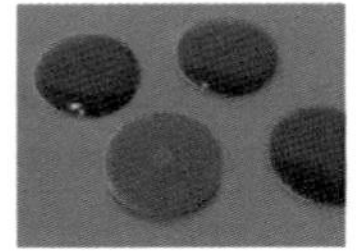

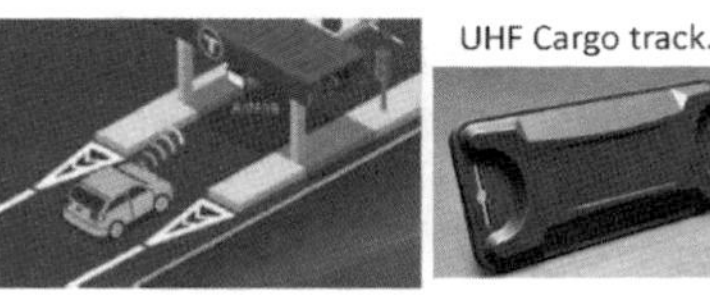

UHF Temperature

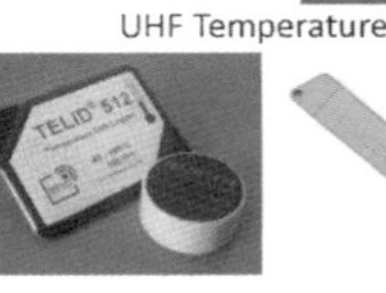

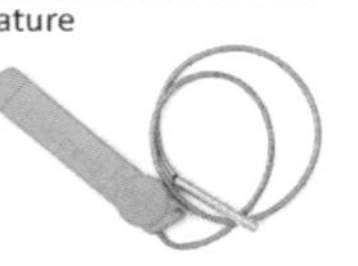

The following slides provide an introduction to semi-passive tags.

A highly spread topology of these tags can be found at high frequency (13.56 MHz). For example, they are used in temperature or pressure data loggers.

Another band used for semi-passive tags is UHF for long-range temperature data loggers. A third example at microwave frequencies are the tags used in highway tolls. In this case, a battery is necessary because the tag must be read very fast and the data rate is also very high.

30 

Backscattering communication

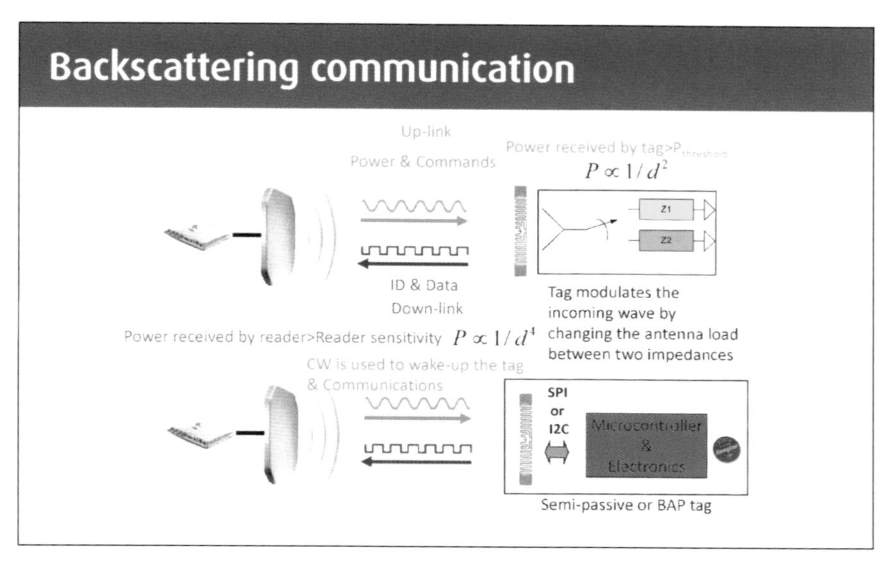

In both cases, in passive and semi-passive technologies, the communication is by backscattering.

The reader transmits an interrogating signal, then the passive tag rectifies the RF signal and obtains DC power to feed the circuits and send the message by modulating the radar cross section. The tag changes the impedance that loads the antenna; therefore, the reflected signal changes as a function of the impedance. For instance, one impedance represents a logical one and the other a logical zero.

The power received at the tag must be higher than the threshold power to allow the tag to wake up. This received power depends on the inverse of the square of distance.

Then, the reflected signal must propagate to the reader; therefore, the received signal at the reader is inversely proportional to the distance to the fourth. This power must be higher than the reader sensitivity. The main differences in semi-passive tags are that the threshold power is considerably lower than that in passive tags and tags can be powered by the battery during the answer period.

31 

Read Range: UHF passive vs BAP tags

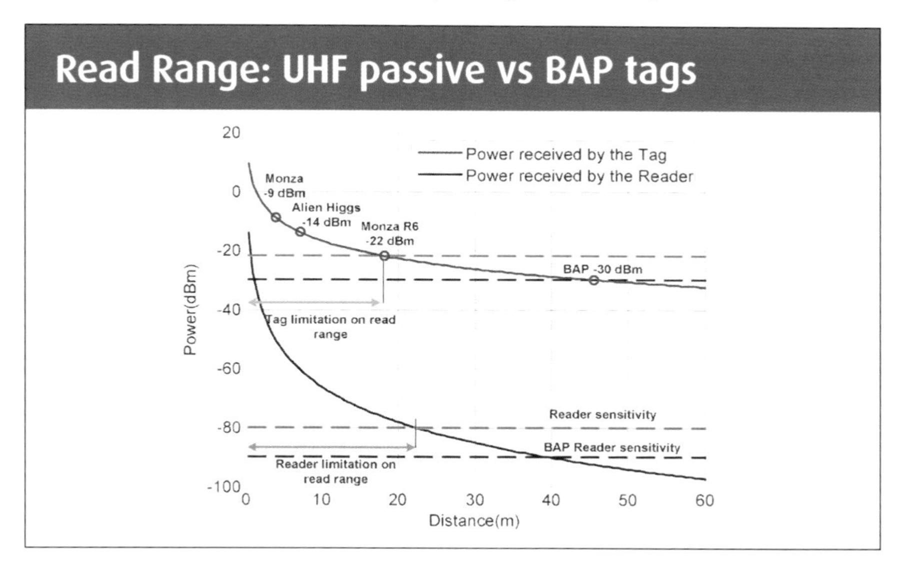

In order to understand the read range limitations, this plot depicts the received power at the tag at UHF (868 MHz). The black continuous line is the backscattered power received at the reader.

The points are the threshold power or tag sensitivity for different passive tags. The first generation of tags (e.g., Monza) have a poor sensitivity; however, the last generation (Monza R6) have sensitivities in the order of -22 dBm. The sensitivity of passive readers is in the order of -80 dBm. Therefore, the passive tags are limited by the tag sensitivity.

In the case of semi-passive tags, the tag sensitivity can be up to -30 dBm or higher. Then, the read range limitation is in the sensitivity of the reader. In addition, often, semi-passive readers have better sensitivity than passive readers. Therefore, read ranges close to 50 m can be obtained with battery-assisted passive tags.

32

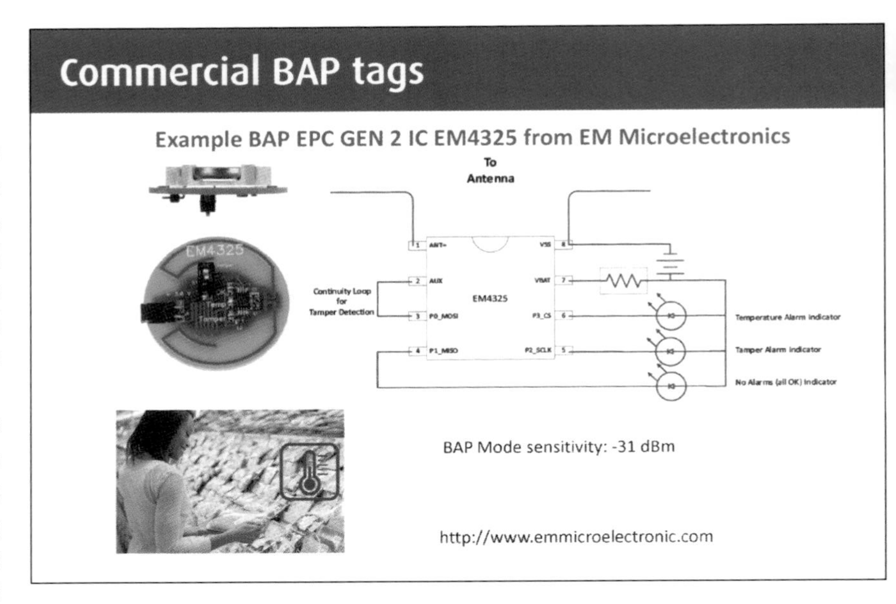

Next slides show some examples of semi-passive tags based on commercial ICs. EM4325 is a UHF Battery-Assisted Passive RFID IC from EM Microelectronics.

The antenna is connected between ANT+ and VS pins, and is a dipole with a matching network. The tag can be used as passive without battery or with battery. Depending on the mode, the sensitivity and the chip impedance change. Therefore, the matching network changes. The sensitivity of this IC is -31dBm in semi-passive mode.

The tag can be configured to read some alarms, for example, a tamper alarm, or using a threshold temperature chip when the temperature surpasses a limit (up or down). These alarms are saved in some bits of the user memory, and this information can be read using a conventional UHF reader.

33

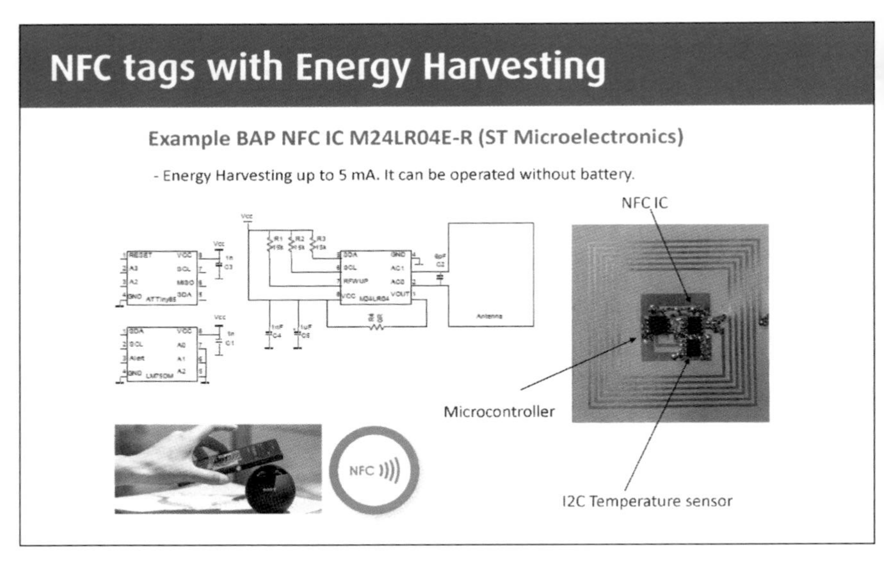

This slide shows another example of semi-passive tag. In this case, it is an NFC-based tag. There are models of NFC chips that can be connected to the battery and have an IC bus to communicate with a microcontroller or sensors. For example, in this case, a chip that has the energy harvesting function can obtain up to 5 mA from the RF signal. For higher currents or higher read distances, the tag must be powered by the battery.

In this picture, it can be seen a custom temperature sensor. It is composed of the NFC chip, the I2C temperature sensor, the pull-up resistance, and an Atmel ATTiny85 microcontroller. By reducing the clock frequency, it is possible to reduce the current consumption below 1 mA, therefore it can read the temperature approaching a mobile phone without any battery. The main attraction of this sensor technology is the low-cost reader (the smartphone) and the compatibility with metal or wearable devices.

34 

Semi-passive UWB RFID (I)

Microcontroller-based semi-passive system (i)

2.45 GHz
Reader
UWB
Tx
Rx
Wake-up interrogation
Tag
UWB Backscattered answer
Narrowband (2.4 GHz ISM) **wake-up** + UWB **backscatterer**
Modulated Signal
On-Off Keying
Reader
2.45 GHz Antenna
2.45 GHz Antenna
DC Block (100 pF)
Matching Network
HSMS-2852
Rectifier
RC Block
Tag
70.93 mm
50.97 mm
UWB Backscatterer
PIN Diode
2.45 GHz Detector
Delay Line
2.45 GHz Antenna
Detector
PIC
UWB Antenna
Program. pins
UWB Antenna
2.45 GHz Antenna
Battery
Programming pins (PIC)

Semi-passive UWB RFID with a time-coded topology is an alternative to time-coded chipless RFID. In this case, some electronics are added, achieving some advantages as introduced in the former slides. The prototype shown here is composed of a time-coded tag (UWB antenna and delay line), in this case loaded with a pin diode. The state of this pin diode is changed (on/off) usign a microcontroller, which has a stored ID code and can also integrate sensors connected. The tag also contains a wake-up circuit made of a rectenna at 2.45 GHz ISM band. The system works as follows: normally it is at a sleep mode, when the wake-up signal is received, the tag wakes up and the communication between tag and reader (UWB radar) is done by backscattering, as in time-coded chipless RFID.

35 

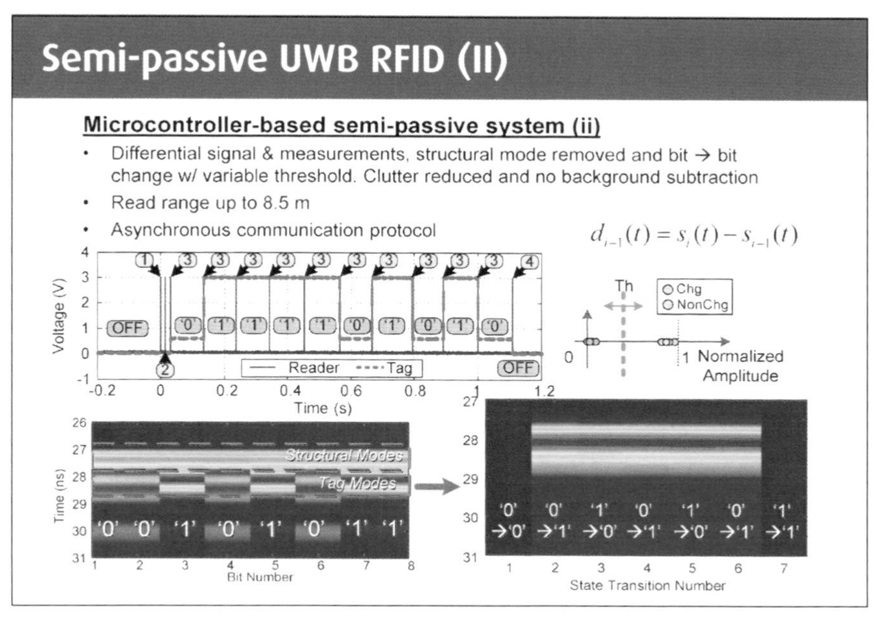

A communication protocol can be defined since a microcontroller is used. On one hand, the wake up of the tag is done by sending a specific code; this is done in order to avoid false wake-ups, which would lead to a shorter lifetime of the tag battery. Then, the communication by backscattering starts between the UWB radar (reader) and the tag, to download all the information from the tag to the reader (ID, sensor...). In order to avoid background subtraction, a differential encoding protocol is programmed: this means that the difference between consecutive bits is detected and the information concerning a 0/1 comes from a change/no-change event. The read range of the tag is 8.5 m.

36 

Semi-passive UWB RFID sensors (I)

Solar-powered temperature sensor (i)

- Energy harvesting enables remote sensors with no need of external power supply and batteries, ideal for long-term monitoring and applications where a vast number of sensors are scattered
- Analog topology
- PIN diode backscatterer
- **NTC** 0805 as temperature sensor
- Powered by solar cell
- 82 µA power consumption

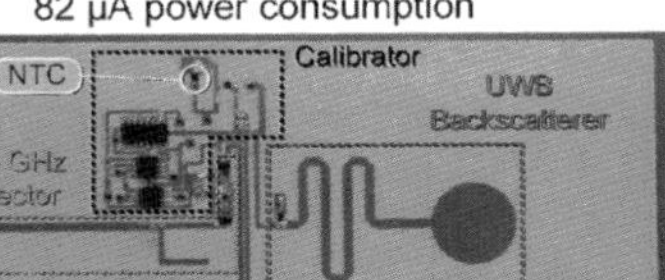

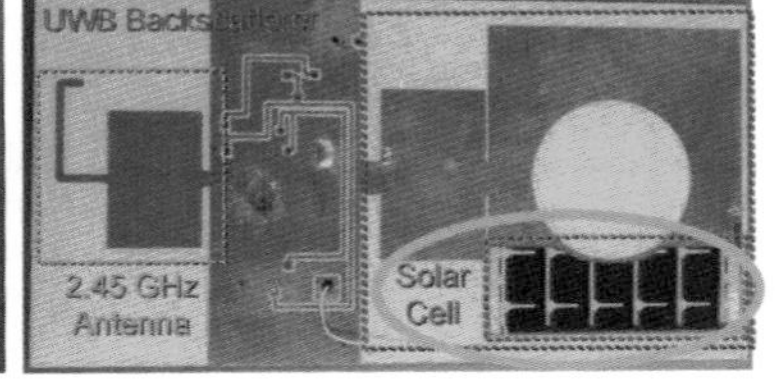

This is a first example of integration of sensors in a semi-passive UWB RFID tag. The tag consists again of a UWB antenna connected to a delay line loaded by a diode. A wake-up circuit made of a rectenna at the 2.45 GHz ISM band is also integrated, with the same performance as former slide. Here, a solar cell is integrated as energy harvester in order to provide energy to power up the tag, so no additional battery is required. No microcontroller is used in this prototype in order to save battery, leading to an energy consumption of 82 mA, able to be powererd from the solar cell, which is placed on the ground plane at the backside of the UWB antenna. The tag detects temperature using an NTC resistor. A calibration circuit is also added, which permits to remotely obtain the temperature from three measurements: temperature measurement, and two calibration measurements in order to avoid background subtraction and calibrate temperature.

37 

Solar-powered temperature sensor (ii)

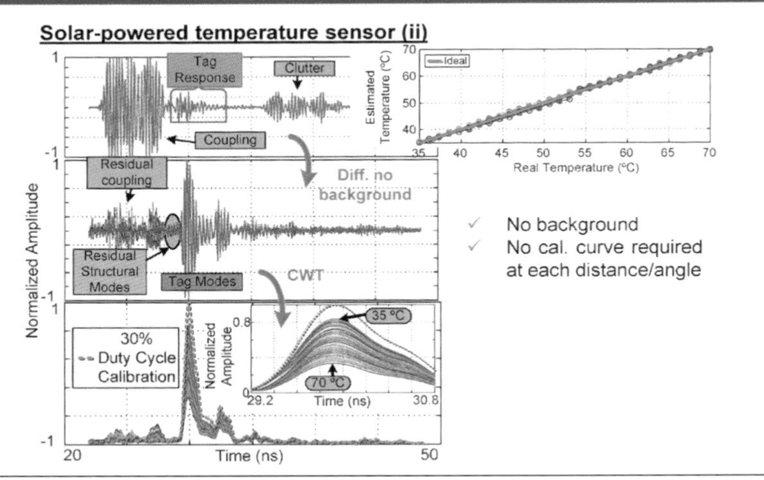

Here, the behavior of the solar-powered temperature sensor tag is shown. It can be shown that by using differential coding both clutter and structural mode are diminished, without the need for background subtraction. Since three consecutive measurements are done every time, a temperature is measured, the value can be calibrated without the need for a calibration curve at all distances/angles that the tag can be measured (as necessary when a chipless topology is used). A measurement of the change of the tag mode amplitude as a function of the temperature is also shown, which comes from the change in the bias of the diode that loads the delay line, directly dependent on the temperature.

38 Semi-passive UWB RFID sensors (III)

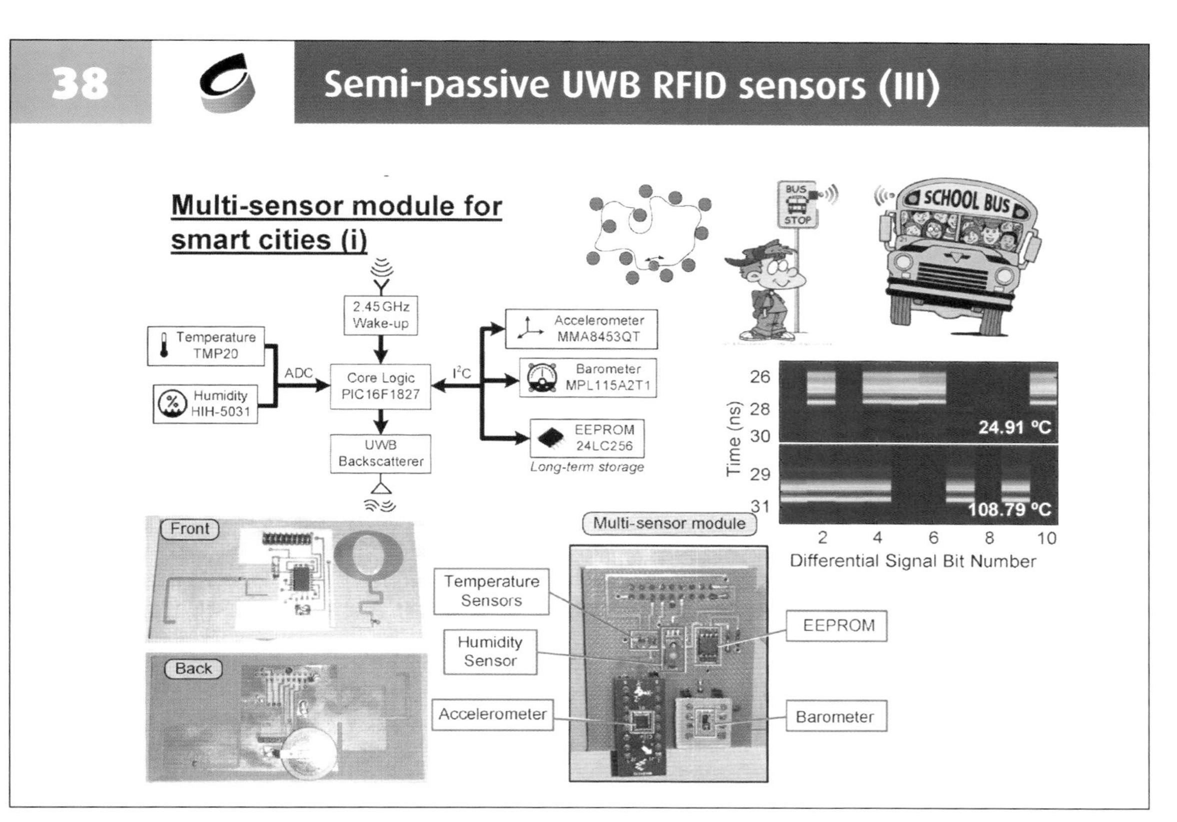

A second example of the UWB RFID can be this multi-sensor module for smart cities. It consists of the microcontroller tag shown before where now, several sensors have been added, a temperature sensor, a humidity sensor, a barometer, and an accelerometer. An EEPROM memory has been integrated too. This sensor can work autonomously and store the measurements in the memory. The information from the memory is downloaded to the reader by backscatering only when the reader wakes up the tag. The concept is to place the sensors at the bus stops and the reader in a bus. In this way, when the bus arrives at the bus stop, it contacts the tag and downloads all the information. This information is, in turn, uploaded to the cloud when the bus arrives at a point prepared for this purpose. Two measurements of the temperature are shown in this slide to demonstrate the system operability.

39  Semi-passive UWB RFID sensors (IV)

The third example of UWB RFID sensors is a sensor based on carbon nanotubes for gas sensing. Carbon nanotubes are an enabling technology for sensing gases. Here, they have been functionalized to detect NO2, and integrated in the RFID tag. An interdigital structure has been designed as carrier of the sensor, which changes their resistance as a function of NO_2 concentration. This structure is placed inside a gas chamber and wire-connected to the tag. It is shown here the resistance variation of the sensor as a function of frequency, and it can be observed that only low-frequency operation is achieved. This is why, they have been integrated in a semi-passive structure and not in a chipless tag.

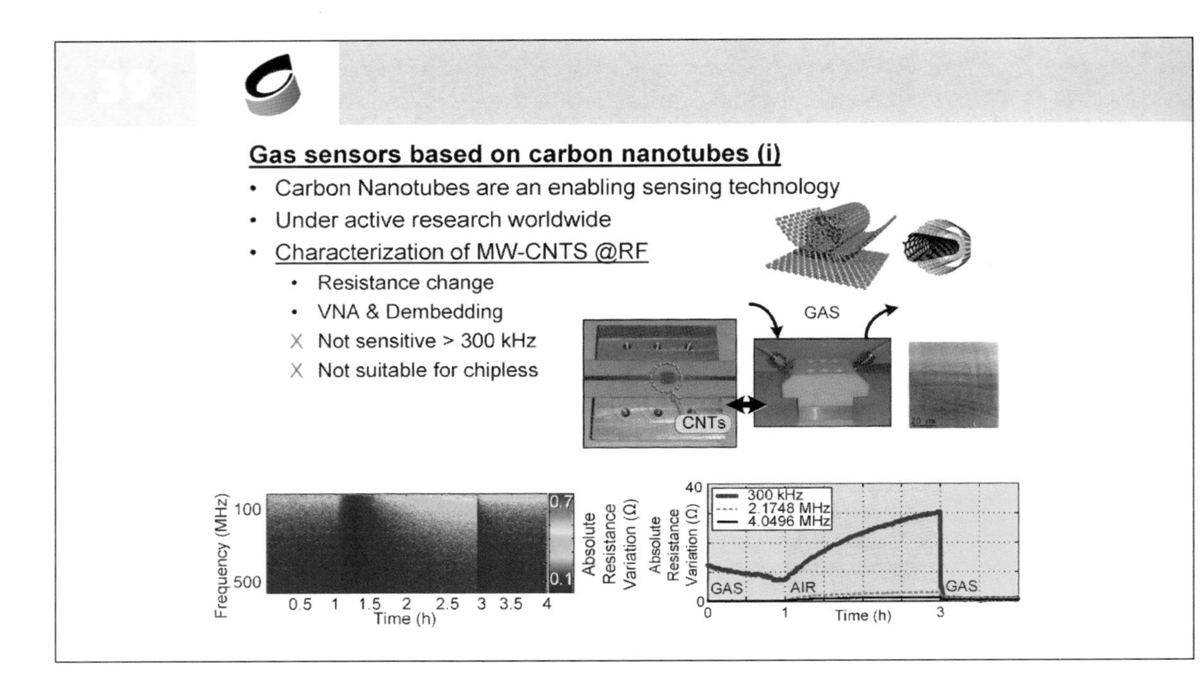

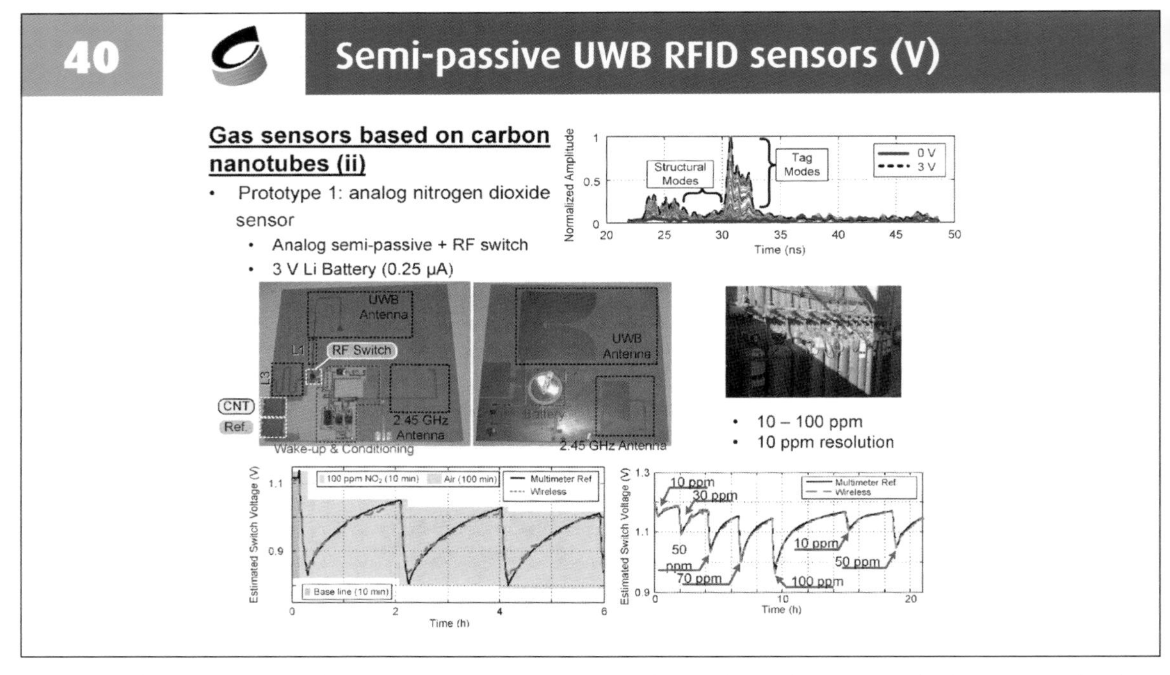

Here, a first prototype of UWB RFID sensor tag based on carbon nanotubes is shown. It is again composed of a UWB antenna and a delay line, connected in this case to an RF switch. This switch commutates between a delay line used for ID of the tag and a circuit that contains the carbon nanotube. This circuit modulates the switch insertion loss depending on the temperature. A rectenna is also used to wake up the circuit, which is powered up by a Li battery and has a small power consumption (only 25 mA). Measurements are shown here, where the nitrogen dioxide concentration is wirelessly detected and compared to a reference value obtained from wire measurements as reference, for calibration purposes and validation.

41

Semi-passive UWB RFID sensors (VI)

Gas sensors based on carbon nanotubes (iii)

- Prototype 2: digital (microcontroller) nitrogen dioxide sensor
  - Read by ADC + conditioning circuit

Gas Chamber

UWB Backscat

Logic

Detected Voltage (V)

Time (h)

10 ppm

100 ppm

30 ppm

50 ppm

70 ppm

Wired

Wireless

A second example of UWB RFID tag with carbon nanotubes for nitrogen dioxide sensing is shown here. The difference between this topology and the former one is that it integrates a microcontroller, and the information coming from the CNT is digitized by a A/D converter. Here, the information between the tag and reader is sent by modulating (on/off) the state of a diode connected at the end of the delay line. Measurements are shown to validate the prototype.

42

Semi-passive UWB RFID sensors (VII)

Smart floor for guidance systems (i)

- Based on using time-Coded UWB RFID and GPR (Ground Penetrating Radar)
- Scattered RFID tags + mobile reader
  - Robot guidance, assistance for blind, etc.
  - Proposed UWB tags: time-domain localization

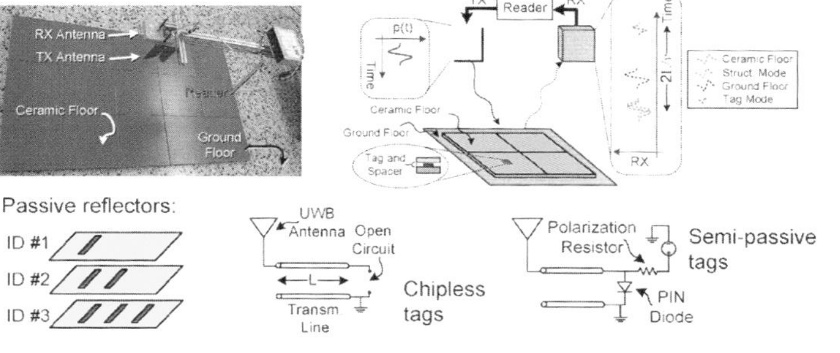

The last application for time-coded chipless RFID and semi-passive UWB RFID is shown in this slide: smart floor guidance systems. The two techniques are suitable to be inserted in floors or walls in order to offer to disabled people or robots information concerning guidance. A third topology is also proposed here, based on Ground Penetrating Radar (GPR) techniques, and tags made of simple metal strips, which, depending on the number, width, and separation of these strips can code information.

43

Semi-passive UWB RFID sensors (VIII)

Smart floor for guidance systems (ii)

- Passive reflectors:
  - Different widths (10/15/20 cm)

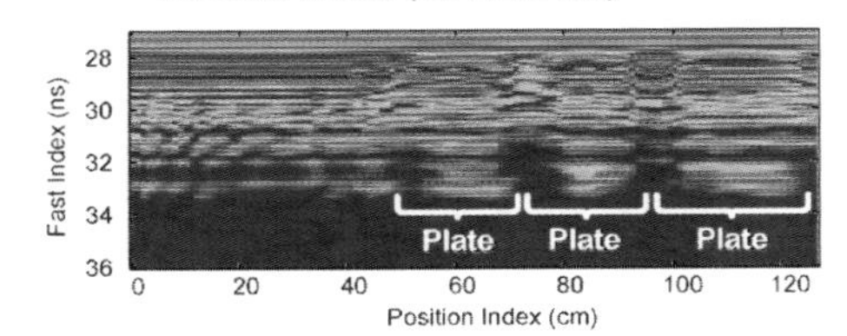

- Semi-Passive tag:

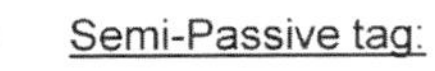

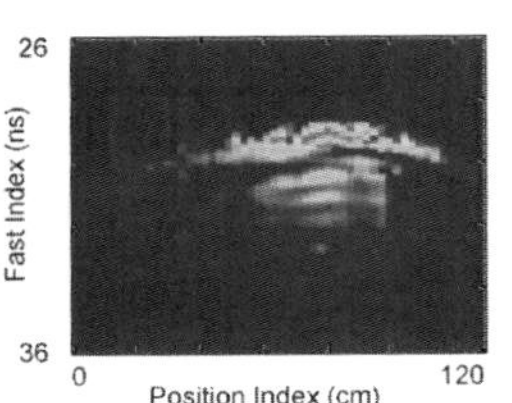

- Chipless tag:

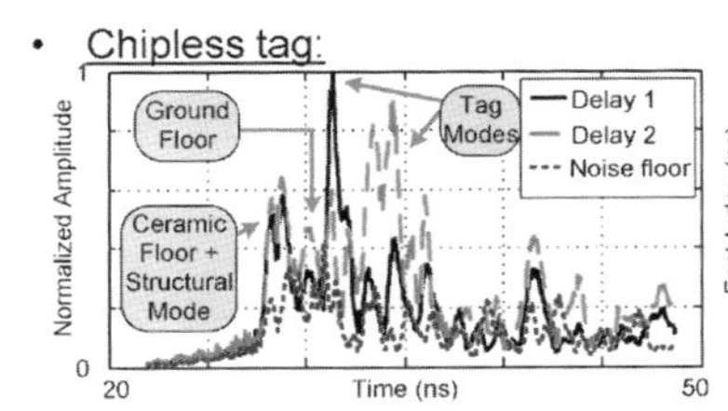

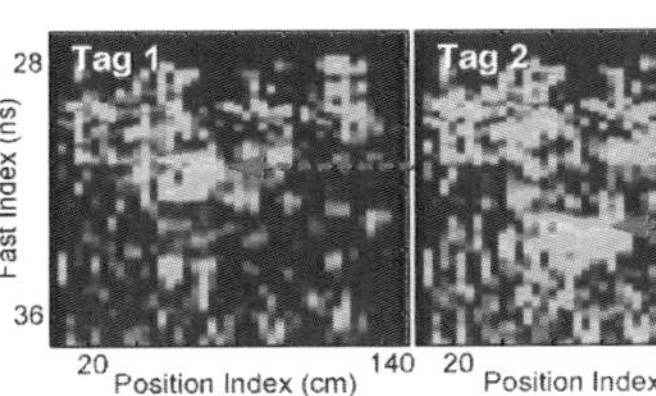

Here, the measurements of the three tags, buried under the smart floor and measured using a UWB radar as reader are shown. It can be observed that both the tags made of metal strips and the semi-passive tag can be well detected, while the chipless tag is detected with some difficulty. It must be noted that this is a complicated scenario for detection, since tags are placed under, but very close to the ground tiles, which generate a very strong reflection, which is very close in time from the expected response of the tags, and with a much larger amplitude

44

Tags based on Modulated FSS

Applications of FSS

- Traditional applications:
  - Radomes
  - Multi-Frequency Reflectors
  - Beam Control Arrays
  - At optical frequencies, FSS can be used as mirrors for solar power applications, and filters for lasers and beam-splitting
- More recent applications:
  - FSS are used in design of Artificial Magnetic Conductors (AMC) and EM Band Gap (EBG) materials
  - RFID tags and transponders

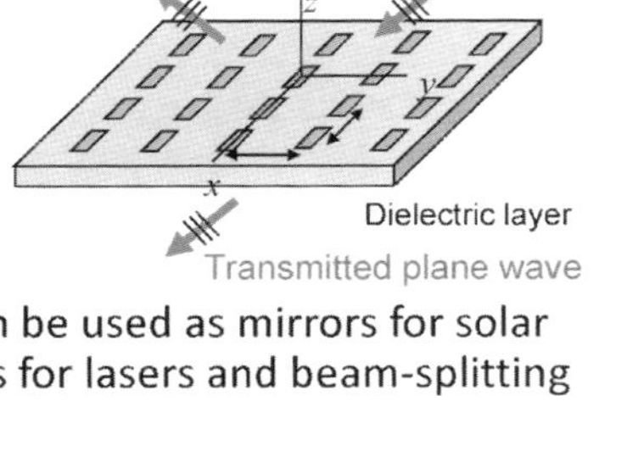

The next part of the chapter is focused on the application of Frequency Selective Surfaces (FSS) to semi-passive RFID.

Frequency Selective Surfaces are periodic structures that act as a filter. Some frequency bands are transmitted and the others are reflected.

Two simple interpretations of an FSS are as an array of antennas/reflectors, or as a grating in optical frequencies. Typical applications of FSS are radomes or as metamaterials for antennas. Here, we will focus on applications of FSSs in RFID tags and transponders.

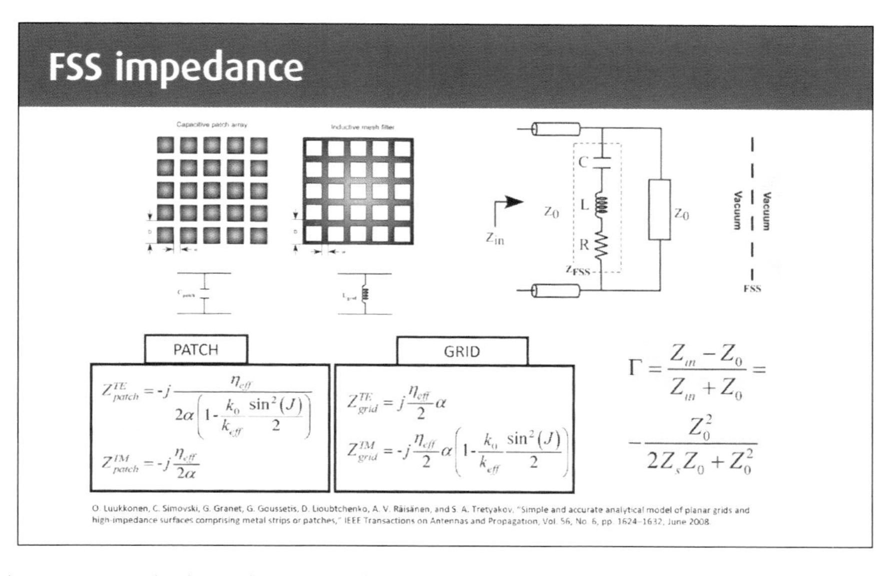

Frequency domain chipless tags based on scattering particles are, in the end, Frequency Selective Surfaces (FSSs). The basic theory of FSS is very useful to engineer these kinds of resonators. FSSs are periodic surfaces comprising a unit cell, with a certain geometry, repeated on a regular grid. The square lattice is the most popular but other lattices, e.g. triangular, are also possible. The single unit cell can be capacitive or inductive. In the first case, the FSS cells are disconnected and the behavior resembles the one of stop band filter. Inductive FSS can be instead described with a parallel LC circuit. Analytical expressions for the sheet impedance are available for simple shapes such as patches or grids, which shows a purely capacitive or purely inductive behavior. The transmission line equivalence is valid in the frequency region where a single harmonic is in propagation, say, when the wavelength is shorter than the FSS periodicity.

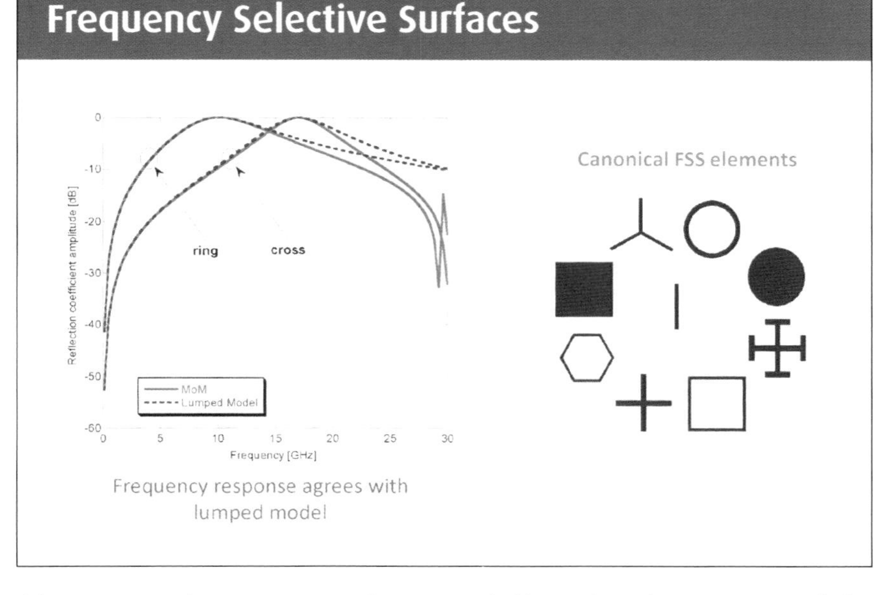

Patches and grids, whose sheet impedance has been reported in the previous slide do not present any resonant behavior. However, it is possible to modify the unit cell geometry in order to achieve a single resonant frequency or even multiple ones. Some FSS topologies are shown in the slide. It is also reported the reflection coefficient of loop type and a cross type FSS. In the former case, the resonance is achieved when the perimeter of the loop becomes a wavelength. In case of the cross element, the resonance is achieved when the length of the cross arm becomes a half-wavelength. Once retrieved, the correct values of the capacitance and the inductance, the reflection response can be accurately predicted by the transmission line model described in the previous slide.

47

Dielectric effects

Capacitive FSS

Inductive FSS

ε_{r1} ε_{r2}

for d>λ/10:

$$f_{r1} = \frac{f_0}{\sqrt{\varepsilon_{eff}}}$$

Where:

$$\varepsilon_{eff} = \frac{\varepsilon_{r1} + \varepsilon_{r2}}{2}$$

f_{r0} f_{r1} D/5

P. Callaghan, E. Parker and R. Langley, "Influence of supporting dielectric layers on the transmission properties of frequency selective surfaces", IEE Proc., Part H: Microwaves, Antennas Propag. 138(5), 448-454, 1991.

If the FSS is printed on a dielectric layer, the presence of supporting dielectric induces a shift in the resonance frequency of the FSS filter toward lower frequencies. This effect is different for capacitive or inductive FSSs. In the former case, the resonance shift is characterized by an exponential behavior whereas, for inductive shapes, there is an oscillation around the final resonance value. For thick dielectrics, the resonance approaches to initial frequency scaled by the square root of the effective permittivity. This is due to the fact that the inductance of the FSS in unaffected by the presence of the dielectric layer, whereas the FSS capacitance is multiplied by the effective permittivity. If the dielectrics of the two sides of the FSS are different but characterized by the same thickness, a mathematical mean of the two permittivity values can be a good approximation of the effective permittivity. In other cases, more complicate formulas are used.

48

Tags based on Modulated FSS (I)

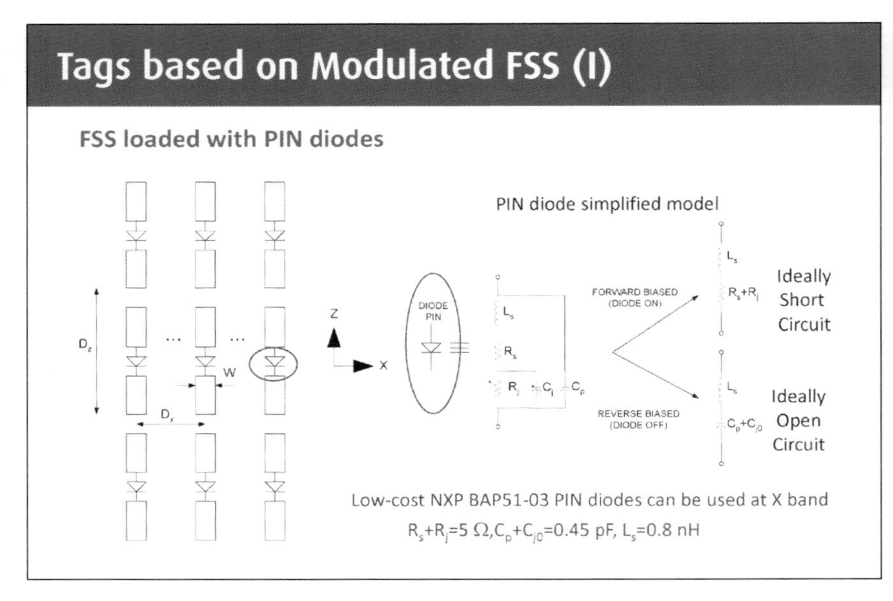

For this application, the FSS is like an array. The simplest FSS is composed of an array of dipoles loaded with diodes.

When the incident wave has a frequency close to the resonance frequency (approximately half of the wavelength), the wave is reflected. When the frequency is outside the resonance, the wave is transmitted and the FSS is transparent. In order to modulate the FSS, the structure is loaded with switching diodes (for example, PIN diodes or varactors). A simplified model used in simulations is shown. When the diodes are in on state, they present a low impedance (short circuit), whereas when they are in off state, they present a high impedance (open circuit). The circuit element values are obtained from the datasheet of the diode manufacturer.

49

Tags based on Modulated FSS (II)

Equivalent circuit model of the FSS for the two diode states for TM waves.

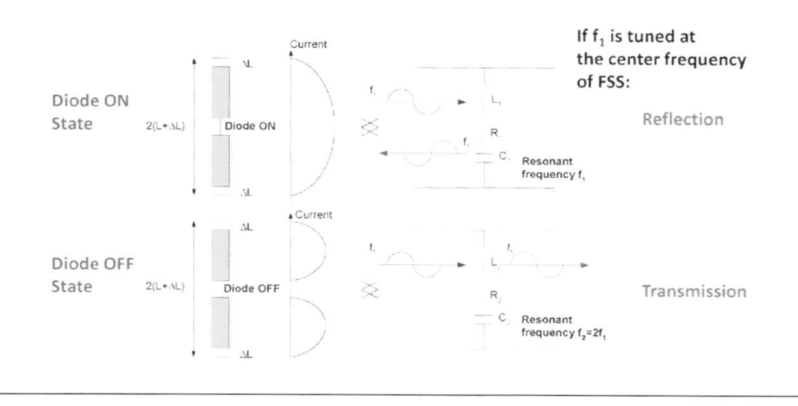

In order to transmit information, the FSS can be modulated. To this end, the dipoles can be loaded with switching diodes. In both cases, when the diodes are in On state, they present a small impedance as a short circuit, whereas when the diodes are off, they present a high impedance. Therefore, the resonance frequency changes between the two states. For ideal short-and-open-circuit cases, the resonance frequency will be double than in the open-state case because the dipole length is the half.

50

Tags based on Modulated FSS (III)

Modulation of backscattered field

The antenna or tag mode can be modulated changing the diodes states at rate or modulating frequency f_{tag}

Γ_{on} Γ_{off}

$1/f_{tag}$ t

$$\Gamma_L(f) = \frac{Z_L - Z_a^*}{Z_L + Z_a} = \sum_{n=-\infty}^{+\infty} c_n \delta(f - (f_c + n f_{tag}))$$

For a square waveform with duty cycle δ, the coefficients c_n are given by*:

$$c_n = \begin{cases} \Gamma_{avg} & , n = 0 \\ \Delta\Gamma\delta\left(\frac{\sin n\pi\delta}{n\pi\delta}\right) & , n \neq 0 \end{cases}$$

* A. Lázaro, et al, "Backscatter Transponder based on Frequency Selective Surface for FMCW Radar Applications," Radioengineering Vol.23,No.2 (2014).

The backscattered or reflected field can be studied using the backscattering antenna theory interpreting the FSS as an antenna array.

The backscattered field can be split into two components: the structural mode, which is the reflection in the structure of the FSS, the objects that support the FSS, and it is constant, and a second term, which is known as antenna mode or tag mode that depends on the load of each element.

The reflection coefficient is changed between the on and off diode states at a rate equal to the modulating tag frequency f_{tag}. Then, the reflection coefficient can be expanded using Fourier series. c_n are the Fourier coefficients at each frequency.

51

Tags based on Modulated FSS (IV)

Modulation of backscattered field

The backscattered field spectrum:

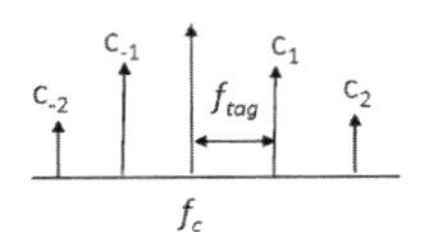

$$\overline{E}_s = \left(\overline{E}_{est} + \overline{E}_m\Gamma_{avg}\right)\delta(f-f_c) + \overline{E}_m \sum_{n\neq 0} c_n\delta(f-(f_c+nf_{tag}))$$

The detected signal is proportional to the Differential RCS:

$$RCS_{dif} = \lim_{d\to\infty} 4\pi d^2 \frac{\left|\overline{E}_{S,ON} - \overline{E}_{S,OFF}\right|^2}{\left|\overline{E}_{in}\right|^2}$$

$$RCS_{dif} = \lim_{d\to\infty} 4\pi d^2 \frac{\left|\overline{E}_m c_1\right|^2}{\left|\overline{E}_{in}\right|^2} = \frac{\lambda^2}{4\pi} G^2 \left|\Delta\Gamma\right|^2 m$$

Level and bandwidth increase with the number of dipoles

The result is that the reflected spectrum is composed of a train of delta functions spaced by the modulating tag frequency around the carrier frequency transmitted by the reader.

The received signal at the frequency f_c+f_{tag} is proportional to the radar cross section, which is proportional, in turn, to the field difference between the two states when the diodes are on and off.

This differential radar cross section (RCS_{dif}) can be expressed as a function of the difference between the reflection coefficients and the gain of the array.

Thus, increasing the number of FSS elements increases the gain and the differential radar cross section compared to the case of using a single dipole antenna. This is one of the main motivations of using FSS instead of using an antenna.

52

Distance measurement with a backscattered FSS using FMCW (I)

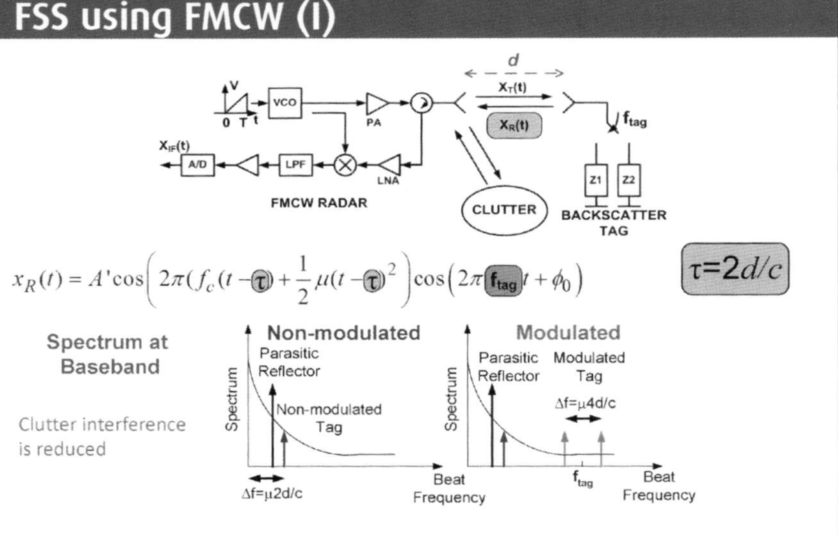

The next slides show an application of modulated FSS for the design of FMCW radar transponders. Briefly, a FMCW radar is a radar that transmits a CW that is frequency-swept in a band. This waveform is called a chirp signal. When this CW is reflected by a target, there is a delay due to the propagation distance between the radar and the target. Therefore, when the reflected wave is received, its frequency is different from the transmission frequency at this instant. The two frequencies are compared using a mixer.

The distance can be obtained from the difference between the transmitted and received frequencies. The parameter m is the sweep slope. The distance resolution increases with the radar bandwidth.

53

Distance measurement with a backscattered FSS using FMCW (II)

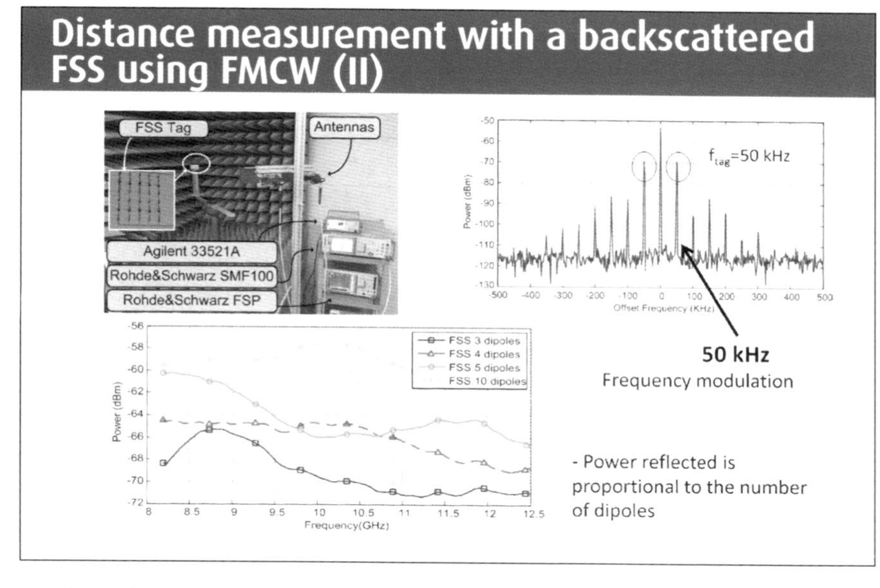

FMCW radars are often used in localization applications. The main problem in indoor environments is that there are reflections from several objects that sometimes have larger radar cross section than the tag. In addition, there is an interference from the phase noise of the radar. Therefore, it is difficult to detect tags close to big objects or close to the radar.

A solution for this problem is to use modulated tags. Then, the output of the IF signal is modulated by the tag modulation frequency. The modulation consists of changing the impedance of an antenna or FSS at rate equal to the tag modulation frequency. Then, two peaks appear in the baseband spectrum centered at frequency, ftag, and whose frequency spacing are proportional to the distance. The modulation of the tag is generated with a low-frequency oscillator connected to the bias of the diodes, generating a square wave between 0 and 3 V, which switches the diodes on or off. The FSS is illuminated with a signal from a microwave generator, and the received signal is received with a spectrum analyzer. It is experimentally demonstrated that the received power increases with the number of FSS elements and also the bandwidth of the FSS. More than 2 GHz of bandwidth can be achieved with a flat response. The bandwidth must be higher than the radar bandwidth, in our case, 1.5 GHz.

54

Distance measurement with a backscattered FSS using FMCW (III)

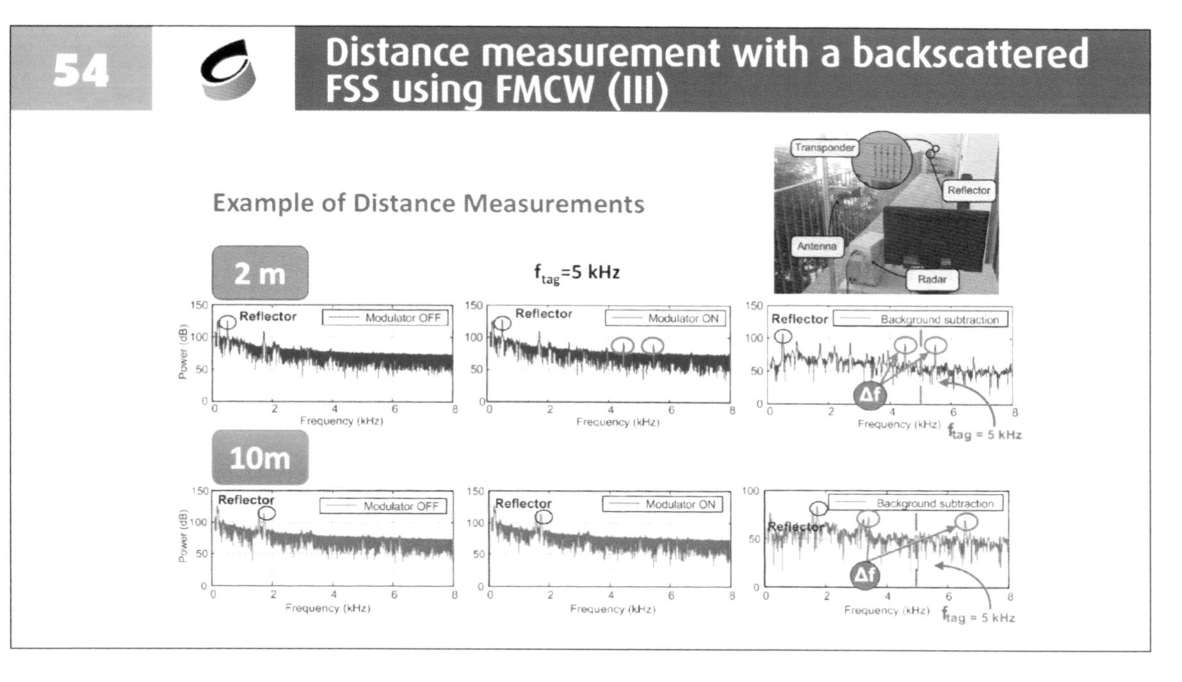

This figure shows the measured baseband spectrum when the tag is illuminated with an FMCW radar. The radar is from Siversima at X band between 9.25 and 10.75 GHz. The radar is connected to a horn antenna with 20 dB of gain. The FSS is modulated, and it is located very close to a large corner that has a large cross section which interferes the tag.

These graphs show some results at 2 and 10 m. Without modulation, the reflector is detected but the tag is not detected. When the modulator is active, two peaks can be observed whose separation depends on the distance. Removing the background (measurement without modulation), the tag is better detected.

55

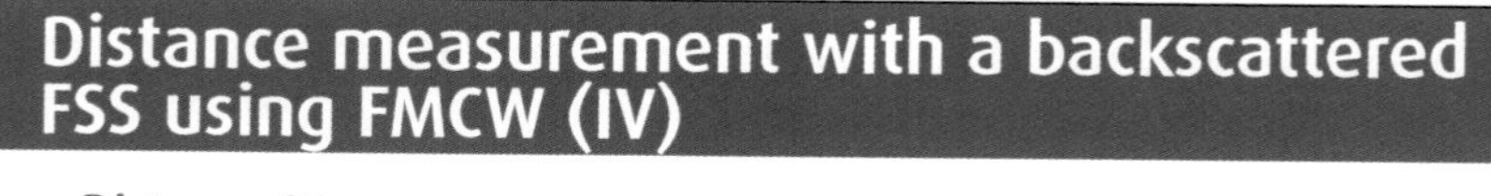

Distance measurement with a backscattered FSS using FMCW (IV)

This plot compares the real distance with the measured distance from the frequency difference in the peaks at baseband spectrum. The average error obtained (12 cm) is close to the theoretical limit given by the radar resolution that depends on the radar bandwidth, ***B***.

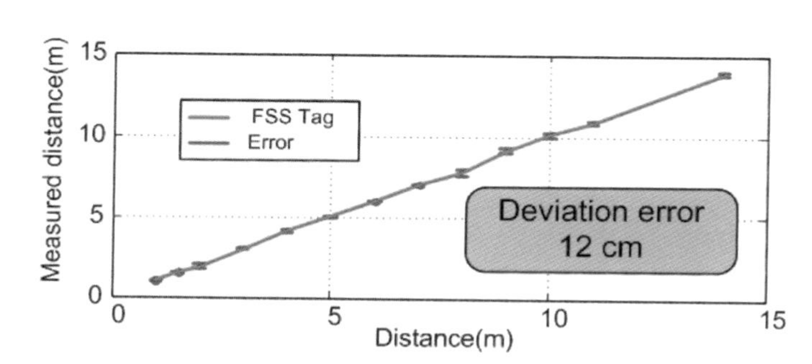

Theoretical FMCW radar resolution: $\Delta d = \frac{c}{2B} = 10cm$

56

Wearable Sensors Based On FSS

Today, wearable devices integrate Bluetooth low-energy transceivers that need about 15 mA in transmission and 7 mA in reception. One motivation of using modulated FSS for wearable devices is the communication by backscattering to save battery, since they do not transmit power. Another motivation to use FSS is that it can be integrated in clothes or textile material, wristbands, or other wearable parts. In addition, some sensors can be easily integrated without additional electronics.

The key is that a modulated FSS can be detected attached on the body. This is an experiment of an FSS on a piece of ham to simulate an arm. The FSS is modulated with a low-frequency oscillator based on a low-power 555 timer. The oscillation frequency depends on a resistance. If the resistance is replaced by a negative temperature resistance, then, the oscillation frequency depends on the temperature. The top plot shows the spectrum at different distances of an FSS on body, modulated at a frequency of 7 kHz at 2.45 GHz (ISM band). This plot shows that it is possible to detect at 3 m. The bottom graph shows the received power from an FSS placed on body, as a function of the interrogating frequency. It is shown that the bandwidth is very large and increases with the number of dipoles. In addition, the response is very close between the FSS on an arm and the FSS on a piece of ham. Therefore, the response is insensitive to changes in the permittivity of the body that can change between people/invidious, or parts of the body.

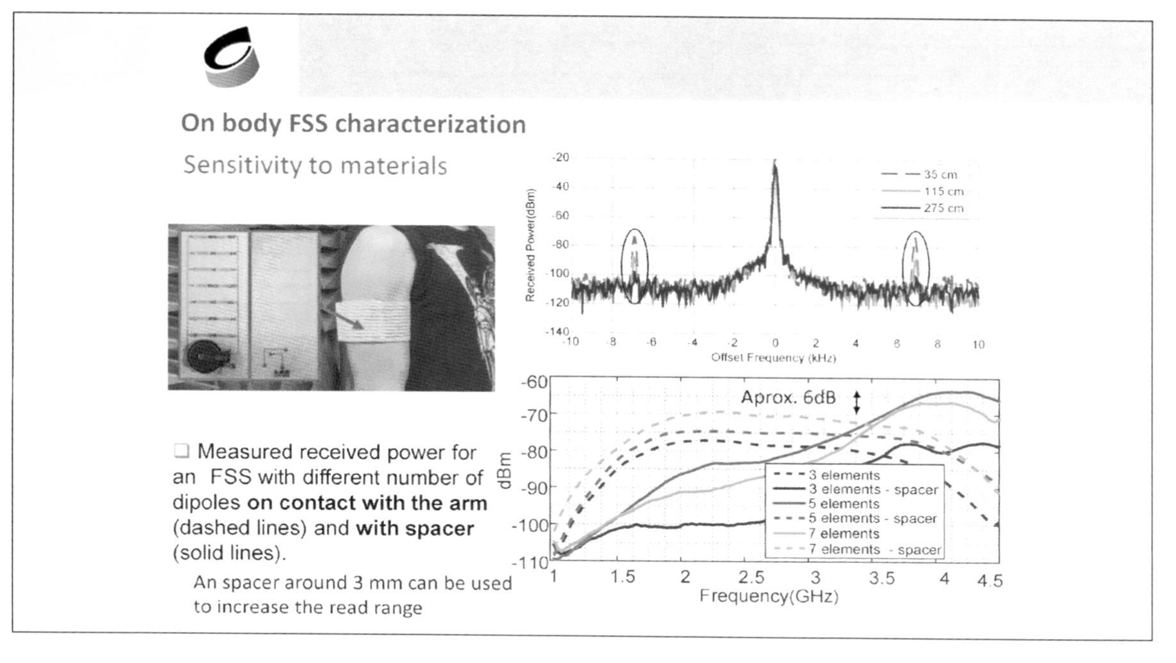

57 

Wearable Sensors Based On FSS (I)

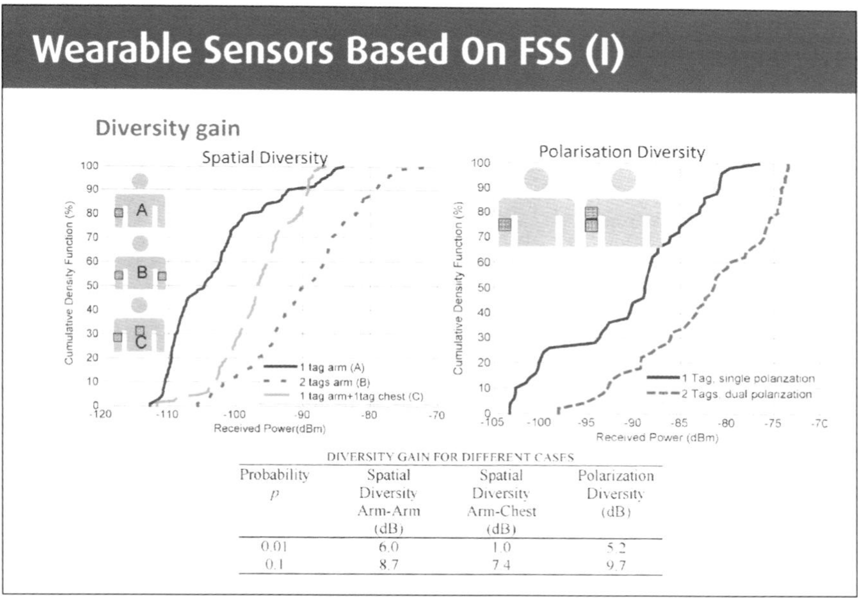

DIVERSITY GAIN FOR DIFFERENT CASES

Probability p	Spatial Diversity Arm-Arm (dB)	Spatial Diversity Arm-Chest (dB)	Polarization Diversity (dB)
0.01	6.0	1.0	5.2
0.1	8.7	7.4	9.7

The main problem in any wireless system is the multipath interference. The received signal can suffer constructive and destructive interferences due to different paths that arrive to our receiver. If a notch falls in our transmission frequency, then communication is not possible.

Diversity techniques are employed to combat multipath. The idea is to use the received signal from two uncorrelated paths. The easiest way is the spatial diversity technique. This is very easy to implement with modulated FSS because we can use two FSS modulated with the same oscillator, connected by a wire and integrated in the clothes. In all cases, the received power is higher using two FSS instead of one. These plots show the cumulative density probability function (CDF), that is the probability that the received power is under a value. The diversity gain is close to 9 dB when the two FSS are more spaced (between the two arm). The gain is smaller when one FSS is on the arm and other is on the chest because the paths are not enough uncorrelated. Another possibility is to use one FSS with the dipoles in one direction and the second FSS with the dipoles in orthogonal direction. Thus, it is a simple polarization diversity technique. The diversity gain is close to 9 dB in this case.

58 

Wearable Sensors Based On FSS (II)

Breathing sensor based on FSS

Airway open during normal breathing (left) and closed airflow in an apnea event (right).

FSS

f_m

Tx

Rx

f_c

To A/D

READER

NTC

f

FSS

NTC sensor

Oscillator + Battery

Oscillator + Coin battery

FSS

NTC sensor

A wearable breathing sensor can be designed with the above concepts. Sleep disorders affect up to 25% of population. The most important are the apneas or obstruction of the air during breathing.

The method for diagnosis normally is the polysomnography, which consists in connecting several sensors on the body. The measurements must be done in the hospital and almost 3 nights are required for the diagnosis.

Therefore, a screening method, noninvasive or less invasive and economic, is required, which can be performed at home. The proposed sensor is based on the temperature sensor based on the modulated FSS presented before. The NTC thermistor is located close to the nose in order to sense the change in the air temperature during the breathing. The frequency of the low-frequency oscillator depends on the resistance and therefore on the air temperature. This oscillator modulates the FSS.

The FSS is integrated in a band around the head and has two orthogonal polarizations, in order to be read from any angle and polarization, exploiting the spatial and polarization diversity studied before.

59 

Wearable Sensors Based On FSS (III)

Temperature Measurement error

The NTC resistance is a function of the air temperature

$$\frac{1}{T} \approx \frac{1}{T_0} + \frac{1}{\beta} \ln \frac{R}{R_0} \xrightarrow{\text{Variation of oscillation frequency}} \frac{\Delta f_m}{f_m} \approx \frac{\beta}{T} \frac{\Delta T}{T}$$

Sensor gain

β=4400 K is the Steinhart–Hart parameter that is taken from the thermistor datasheet

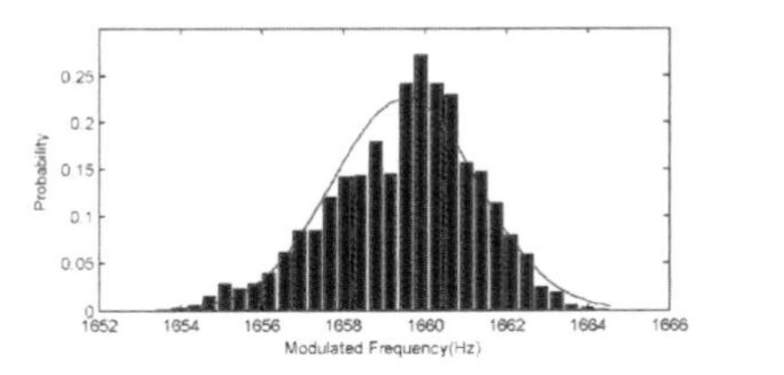

Δf_m deviation<1Hz

ΔT deviation<0.01 K

The variation of the NTC resistance as a function of temperature is given by the Steinhart-Hart equation. In commercial NTC, the parameter b is very high. The result is that the sensitivity in the shift of the oscillation frequency is very high. This factor acts as a sensor gain.

From the histogram obtained experimentally, a frequency resolution or deviation of 1 Hz is obtained; therefore, the temperature measurement deviation is about 0.01 K. This temperature resolution is enough for most applications.

60

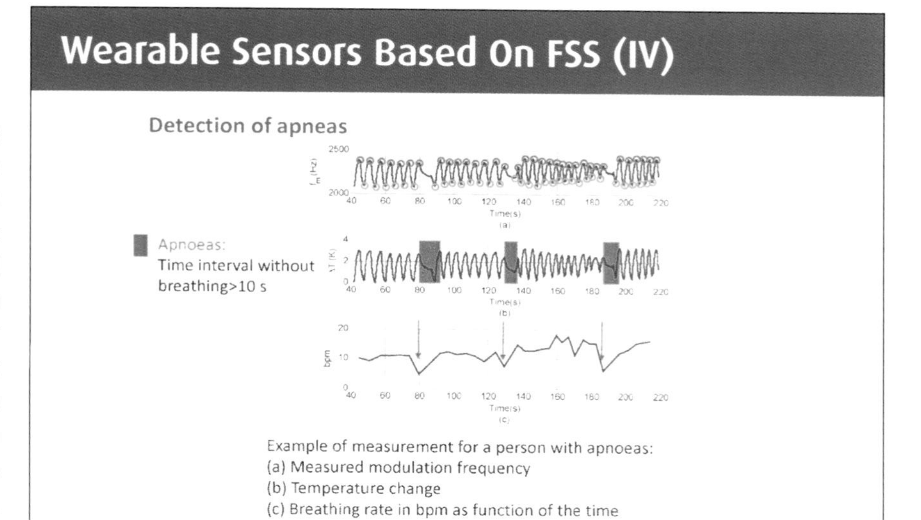

Example of measurement for a person with apnoeas:
(a) Measured modulation frequency
(b) Temperature change
(c) Breathing rate in bpm as function of the time

In order to obtain the breathing rate, some signal processing is required. At the output of the mixer, the signal is sampled during a period, then the Fourier transform experiment is performed to obtain the modulation frequency that depends on the temperature. Then, the variation of the temperature as a function of time is obtained. Then, a peak algorithm is used to detect the breathing interval and the apnea intervals.

The first plot shows the modulating frequency and the corresponding temperature variation as a function of time, during a normal breathing and with 3 apneas of obstruction of the breathing for more than 10 s. The breathing rate by minute (bpm) is shown at the bottom figure. The average breathing rate falls if apneas are present.

61

FD chipless RFID tags

- Based on chipless RFID tag scattering
- Based on chipless RFID tag retransmission
- Based on quantization of amplitude/ phase response, polarization

As already mentioned, another possible approach to design chipless RFID tags consists in encoding the information into the spectrum response of the identification tag by exploiting the properties of resonant structures. The binary information can be associated to the presence ('1') or absence ('0') of a resonant peak in the backscattered field at a fixed frequency within the available spectrum. These chipless RFID tags can be realized by adopting several technologies, from standard planar microstrip or waveguides up to additive-manufacturing processes. They can be realized on dielectric substrates, printed on flexible laminates or manufactures by using polymers.

62 

Frequency domain chipless tags

Scattering

Retransmission

E_{inc}

E_{scatt}

Frequency domain chipless tags can be subdivided into two main categories. The scattering-based ones are composed of an array of particles with different sizes, which introduce several resonances in the backscattered response. The other typical configuration comprises a couple of orthogonally placed wideband monopole antennas separated by a transmission line with multiple filters in series. The scattering, based tags can be copular or cross-polar. In the first case, the copolar backscattered frequency response is characterized by multiple frequency peaks. The main problem of this encoding strategy is that the copular response is easily affected by the multipath. On the contrary, if cross-polar tags (or depolarizing tags) are designed, then the isolation of the tag response from the response of the environment is greatly improved.

63 

FD chipless RFID tags (I)

Based on chipless RFID tag scattering

- The chipless tag consists of a number of capacitively-tuned dipole antennas, which resonate at different frequencies. When the tag is illuminated by a frequency sweep signal, the tag responds with a spectrum with dips in correspondence of a '1' bit. Each dipole has a 1:1 correspondence to a data bit.
- Tag size is an issue (lower frequency longer dipole—half wavelength) as well as mutual coupling between dipoles. High-gain antenna are required.

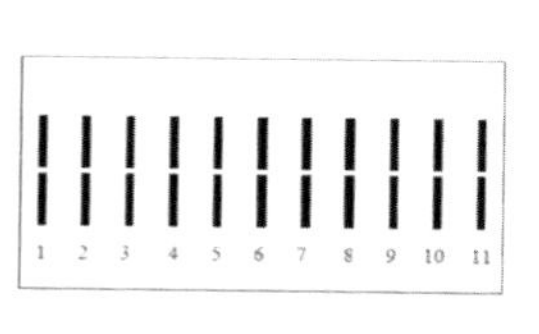

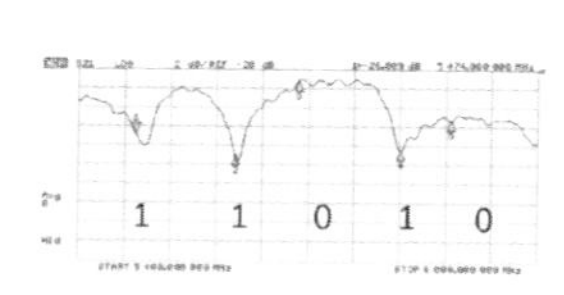

More details in: I. Jalaly and I. D. Robertson, "Capacitively-tuned split microstrip resonators for RFID barcodes," in 2005 European Microwave Conference, 2005, vol. 2.

A first example resort to chipless tag comprising several capacitively tuned dipole antennas, which resonate at different frequencies. When the tag is illuminated by a frequency- swept signal, the tag responds with a spectrum with dips. Each one of these peaks encodes a '1' bit and it is determined by the presence of a dipole; therefore, each dipole has a 1:1 correspondence to a data bit. To make the chipless RFID tag more compact, space-filling curves have been used instead of dipoles. The space-filling curve allows the resonance at a frequency that has a wavelength much greater than its footprint. This is an advantage since it allows the development of small footprint tags at UHF ranges. This shrinking process allows tag of compact size but requires significant modifications of the tag layout in order to encode the data. More details on this topic are available in: J. McVay, N. Engheta, and A. Hoorfar, "High impedance metamaterial surfaces using Hilbert-curve inclusions," *IEEE Microwave and Wireless Components Letters*, vol. 14, no. 3, pp. 130–132, Mar. 2004.

64

FD chipless RFID tags (II)

Based on chipless RFID tag retransmission

- Two UWB antennas are necessary to collect the interrogating signal and to retransmit the encoded information to the reader.
- The Rx monopole antenna collects the interrogating signal and then it propagates towards the multiresonating circuit. The multiresonating circuit encodes data bits using cascaded the spiral resonators (unique spectral signature).
- The encoded information is then transmitted back to the reader by the Tx monopole tag antenna.

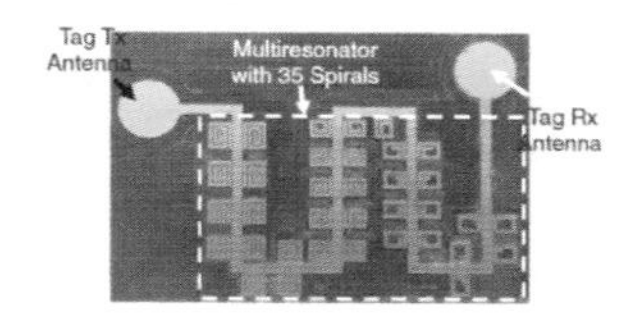

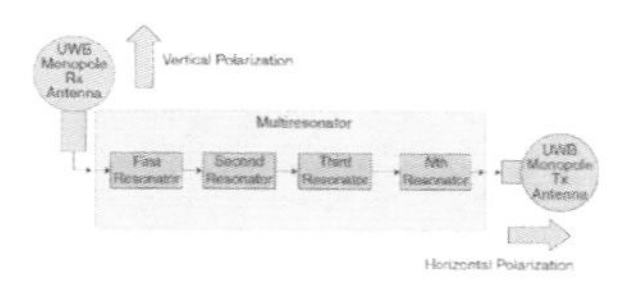

Another example is provided by chipless RFDI tag based on the tag retransmission. In this kind of tag, the signal is "elaborated" by a multiresonator (a filter) that interacts with the signal traveling from the receiving antenna to the re-transmitting one. The signal collected by the receiving antenna tag during its propagation toward the tag TX antenna undergoes an amplitude and phase modulation determined by the number and shape of the resonator placed along the transmitting line. In the proposed example, all spiral resonators are present, but they can be shorted to remove the correspondent resonance.

65

FD chipless RFID tags (III)

Based on chipless RFID tag retransmission

- The Rx and Tx tag antennas are cross-polarized in order to minimize interference between the interrogation signal and the retransmitted encoded signal which contains the information.
- The tag responses are not based on RCS backscattering as in the previous cases but on the retransmission of the cross-polarized interrogation signal with the unique spectral ID encoded by the multiresonator (both amplitude and phase).

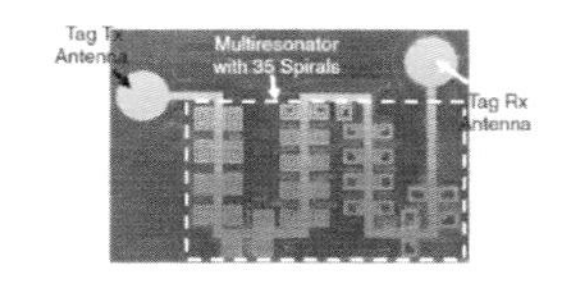

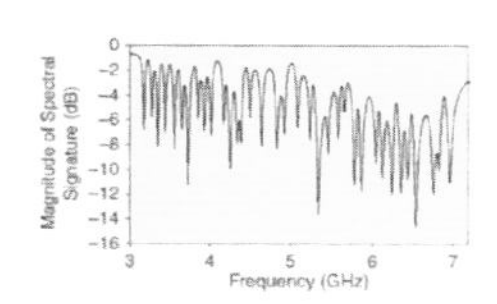

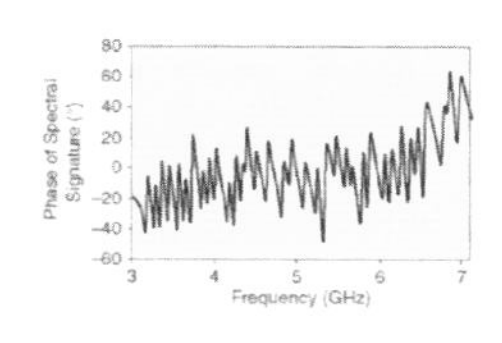

More in detail, two UWB antennas are necessary to collect the interrogating single and to retransmit the encoded information to the reader. The Rx monopole antenna collects the interrogating signal and then it propagates toward the multiresonating circuit. The multiresonating circuit encodes data bits using cascaded the spiral resonators (unique spectral signature). The encoded information is then transmitted back to the reader by the Tx monopole tag antenna. The Rx and Tx tag antennas are cross-polarized in order to minimize interference between the interrogation signal and the retransmitted encoded signal which contains the information. The tag responses are not based on RCS backscattering as in the previous cases but on the retransmission of the cross-polarized interrogation signal with the unique spectral ID encoded by the multiresonator (both amplitude and phase).

66

Artificial Impedance Surfaces (AIS) (I)

HIS as an L_S-LC parallel circuit

Z

C L d ζ_1 SC

C L L_s

$$Z = \frac{j\omega L_s\left(1-\omega^2 LC\right)}{1-\omega^2\left(L+L_s\right)C}$$

Impedance

$$\omega = \frac{1}{\sqrt{(L+L_s)C}}$$

Frequency ω

Reflection coefficient phase [deg]

180 135 90 45 0 -45 -90 -135 -180

HIS band

Frequency

A chipless tag configuration more robust to the presence of nearby objects is the so-called Artificial Impedance Surface (AIS). The AIS comprises an FSS placed in the close proximity of a metallic ground plane. The FSS is printed on a lossy dielectric. In the absence of losses, the reflection amplitude of this surface is equal to one, while the phase is characterized by a transition though zero degrees. If a suitable amount of losses is added in the substrate, a deep amplitude absorption peak is achieved. If a multiresonant FSS is used instead of a single resonant one, then multiple absorption peaks are synthesized. Due to the presence of the ground plane, the structure is much more robust to the presence of nearby objects. It is not necessary to print the chipless resonator with the fround plane since the FSS, printed on the lossy substrate, can be directly applied to a metallic structure so as to form the AIS resonator.

67

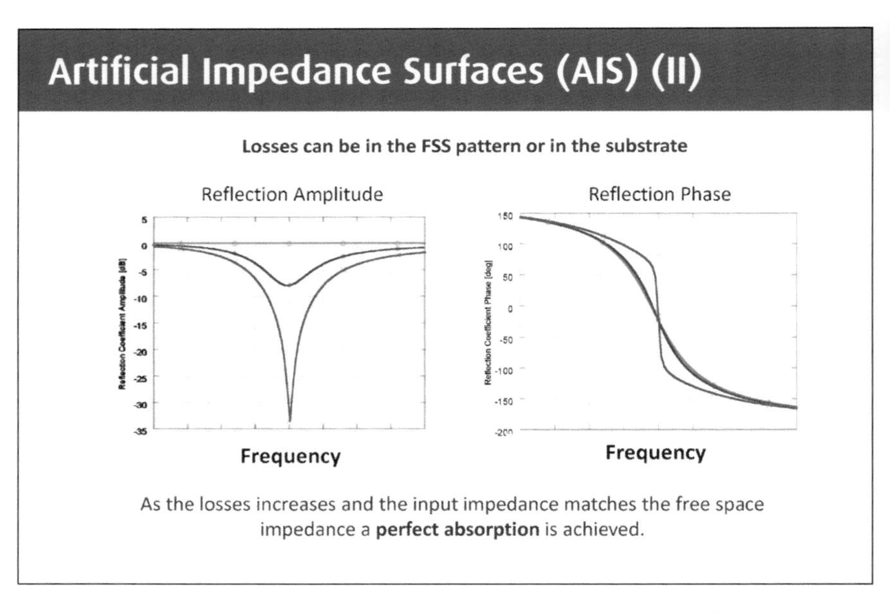

This slide reports the amplitude and phase reflection coefficient of a single resonant AIS. If the losses are neglected, a total reflection amplitude and a smooth phase transition are achieved. If the amount of losses increases, then a deep reflection peak due to power absorption and a steeper phase profile is achieved. The resonant peak is preserved even in the presence of high lossy substrates such us FR4. The total absorption is achieved when the input impedance of the AIS resonator is matched with the free space impedance. If the losses of the substrate of in the periodic surface are further increased with respect to the matching condition, then the absorption decreases, and the phase profile shows an unconventional reverse transition.

68

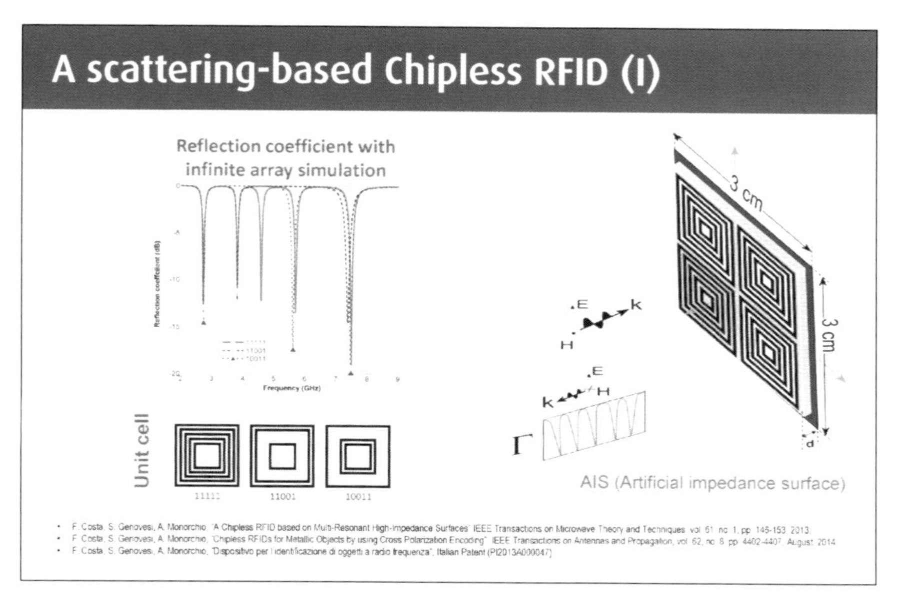

The slide reports the chipless tag configuration proposed by our group. It comprises a multiresonant frequency-selective surface printed on a grounded FR4 dielectric slab. The presence of the loop induces the presence of an absorption peak at a predetermined frequency and this codifies the bit 1. When one of the loops is removed from the metallic pattern, its corresponding frequency peak disappears as well codifying the state 1. The number of coded bits is proportional to the number of loops. There exists a limited frequency shift of the remaining frequency peaks due to mutual coupling when a resonant loop is removed from the metallic pattern. The undesired effect can be adjusting the length of the remaining loop.

69

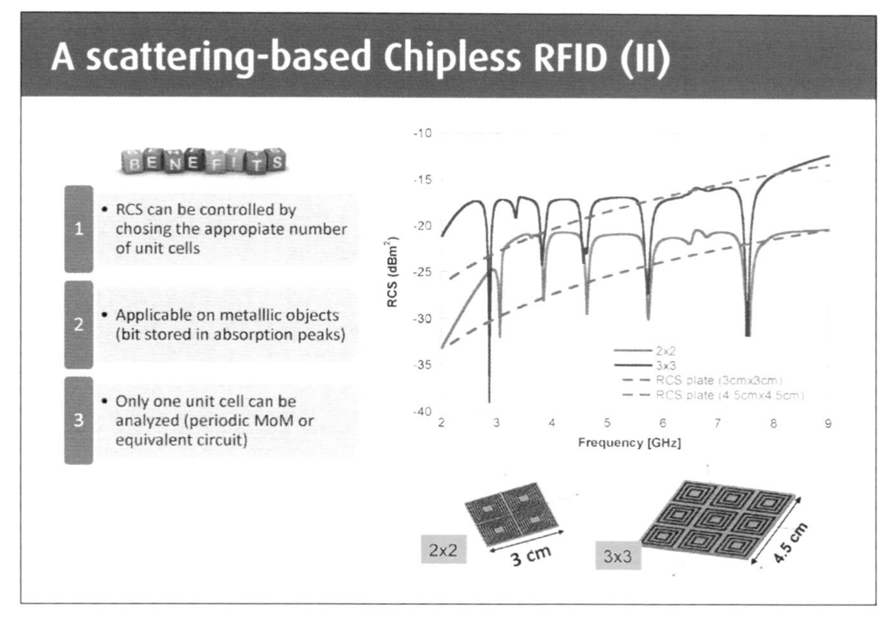

The results of the previous simulations were about the infinite periodic configuration. The simulations were performed with a periodic method of moments (PMM) in which a single unit cell is considered. However, a real structure has a finite size and the periodic surface has to be cut considering a certain number of unit cells. The use of a large surface with a high number of unit cell is not advantageous in terms of amount of information encoded. However, the increase of the number of unit cells allows to increase the amount of backscattered energy. Indeed, the radar cross section (RCS) is proportional to the size of the panel.

70

Exploiting amp/phase/pol (I)

Delta-Phase quantization encoding

- The rectangular unit cell of the considered periodic structure structure comprises a grounded dielectric substrate (FR4 with ε_r = 4.4-j0.088, thickness h = 3.6 mm) with a rectangular loop printed on the top face.
- The periodicity of the unit cell is equal to T_x = 1.5 cm and T_y = 2 cm along x and y axis, respectively.
- The dimension of the rectangular loop is equal to D_x in x-direction and D_y in y-direction.
- A stub of length S is attached in correspondence of each loop corner.
- The unit cell is defined by using a 64 x 64 pixel matrix for the analysis with a Periodic Method of Moments.

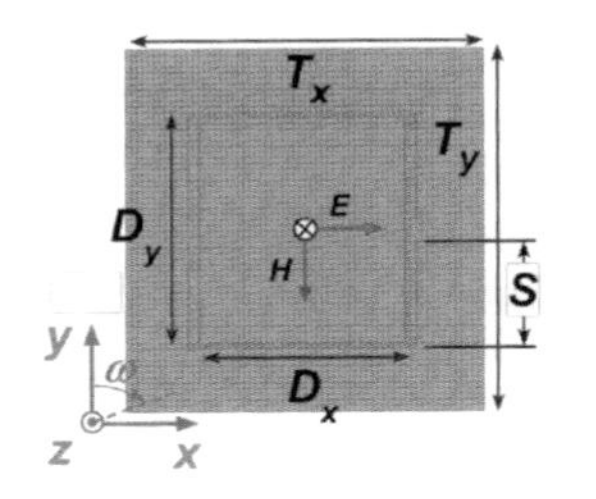

The exploitation of amplitude, phase, or polarization can provide chipless RFID tags, which may require narrower operational bandwidths with respect to the previous solutions. An example is offered by the "delta-phase" quantization encoding scheme. To introduce this concept, let us consider the phase response of a periodic surface (whose rectangular unit cell is shown in the figure) when it is illuminated at normal incidence,both with a TE plane (E-field parallel to x–axis) wave and a TM plane wave (H-field parallel to x-axis).

71

Exploiting amp/phase/pol (II)

Delta-Phase quantization encoding

- The width of the rectangular ring and the stub is one pixel, as well as the space between the ring side and the stub.
- Let us fix the dimension D_x = 58 pixel and D_y = 62 pixel and look the phase response for two different values of stub length, S_1 = 12 pixel and S_2 = 16 pixel.
- Phase response of the periodic when illuminated both with a TE plane wave and a TM plane wave. The plane wave is normally impinging on the surface. Electric field E is parallel to x axis.
- Even a small change in the stub length determines an apparent shift of the phase response for a TE incident plane wave whereas the TM response is almost unaffected.

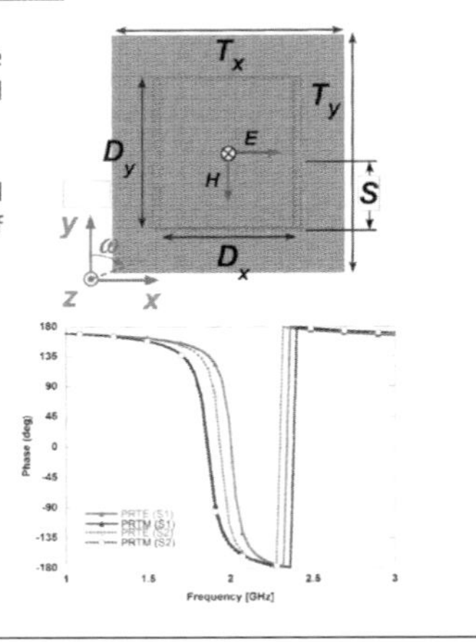

The structure comprises a grounded dielectric substrate with a rectangular loop printed on the top face. A stub of length S is attached in correspondence of each loop corner. Looking at the phase response, it can be noticed that even a small change in the stub length determines a shift of the phase response for a TE incident plane wave, whereas the TM response is almost unchanged. The unit cell is discretized into 64 x 64 pixel matrix for the analysis with a Periodic Method of Moments (PMM).

72  Exploiting amp/phase/pol (III)

Delta-Phase quantization encoding

- It can be seen that the delta-phase value spans within the interval (- 25°, 250°), with short stubs exhibiting the highest differential phase values.
- A stub can encode a ***multi-value*** bit with more than two states.
- The set of stub lengths employed in the codification depends on the criterion used for ***quantizing*** and ***discriminating*** two phase states. The stubs whose delta-phase differs at least Δ degrees is adopted.
- Δ = 10° → 10 stub lengths will be available ;
- Δ = 20°/ 30° → 8 / 6 different selectable states.
- ***One*** stubbed ring equals ***three*** bits if Δ = 20°.

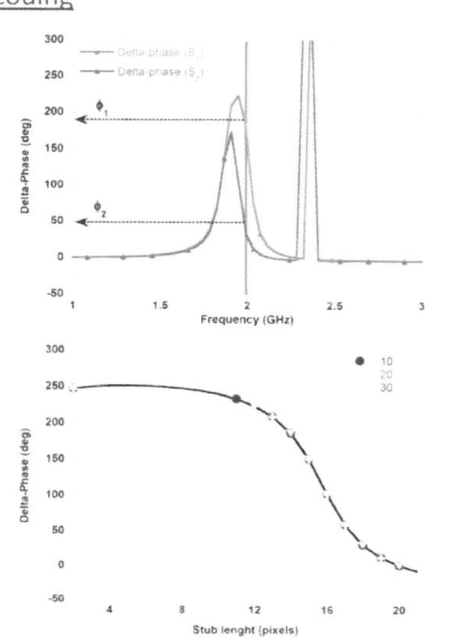

Looking at the phase response previously reported, it is possible to calculate the difference between the TE and TM response. As highlighted in the plot, it is possible to exploit the delta phase associated with a particular stub length as a bit codification. For example, the differential phases at frequency f_1 = 2 GHz for the two stub lengths S_1 and S_2 are equal to $\mathbf{f}_1$ and $\mathbf{f}_2$ and more values can be obtained with different stub lengths. Let us change the stub length and look at the delta-phase value exhibited at frequency f_1. The length is expressed by using the number of pixels composing the stub. It can be seen that the delta-phase value spans within the interval (- 25°, 250°), with short stubs exhibiting the highest differential phase values. Therefore, a stub can encode a multi-value bit with more than two states. The set of stub lengths employed in the codification depends on the criterion used for quantizing and discriminating two phase states. The stubs whose delta-phase differs at least D degrees is adopted. It is apparent that 10 stub lengths will be available if D = 10° whereas keeping D = 20° or 30° the different selectable states will be 8 and 6, respectively. This means that one stubbed ring allows codifying 3 bits if D = 20° is chosen.

73  Exploiting amp/phase/pol (IV)

It is then necessary to define a decoding procedure. Let us consider again the delta phase f at frequency f_1. for D = 20°. This choice individuates a set of delta phases within the interval [f - D/2, f + D/2], where D/2 is the accepted phase deviation. The individuated intervals do not intersect; thus, there is no ambiguity in the reading process. Finally, it is also interesting to assess the effect of the incident wave angle on the delta-phase behavior. A numerical study considering several incidence angles proved that the proposed codification can be employed up to w = 25° and q = ±30°.

Delta-Phase quantization encoding

- A decoding procedure has to be defined in order to avoid ambiguity and make a simple but robust information extraction.
- Let us consider again the delta-phase ϕ at frequency f_1 for Δ = 20°. This choice individuates a set of delta-phases within the interval [ϕ - Δ /2, ϕ + Δ /2], where Δ/2 is the accepted phase deviation.
- The individuated intervals do not intersect thus there is no ambiguity in the reading process.
- Example of stub length = 16 pixels (differential phase of 100°). The information is considered correctly retrieved if the measures delta-phase response is within [100°- Δ /2,100 + Δ /2].
- Stability of the delta-phase quantization with respect to the incidence angle. The proposed codification can be employed up to ω = 25° and θ = ±30°.

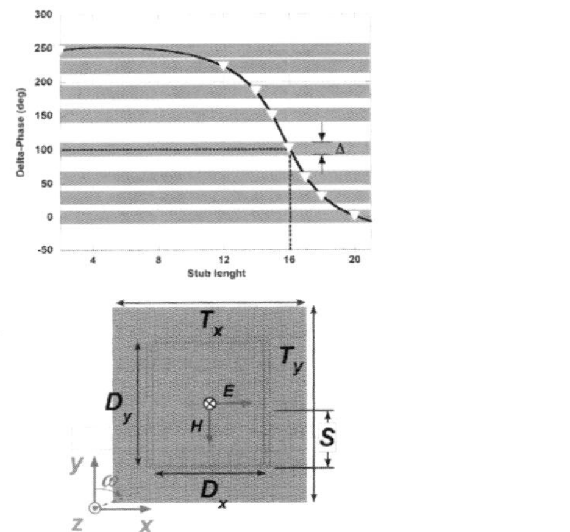

74

Exploiting amp/phase/pol (V)

Delta-Phase quantization encoding

- A finite-size tag comprising 3x3 unit cells has been realized and measured. The unit cells number employed does not modify the presented concepts but it has an impact only on the read-range. By using a 3x3 unit cells tag it is possible to obtain a read range of almost 10 m with conventional readers.
- A dual-polarized wideband horn antenna (Flann DP280) and an Agilent E5071C VNA have been employed for the measurements in a non-anechoic environment.
- Measured frequency response with the occupied frequency band highlighted with blue vertical bars.
- Performance has been observed in all the manufactured tags and suggest that a Λ = 20° can be considered a good choice able to guarantee the trade-off between encoding capacity and correct recovering of the information.

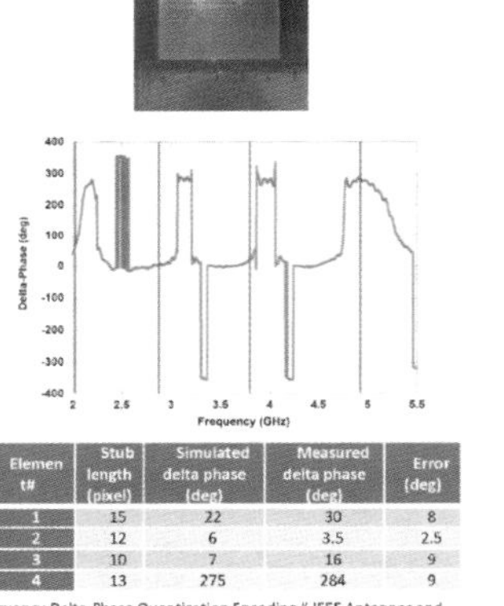

Element#	Stub length (pixel)	Simulated delta phase (deg)	Measured delta phase (deg)	Error (deg)
1	15	22	30	8
2	12	6	3.5	2.5
3	10	7	16	9
4	13	275	284	9

More details in: S. Genovesi, F. Costa, A. Monorchio, and G. Manara, "Chipless RFID Tag Exploiting Multifrequency Delta-Phase Quantization Encoding," IEEE Antennas and Wireless Propagation Letters, vol. 15, pp. 738–741, 2016.

In order to increase the quantity of information stored in the chipless tag, more nested rings can be employed. Each ring J has its own stubs S^J of equal length attached to the corner and obviously different D_{xJ} and D_{yJ}. In this case, the codification of the information is related to the differential phase exhibited by the tag at a fixed frequency. In the case of four nested rings, we have four reference frequencies f_i (i = 1,2,3,4). It is important to highlight that the change in the delta phase exhibited at the frequencies fi is mostly related to the stub length of the corresponding ring and it is weakly related to the adjacent elements. Let us now calculate the number of states encoded by the described structure. Considering D = 10,° the total number of combinations is equal to 13104, that is 13.67 bits, whereas by choosing D = 30° the bit number is 10.49. Contrarily to many encoding schemes that require an ultra-wide or wide band occupation, the proposed codification paradigm requires the chipless RFID tag phase response at 4 fixed frequencies only. A finite-size tag comprising 3x3 unit cells has been manufactured and tested as a representative example. The observed results suggest that a D = 20°can be considered a good choice able to guarantee the trade-off between encoding capacity and correct recovering of the information.

75

Exploiting amp/phase/pol (VI)

Exploiting Cross-polarization

- If a chipless RFID is placed on a metallic structure, the radar cross section of the backing metallic structure becomes more and more strong The increase of the metallic platform tends to conceal the information encoded within the structural RCS of the chipless RFID tag.
- However, the detection of the presence of the two narrowband peaks is straightforward by using the cross-polar reflection. The reason is that the metallic platform does not de-polarize the reflected signal as the tag does.

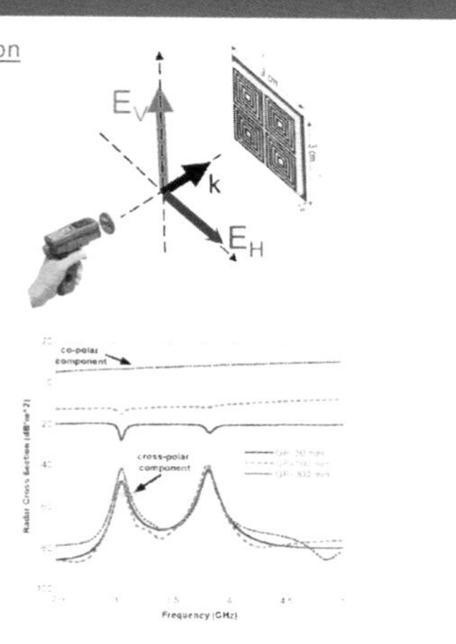

If a chipless RFID based on the frequency signature concept is placed on a metallic structure, the radar cross section of the backing metallic structure becomes more and more strong and conceal the absorptive peaks of the resonator where the information is encoded. In order to discriminate the signal from the tag with respect to the large echo coming from the surrounding big metallic platform, it can be useful to associate the information with the cross-polar reflection scattered field. This solution allows the detectability of the tag backscattering, no matter how large the metallic platform is, since the metallic platform does not de-polarize the reflected signal as the tag does.

76

Exploiting amp/phase/pol (VII)

Exploiting Cross-polarization

- The highest cross-polar is excited when the E filed is impinging along the diagonal of the nested rings.
- Measurements confirm the estimated results.

More details in: F. Costa, S. Genovesi, and A. Monorchio, "Chipless RFIDs for Metallic Objects by Using Cross Polarization Encoding," IEEE Transactions on Antennas and Propagation, vol. 62, no. 8, pp. 4402-4407, Aug. 2014.

If we consider a loop as the basic element of the unit cell, the highest cross-polar is excited when the E filed is impinging along the diagonal of the ring. Also in this case, more rings can be nested in order to have a multi-bit codification.

In order to verify the reliability of the proposed chipless design, a set of prototypes have been manufactured with a standard photolithographic technology on a commercial FR4 substrate. The measured reflection of a 10-bit tag glued on the door is also reported in the slide. The 10 peaks are visible at a distance of 40 cm from the door.

77 

Cross-pol generation

ϕ=45°, θ=0°

$E_{inc}^{H} \Rightarrow \Gamma = -1$

$E_{inc}^{V} \Rightarrow \Gamma = 1$

The previous tag comprising loop resonators radiates some electric field in cross-polarization, but the tag is not intentionally designed to work as a depolarizing tag. It is also possible to design a chipless tag which converts the electric field more efficiently. For doing that, an asymmetric FSS unit cell has to be employed. A simple example is the dipole FSS. Let us suppose having a dipole unit cell printed on a grounded dielectric slab. When the field is polarized along the dipole, a near unity reflection profile is obtained provided that the substrate is characterized by small losses. The phase is instead characterized by a transition through zero. The reflection coefficient is therefore +1. If the impinging field is polarized orthogonally to the dipole, a reflection coefficient equal to -1 is obtained. If the field is oriented toward phi=45°, the field can be decomposed into two identical vertical and horizontal vector fields. Since only one of the two vector fields is reversed, the reflected polarization is purely orthogonal with respect to the impinging one.

78 

Limitations and current challenges (I)

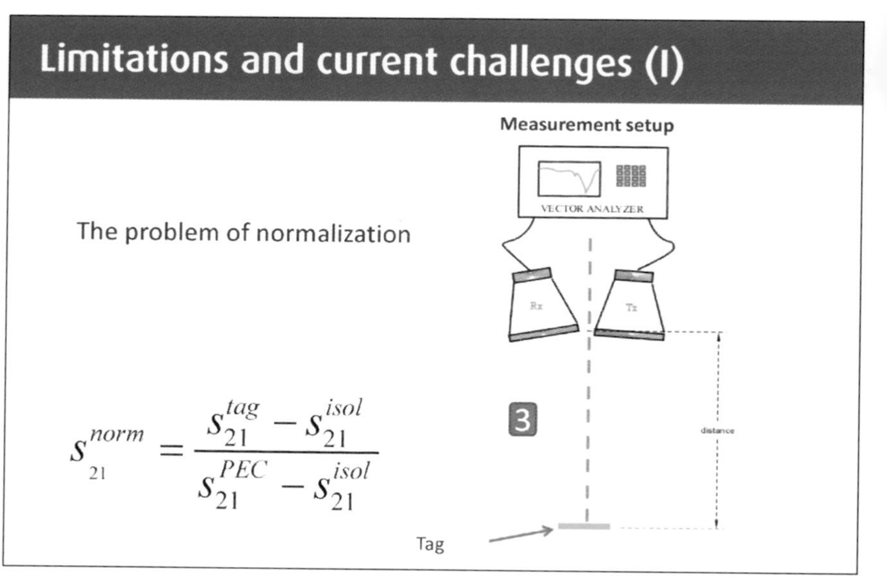

All the results shown for the frequency-encoded chipless RFID tag requires a normalization procedure for the correct recovery of the encoded data. More in detail, three different measurements are required. In the first one (S_{21}1isol), in order to cancel out the mutual coupling effect between the two ports of the antenna and the undesired reflections due to multipath propagation, a reference measurement of the environment in the absence of the tag has been preliminary performed for each polarization. Next, a reference measurement of a metallic object of the same size of the tag must be performed (S_{21}isol). Then, the tag response (S_{21}tag). can be collected. Finally, the normalized measure (S_{21}norm) can be evaluated and the information can be recovered.

79

Limitations and current challenges (II)

- The problem of normalization: a suitably tailored chipless RFID tag and additional signal processing may mitigate this problem.

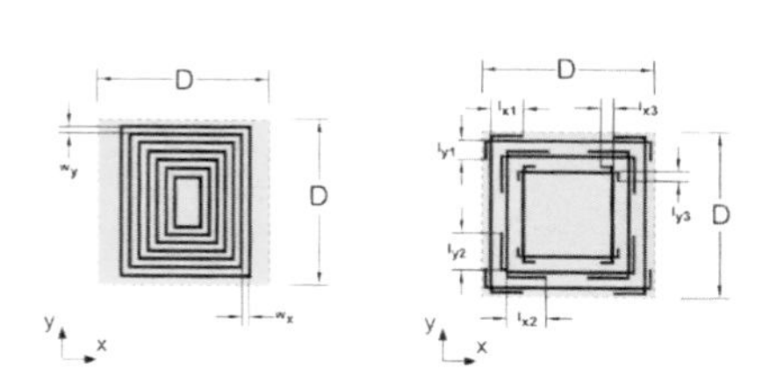

unit cell 1

The different degree of stretch (w_x and w_y) provides the amount of the shift between the two spectral responses

unit cell 2

The length of the stubs is different towards x and y directions in to provide the desired shifted frequency response.

The previously described procedure, based on two or three independent measurements performed on the same scenario (tag, background, and eventually ground plane) is not feasible in a realistic scenario. To cope with this limitation, a new encoding/decoding scheme has been proposed. It requires two measurements along two orthogonal planes of incidence that are further elaborated with post-processing algorithms. This approach can be carried out in a realistic scenario by using a reader with a dual-polarized antenna. To exploit the new calibration scheme, it is necessary that the tag exhibits a shift in the frequency response to the two orthogonally polarized probing waves. As it will be clarified later, this condition provides sharp peaks in the frequency domain when the two orthogonal responses are substracted. This behavior can be achieved by asymmetric resonators such as rectangular loops or by square loops loaded with stubs of different length along the two main planar directions.

80

Limitations and current challenges (III)

Dual-polarized interrogation: reading without normalization

START
Measurement of the s_{11} and s_{22} through a dual polarized antenna
Antenna scattering parameters in free space ($s_{11\_antenna}$, $s_{22\_antenna}$)
$V(f)=s_{11\_with\_tag}-s_{11\_antenna}$
$H(f)=s_{22\_with\_tag}-s_{22\_antenna}$
IFFT of V(f) and H(f): V(t), H(t)
TD gating of V(t) and H(t): $V(t)_{TG}$, $H(t)_{TG}$
FFT of $V(t)_{TG}$, $H(t)_{TG}$: $V(f)_{TG}$, $H(f)_{TG}$
$V(f)_{TG}-H(f)_{TG}$
END

1. s_{11} and the s_{22} of the reader are recorded;
2. Complex subtraction of the s_{11} and the s_{22} of the antennas measured in free space → V(f), H(f);
3. *V(f)* and *H(f)* anti-transformed in the time-domain;
4. TD window (removing of coupling and multipath);
5. The filtered responses are then transformed again into the frequency domain;
6. Amplitude (in decibel) subtraction.

To define a measurement procedure that does not require every time to perform three separate measurements, we take advantage of three strategies: dual-polarization interrogation, time domain gating, and free-space antenna response subtraction. First of all, the information must be encoded in the difference between the reflection coefficients of the tag measured with respect, vertical and horizontal polarizations. In this way, the information is associated with the differential response instead of the absolute value of the backscattered field. The decoding steps are summarized in the reported flowchart and the effects on the received signals will be illustrated in the next slides.

81

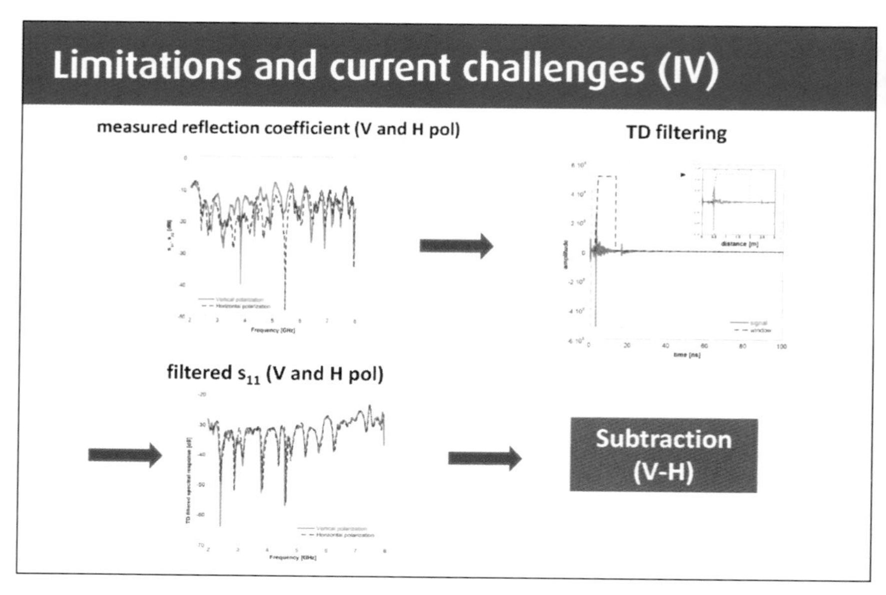

First of all, the tag is interrogated with a dual-polarized antenna and the two responses (V and H) are collected and stored. Next, it is necessary to subtract the reflection coefficients of the reader antenna operating in free space (not in the operative scenario) from the reflection coefficients measured in the presence of the tag (operative scenario). This task must be performed because dual-polarized antennas intrinsically exhibit two different reflection coefficients at its ports and this difference, if not removed, could invalidate the decoding procedure. The unloaded reflection coefficients of the antenna are independent of the scenario and they can be considered as known parameters that can be stored and not measured every time we want to read a tag.

82

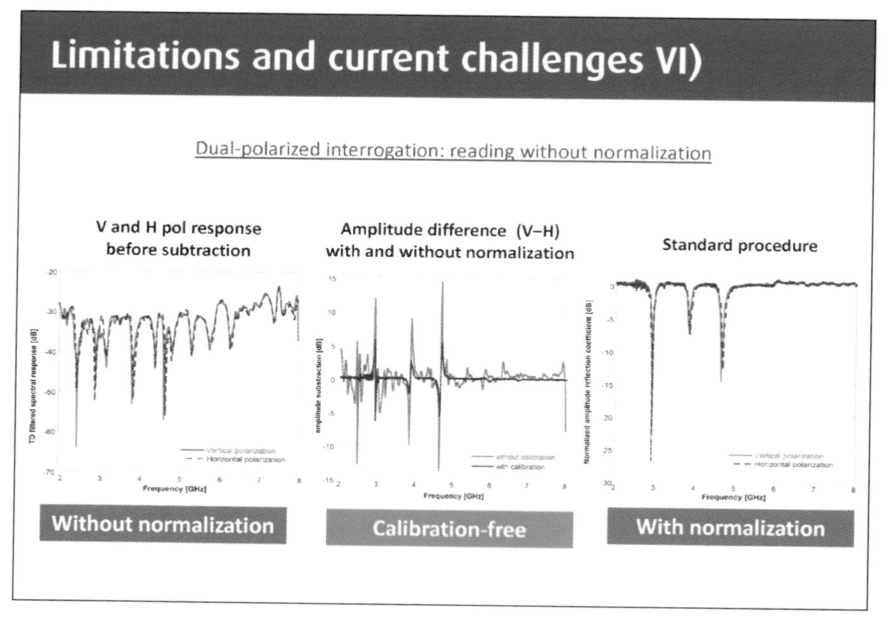

In the final step, a time-domain gating allows removing some of the harmful effects due to the antenna coupling and to the multipath phenomena. The distance from the tag is necessary to perform this last elaboration and it is estimated by tracking the first structural RCS peak.

It is important to point out the importance of the joint application of differential encoding and time gating since none of this two strategies, if separately applied, can be successful.

It is also interesting to observe that even after the subtraction of the free-space antenna parameters and time gating, the V and H signal are not yet intelligible. In fact, it is their difference that, thanks to the imposed differential encoding, reveals the information.

83

Chipless RFID sensor based on HIS (I)

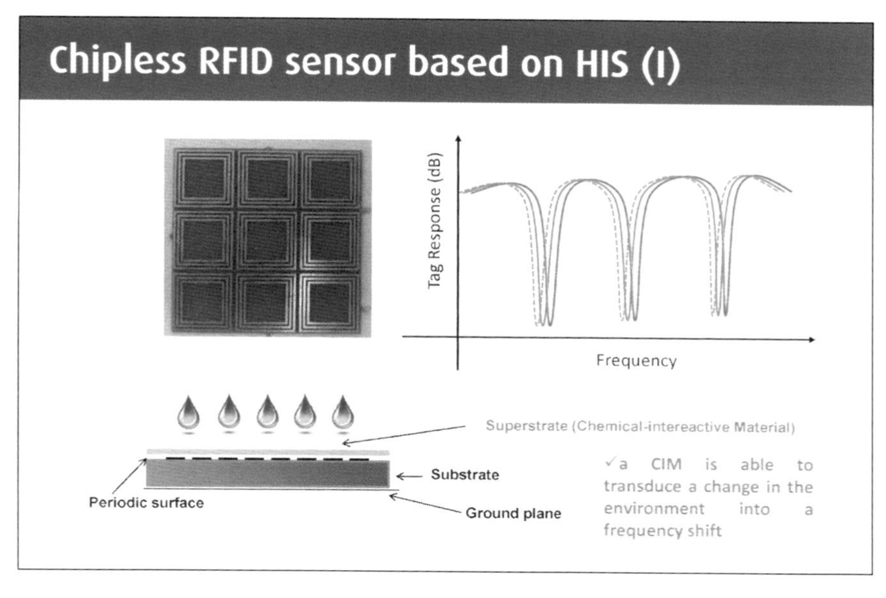

It could be interesting to investigate if a chipless RFID tag can be transformed into a sensor. The sensing function can be added to a properly designed chipless RFID tag by exploiting materials that are susceptible to external environment changes such as pressure, temperature, humidity, and gas concentration. The physical basis grounds on the capability of chemical interactive materials (CIM) of changing their dielectric properties through the interaction with target molecules or by environmental changes. This change can be advantageously exploited to modify the field scattered by a chipless RFID tag equipped with a CIM. The observed variation of the scattered field (*i.e.* frequency shift) contains the information collected by the sensor. A CIM (Chemical-Interactive Material) can be added to a chipless RFID tag in order to perform sensing or even sensing and ID capability. We will consider the case of a CIM that changes its own permittivity as a function of the relative humidty (RH) to which it is exposed.

84

Chipless RFID sensor based on HIS (II)

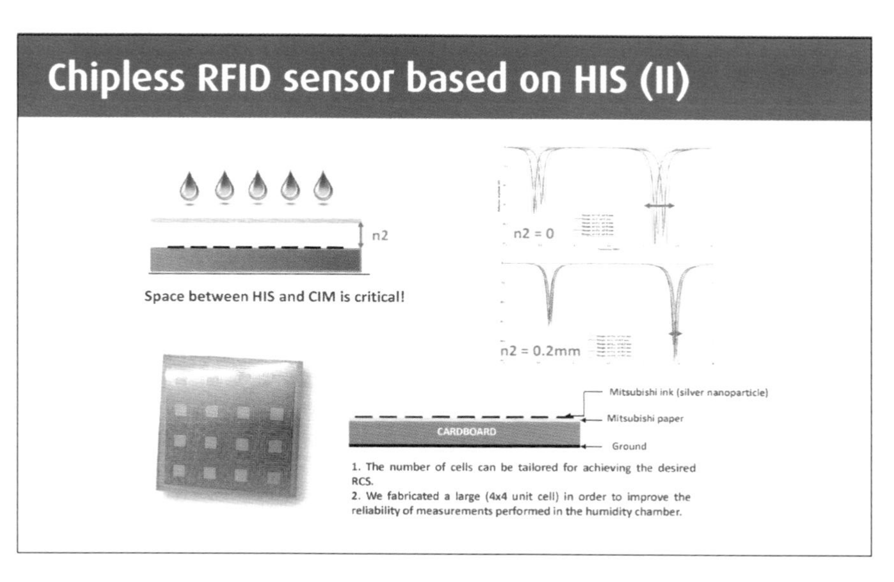

If the CIM is directly applied to the chipless RFID tag and therefore there is no air gap in between, then the observed frequency shift is maximum. On the contrary, it can be proved that even a small air gap can deteriorate the sensor response since it limits the amount of shift achieved with the permittivity change of the CIM.

To this aim, in order to avoid any possible air gap between the resonators comprising the chipless RFID and the CIM, the periodic structure has been directly printed on the sensing material. More in detail, a silver nanoparticle ink has been used to print the resonators on photographic paper by using standard desktop printer. This film has then been applied to a grounded cardboard support.

85

Chipless RFID sensor based on HIS (III)

Design of automatic climatic chamber

Prototype of a humidity-controlled small chamber comprising:

- A Rele;
- Humidity generator controlled by the rele;
- Fans controlled by the rele;
- Sensor for reading RH and Temperature;

The system is currently controlled by Matlab which realizes a feedback action to maintain constant RH.

To assess the change in the electromagnetic response of the tag as a function of the level of humidity of the environment, a high number of measurements is required. Therefore, it is important to use a reliable and controllable setup that allows an automatic testing of the sensor during a certain interval of time. A prototype of a humidity-controlled chamber has been realized. A dual-polarized horn antenna is placed in front of the sensor, both inside the small chamber. A vector network analyzer (VNA) is connected to a laptop via an USB cable. The electromagnetic response of the tag is measured at the desired time intervals by using a Matlab code. At the same time, it is possible to regulate the level of humidity inside the box with a feedback control system control.

86

Chipless RFID sensor based on HIS (IV)

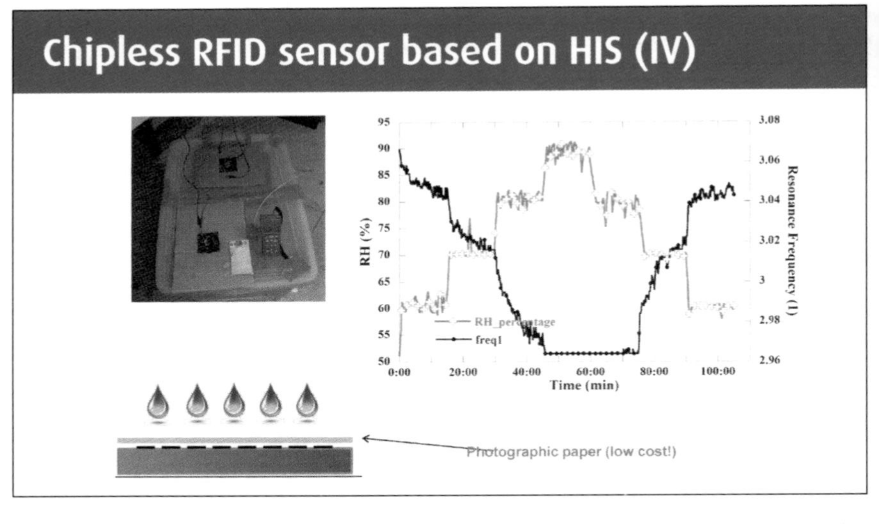

This system is able to automatically control the RH level inside the box according to a humidity profile chosen by the user. During the RH-controlled cycle, the electromagnetic response of the chipless tag is collected at the predetermined time steps. Next, the collected EM signal is correlated to the RH level with a post-processing algorithm. In fact, in order to assess the sensitivity of the chipless sensor, it is necessary to correlate the position of the resonance peaks to the RH level. To perform this task, the frequency response of the tag was monitored when the RH level within the climate chamber was changed from 60% to 90% and back in steps of 10%. To test the moisture absorption time of the tag, each humidity level was kept constant for 15 minutes. With the aim of better displaying the shift of the resonance peaks with the variation of the RH level, one of the resonant peaks has been plotted as a function of the observation time, together with the variation of the RH level. It is evident that in correspondence to a rapid variation of the RH curve, the resonance frequency varies rapidly, whereas it is almost constant when the RH level is constant. In addition, these graphs confirm that the moisture absorption of the tag is a reversible phenomenon.

This research has been founded by the Emergent Marie-Curie RISE project GA n. 645751 and the Spanish Government Project TEC2015-67883-R.

RISE program funds short-term exchanges for staff to develop careers combining scientific excellence with exposure to other countries and sectors and encourages the organization of trainings, networking activities, and liaisons with other EU projects.

Three universities, namely University of Pisa, University Rovira i Virgili, and Université Grenoble-Alpes, and three companies, namely Cubit, Generation RFID, and Ardeje are the partners involved this project.

CHAPTER 06

Near-Field Focused Antennas for Short-Range Identification and Communication Systems

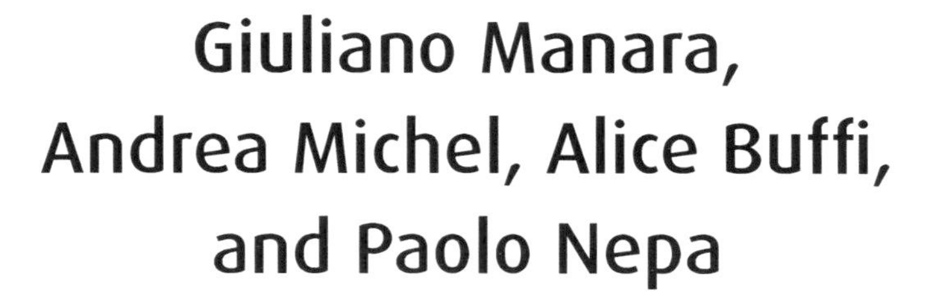

Giuliano Manara, Andrea Michel, Alice Buffi, and Paolo Nepa

University of Pisa, Italy

1. Agenda 162
2. Reactive near-field RFID systems (I) 162
3. Reactive near-field RFID systems (II) 163
4. Near-field UHF RFID systems 163
5. Antenna Requirements 164
6. Different kinds of near-field RFID Antennas 164
7. Loop Antennas (I) 165
8. Loop Antennas (II) 165
9. Resonant antennas and arrays 166
10. Oppositely directed Currents (ODCs) 166
11. Transmission Line (TL) Antennas (I) 167
12. Transmission Line (TL) Antennas (II) 167
13. Antenna layouts and prototypes 168
14. Transmission Line (TL) Antennas (I) 169
15. Transmission Line (TL) Antennas (II) 169
16. TWA Array – Prototype and measured results 170
17. Measured tag detection 170
18. System-level performance 171
19. Stacked Tag Read Range 171
20. Modular antenna (I) 172
21. Modular antenna (II) 172
22. Modular antenna (III) 173
23. Near-field distribution 173
24. System-level antenna performance 174
25. Stacked tag readability 174
26. A multi-function antenna 175
27. Writing tests 175
28. Reconfigurable Modular Antenna Layout 176
29. Two States of the Reconfigurable Antenna 176
30. Modular Antenna Prototype 177
31. Electric field components at 865 MHz 178
32. Tag detection measurement setup 178
33. System-Level Performance 179
34. Future trend 179

In this chapter, near-field antennas for short-range identification and communication systems are described. Specifically, ultra-high frequency (UHF) (840–960 MHz) near-field (NF) Radio Frequency Identification (RFID) systems have attracted increasing attention because of the possibility of achieving much higher reading speeds and capability to detect a larger number of tags (bulk reading). A UHF NF RFID system is a valuable solution to implement a reliable short-range wireless link (up to a few tens of centimeters) for ILT applications. Since the tags can be made smaller, RFID-based applications can be extended to extremely small items (e.g., retail apparel, jewelry, drugs, rented apparel) as well as to successful implementations of RFID-based storage spaces, smart conveyor belts, and shopping carts. This course aims at introducing the RFID technology, focusing on near-field systems and applications. A detailed overview of ad hoc NF reader antennas is proposed, highlighting their main characteristics and achievable performance. The future trends of these systems are also discussed together with their potentialities and advantages.

1  **Agenda**

- **Overview on UHF RFID Near-field Systems**
- **State of the Art on Near-field antennas**
  - Loop-like antennas
  - Traveling Wave Antennas
  - Resonating Antennas
- **Reconfigurable Near-field UHF RFID Antennas**
  - Modular Antenna Concept
  - Example of reconfigurable Near-field UHF RFID antennas
- **Conclusion and Future trends**

In this short course, near-field antennas for short-range identification and communication systems are described, highlighting their main characteristics and achievable performance.

After a brief overview on UHF RFID Near-field systems and their applications, a detailed state-of-the-art review is discussed. Furthermore, reconfigurable near-field antennas are introduced, showing two examples. Finally, conclusions are drawn, focusing on future trends of this kind of applications.

2 

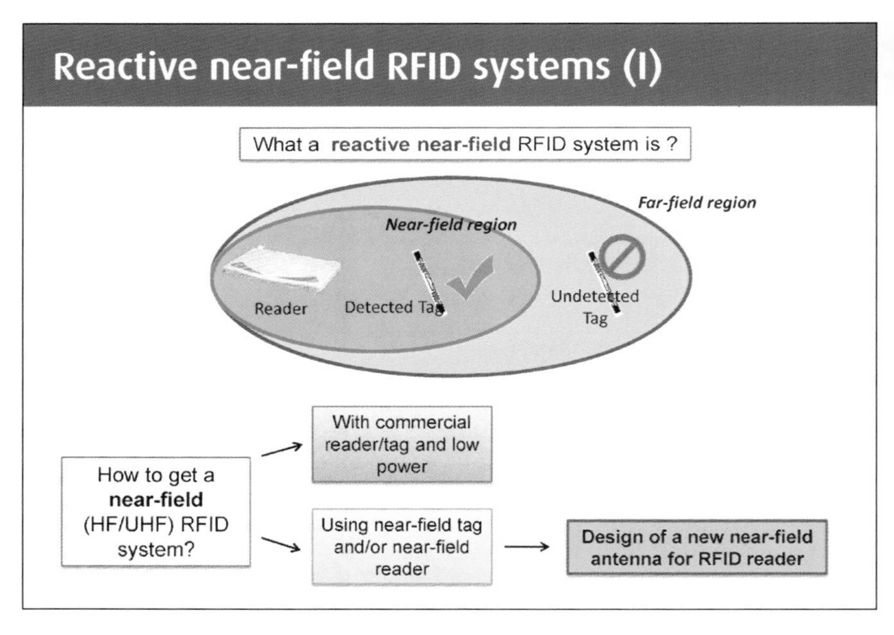

A UHF NF RFID system is a valuable solution to implement a reliable short-range wireless link (up to a few tens of cm) Item Level Tagging (ILT) applications. Since the tags can be made smaller, RFID-based applications can be extended to extremely small items (retail apparel, jewelry, drugs, rented apparel), as well as to a successful implementation of RFID-based storage spaces, smart conveyor belts, and shopping carts.

A continuous effort has been made by researchers to improve the performance of the UHF NF RFID systems. In this context, ad-hoc NF reader antennas have been investigated to enhance the UHF NF RFID system performance, while confining the electromagnetic field in an assigned limited volume close to the reader antenna.

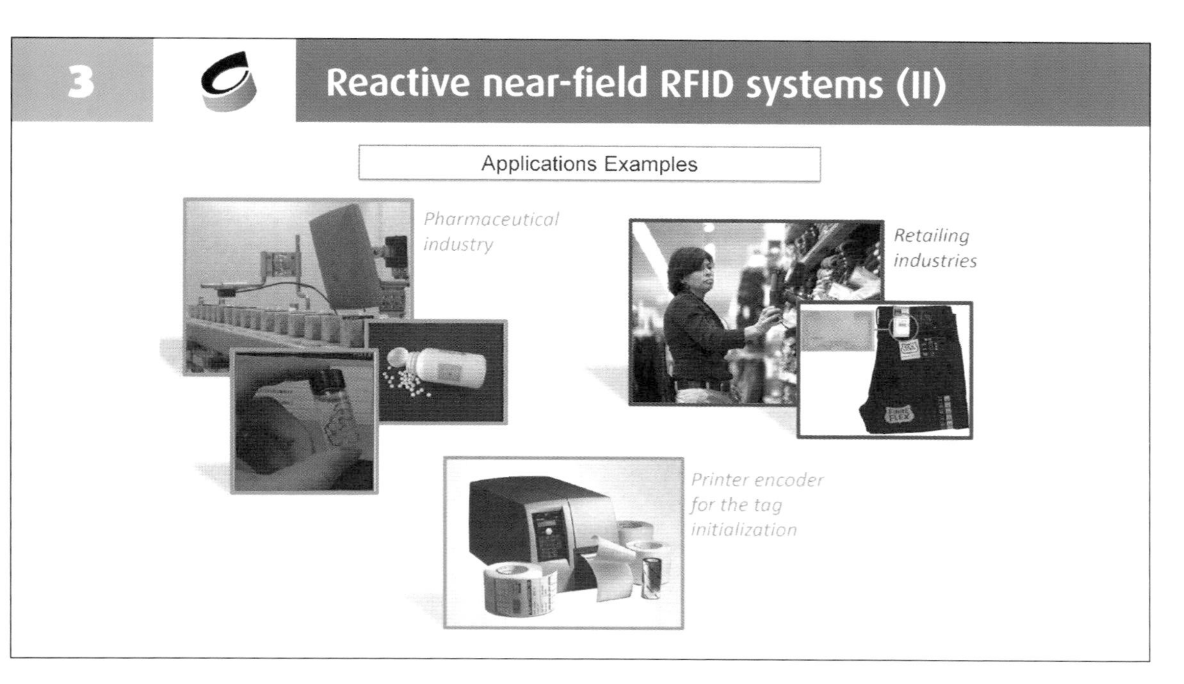

RFID systems are employed in a large number of applications. Specifically, near-field UHF RFID systems are typically used in Item-Level Tagging applications in pharmaceutical and retailing industry. To initialize a tag, specific devices are used, named printer encoders. The RFID printer-encoder represents a specific scenario in which near-field coupling between reader and transponder antennas is involved. The RFID printer-encoder manages smart labels composed of barcode and human-readable text (typically printed on paper or plastic substrates), combined with a UHF-RFID transponder that has to be encoded.

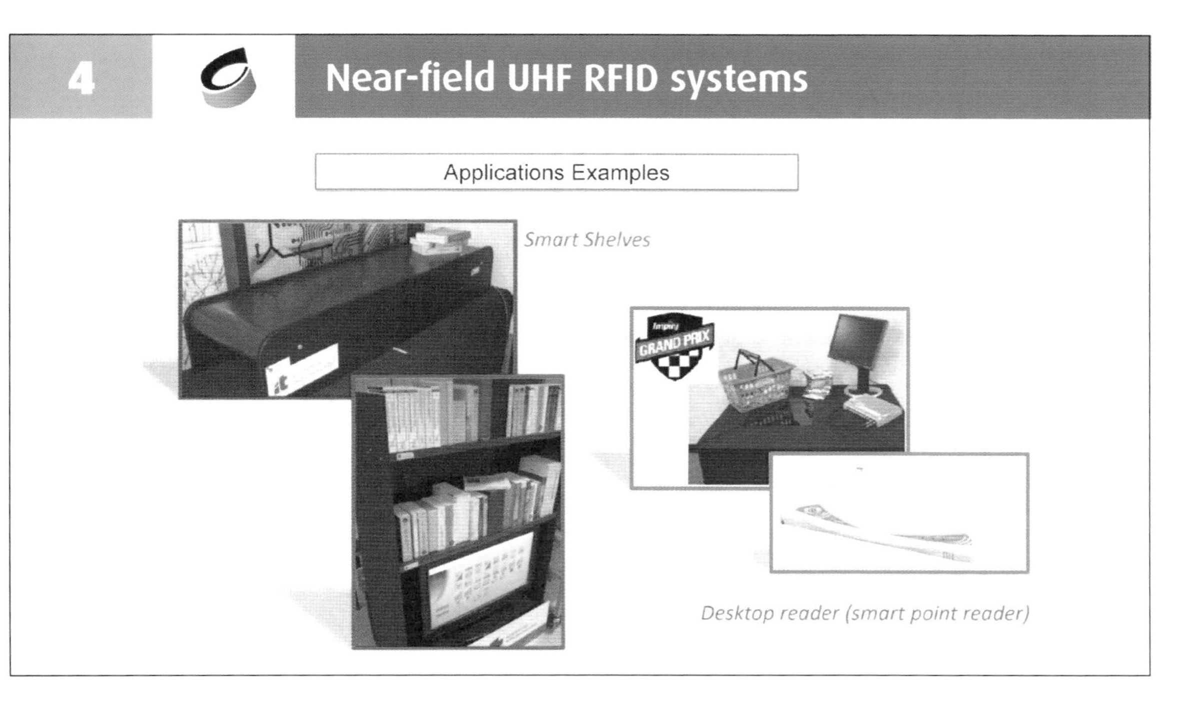

Near-field RFID systems are also used to create Smart Shelves, Smart Point Reader, Smart Drawers.

5

Antenna Requirements

- Low Cost
- Scalable
- Low Profile
- Wideband
- Near-field Application Oriented
- Low Tag Orientation Sensitivity

Electric and magnetic fields should be strong and uniform in a confined region around the reader case and up to few decimeters.

A UHF NF RFID reader antenna is required to generate an as uniform/strong field distribution as possible in a confined interrogation zone to avoid tag detection failures. The required read range can vary from a few millimeters up to a few tens of centimeters, with an assigned reading rate, such as 100%. Moreover, the vector magnetic/electric field has not to exhibit a dominant component, as in most applications, the tag orientation with respect to the reader antenna is unknown. Besides the issues of shape and size of the interrogation zone, field intensity, and distribution, a UHF NF reader antenna should be cost-effective and easy for system implementation as well. For example, the antenna for an RFID smart-shelf should be easily adaptable to different shelf sizes and types, and the field distribution must be controlled carefully to suppress the interference between the antennas in adjacent tiers of the shelves. Furthermore, the antenna for UHF NF RFID readers is usually required to be planar and low-profile design with smaller thickness as well.

6

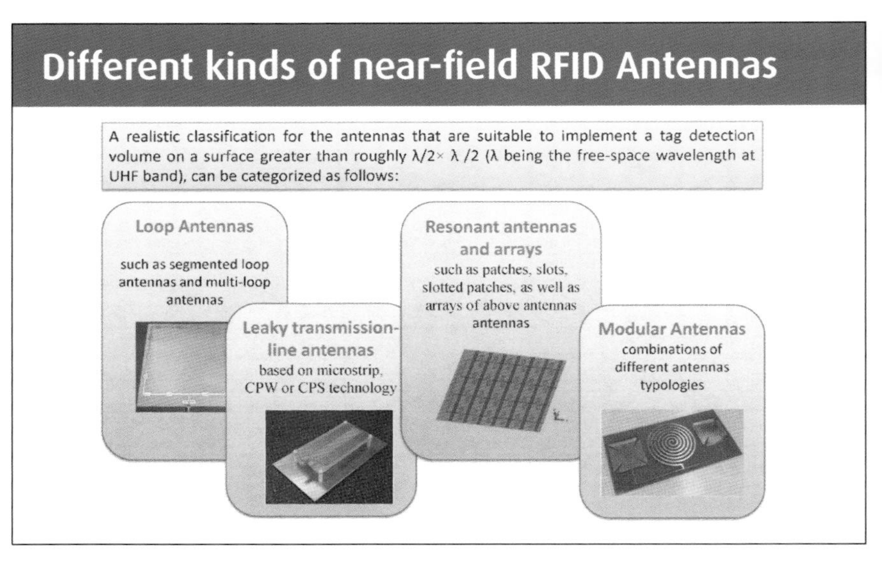

Antennas that are suitable to implement an interrogation zone with surface area greater than $\lambda/2\times\lambda/2$ (λ being the free-space wavelength at UHF band) can be categorized as follows:

- Loop antennas, such as segmented loop antennas and multi-loop antennas;
- Leaky transmission-line antennas, based on microstrip, coplanar waveguide (CPW), or coplanar stripline (CPS) technology;
- Resonant antennas and arrays, such as patches, slots, slotted patches, as well as arrays of above antennas.

Proper combinations of different antenna typologies have also been proposed to make a reader antenna suitable for both near-field and far-field applications. That is, the antenna is able to generate a strong and uniform field in proximity of its surface, while offer, a non-negligible far-field gain for farther tag detection.

7 

Loop Antennas (I)

Some loops made of segmented lines (named as Segmented Loops) have been presented, where the current is kept almost constant and in-phase along the loop, even though the loop perimeter is larger than the operating wavelength.

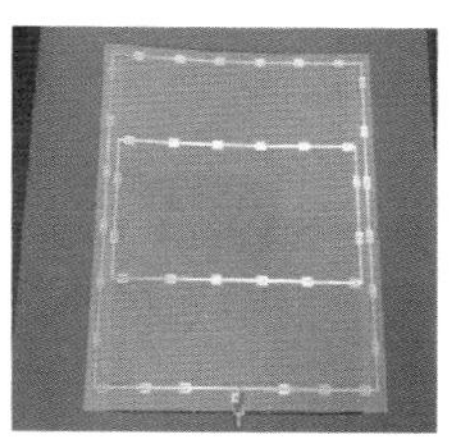

X. Qing and Z. N. Chen, "UHF near-field segmented loop antennas with enlarged interrogation zone," in 2012 IEEE International Workshop on Antenna Technology (iWAT), 2012, pp. 132–135.

J. Shi, X. Qing, Z. N. Chen, and C. K. Goh, "Electrically Large Dual-Loop Antenna for UHF Near-Field RFID Reader," IEEE Trans. Antennas Propag., vol. 61, no. 3, pp. 1019–1025, Mar. 2013

X. Qing, C. K. Goh, and Z. N. Chen, "Segmented loop antenna for UHF near-field RFID applications," *Electron. Lett.*, vol. 45, no. 17, pp. 872–873, Aug. 2009.

Single- and multi-turn solid-line loop antennas are most commonly used in HF RFID readers because of their ability to generate strong magnetic field. However, at UHF band, a physically large loop required to offer an extended interrogation zone exhibits a weak field in its central portion, since the current along the loop experiences phase inversions and current nulls. Some loops made of segmented lines (named as segmented loops) have been presented, where the current is kept almost constant and in-phase along the loop, even though the loop perimeter is larger than λ.

8 

Loop Antennas (II)

Segmented loop antenna made of four large curved strips separated by four pairs of coupled stubs

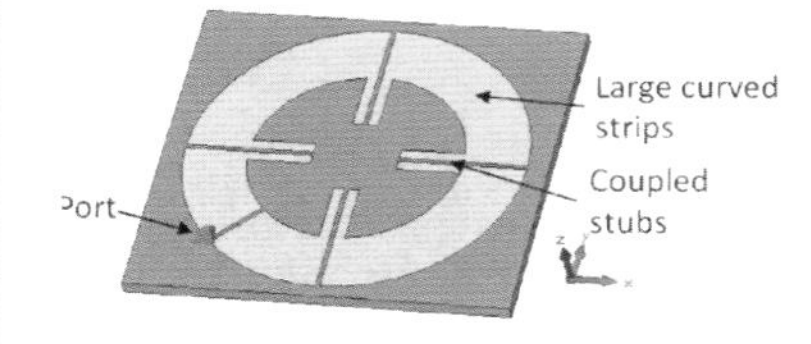

Y.-W. Liu, K.-H. Wu, and C.-F. Yang, "UHF reader loop antenna for near-field RFID applications," Electron. Lett., vol. 46, no. 1, pp. 10–11, Jan. 2010.

Grid array of coupled dash-line segmented loop antenna

J. Shi, X. Qing, and Z. N. Chen, "Electrically Large Zero-Phase-Shift Line Grid-Array UHF Near-Field RFID Reader Antenna," IEEE Trans. Antennas Propag., vol. 62, no. 4, pp. 2201–2208, Apr. 2014.

Several techniques have been presented to design electrically large segmented loop antennas for generating a strong and uniform magnetic field distribution. Specifically, segmented loops can be configured by using segment lines with lumped capacitive elements, distributed capacitors, or coupled lines, dash lines, or embedding phase-shifters into solid-line loop. Moreover, a segmented loop can also be configured by using dual dipoles or dual open loops.

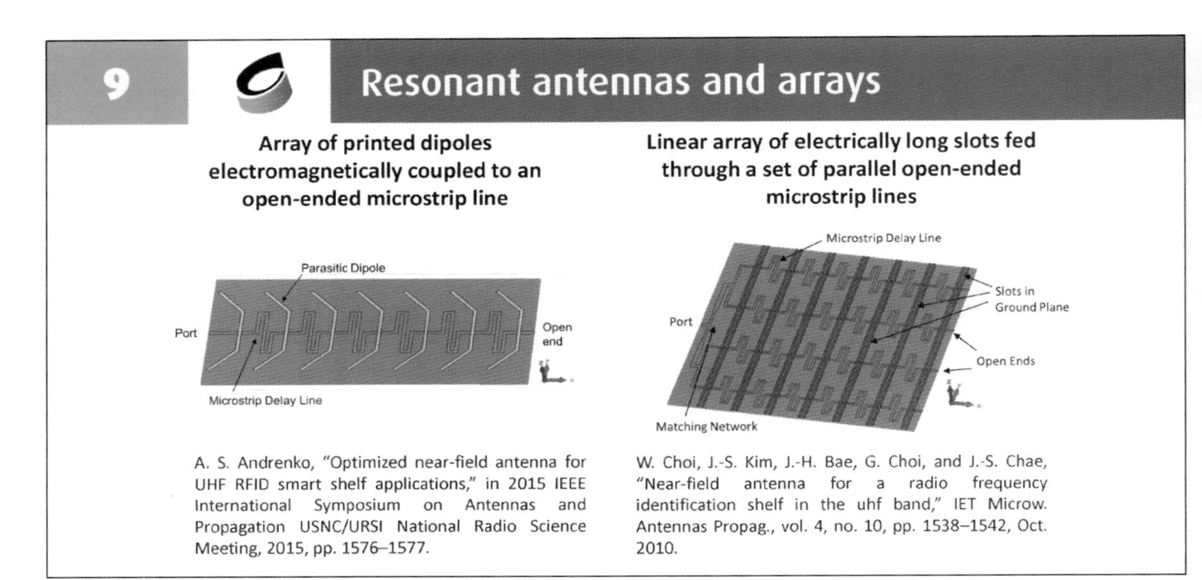

Resonant antennas are characterized by a quite high gain and their size is strictly related to the operating frequency (i.e. a square patch usually fits an area of $\lambda_g/2\times\lambda_g/2$). Miniaturization techniques have been employed to reduce the antenna size, allowing the radiating element to be embedded in commercial UHF RFID readers. To cover large detection areas with resonant antennas, an antenna array is a mandatory solution.

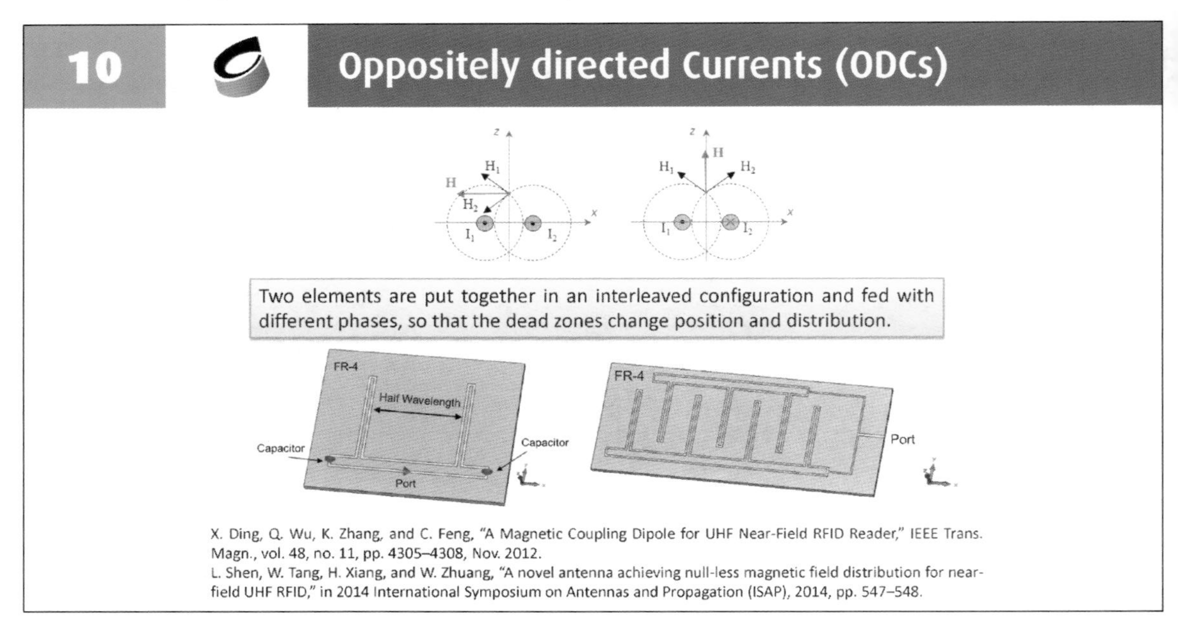

Two closely spaced oppositely directed (*i.e.* out-of-phase) currents (ODCs) are able to generate a strong and uniform magnetic near field over a relatively large detection region, in particular, for the *H*-field component perpendicular to the antenna surface.

It is worth noting that for antennas based on the ODCs concept, the magnetic field component perpendicular to the antenna surface may experience dead zones since the magnetic field right above the current is in parallel with the antenna surface. Thus, two elements are put together in an interleaved configuration and fed with different phases, so that the dead zones change position and distribution.

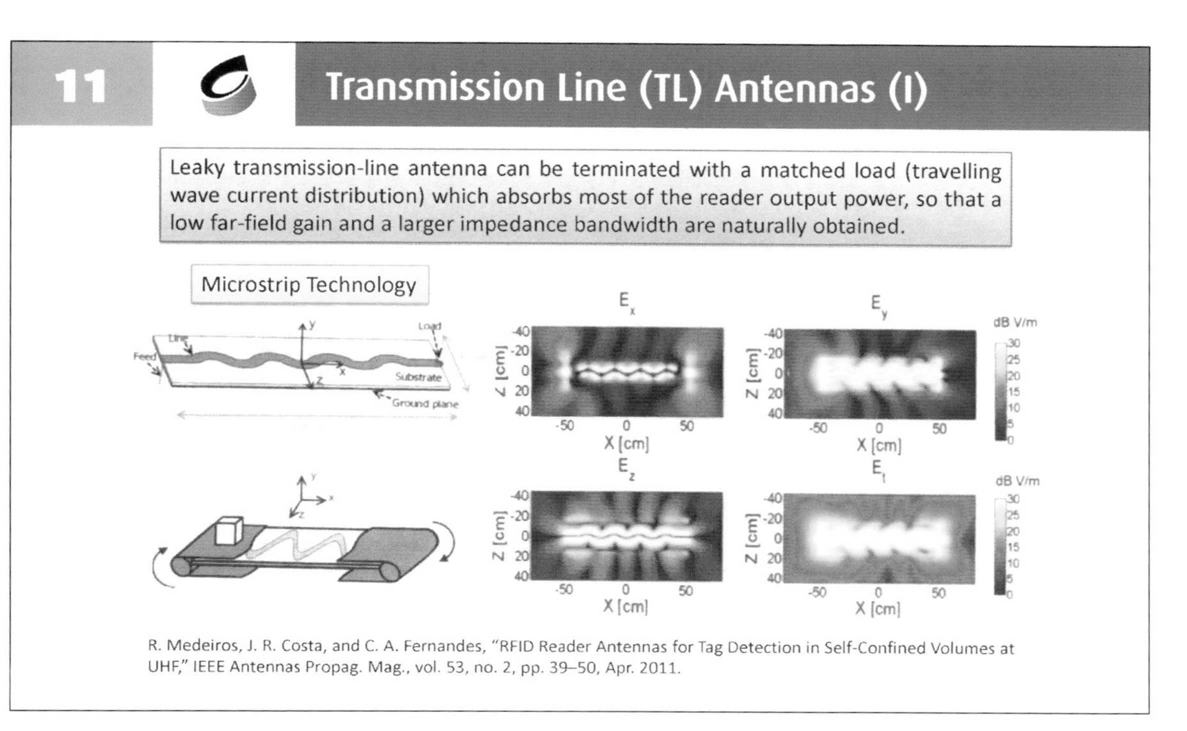

Leaky transmission-line antenna is another type of antenna which has been intensively studied for UHF NF RFID readers. The transmission line can be terminated with either a matched load (traveling wave current distribution) or a resistive load that can be varied to control the amount of energy of the reflected wave (stationary wave current pattern). Different transmission lines have been exploited to design the leaky transmission line antenna: microstrip, Coplanar Stripline, CPS, and Coplanar Waveguide, CPW. If the transmission line is terminated with a matched load, the latter usually absorbs most of the reader output power, so that a low far-field gain is naturally obtained. Moreover, the losses result in a larger impedance bandwidth and make the input impedance matching less sensitive to the presence of the tagged items in the antenna near-field region.

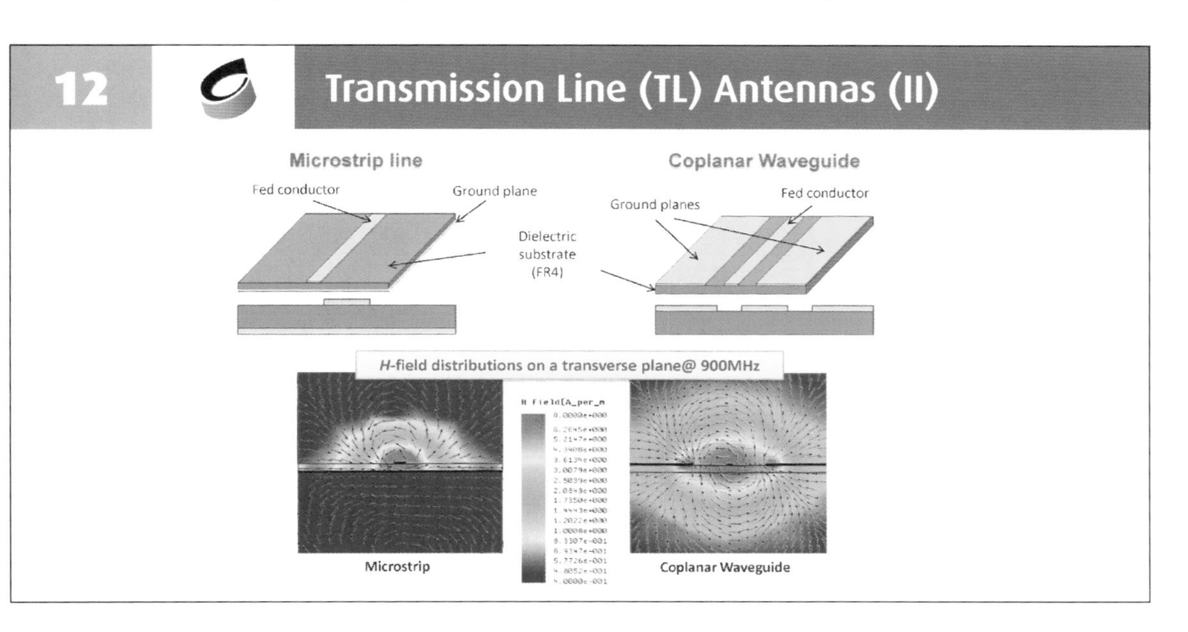

In a microstrip line, the field-force lines are such that the region with the strongest electromagnetic field intensity is concentrated inside the dielectric substrate, between the printed line and the ground plane. Conversely, using a CPW it is possible to increase field intensity outside the dielectric because a strong electromagnetic field is generated into the two slots between the internal conductor and the lateral ground planes. Thus, the electromagnetic field above a CPW line is expected to be stronger than the field above a microstrip line, at the same operative feeding conditions, and this can improve the antenna performance in near-field applications.

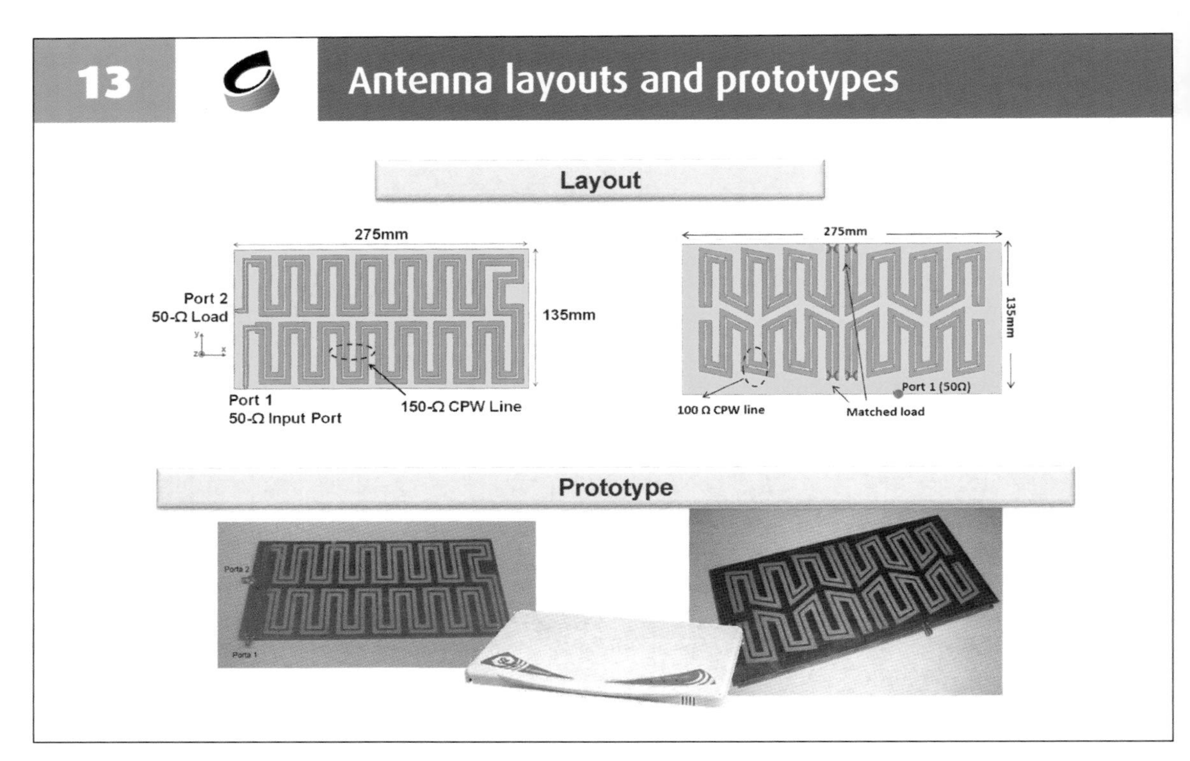

To increase the field intensity above the antenna surface, a high characteristic impedance CPW line is designed. The meandered line is terminated with a matched load. The need for a matched load is twofold: it allows a traveling wave with no-stationary wave field distribution along the transmission line, and also reduces the antenna efficiency. Indeed, low efficiency diminishes the far-field antenna gain, which is mandatory to avoid tag cross readings outside the assigned reader interrogation zone. Moreover, a Traveling Wave Antennas array can be designed, getting a confined and uniform field distribution close to the antenna (up to few centimeters), where tagged items are supposed to lie on. In particular, the presence of 2×2 radiating elements allows maximizing both electric and magnetic fields within a confined volume above the reader antenna surface (10 cm), reducing the false positives issue in the item level tagging applications (e.g. smart point readers). The meandered layout of each radiating element allows exciting field components in all directions, making the tag detection almost independent on the particular tag orientation.

14  Transmission Line (TL) Antennas (I)

The meanders number has been set in order to fully cover the tag detection area, keeping constant the distance (D) among them, along both the x-axis and the y-axis. The overall CPW line length results longer than 10 λ_g (λ_g being the CPW-guided wavelength). The length of a couple of meanders corresponds to around one CPW-guided wavelength (as apparent in the pictures where simulated results for the surface current are shown for a particular phase value, at 900 MHz).

Moreover, a minimum distance between two adjacent meanders (D) was guaranteed in order to limit mutual coupling effects that complicate achieving a wideband impedance matching.

15  Transmission Line (TL) Antennas (II)

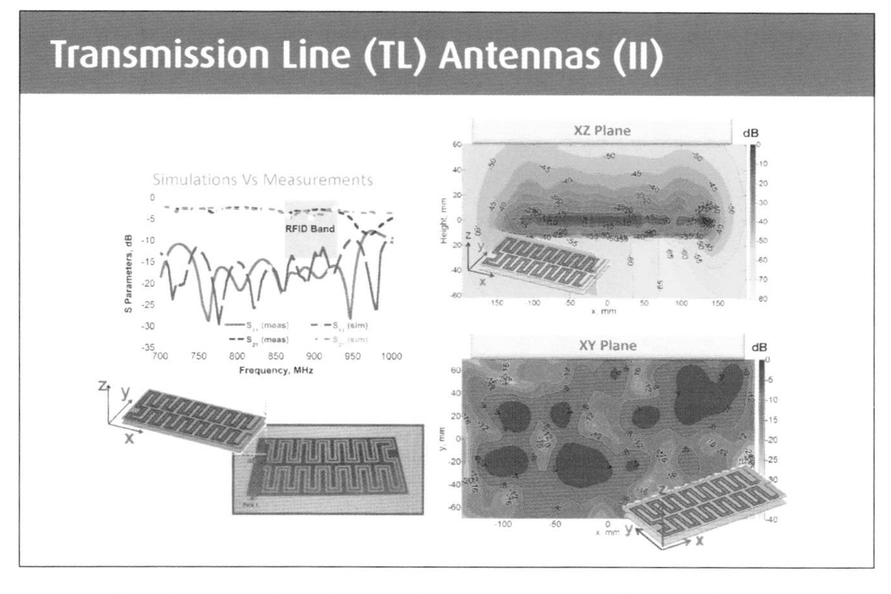

To limit the antenna thickness, in agreement with typical aesthetic specifications for desktop readers, the metallic reflector is placed at a distance of only 10 mm from the FR4 laminate. It has been verified that the presence of such a metallic reflector does not complicate impedance matching, even if it is very close to the CPW line (less than around $\lambda/30$, λ being the free-space wavelength). Since the proposed antenna is a non-resonating structure, it is robust to the presence of the reflector plane as well as to the dielectric and mechanical tolerances, which are attractive features for a simple and cheap production process.

In the HF band, most of the tags are loop-like tags that require a magnetic coupling mechanism. On the other hand, in the NF UHF-RFID systems, both electric and magnetic coupling, are important, since both dipole-like and loop-like tags may be used. Since at UHF band it is expected that the homogeneity of the electric field also implies the homogeneity of the magnetic field, simulation results only for the magnetic field behavior are considered in the following.

16

TWA Array – Prototype and measured results

TWA Array Prototype

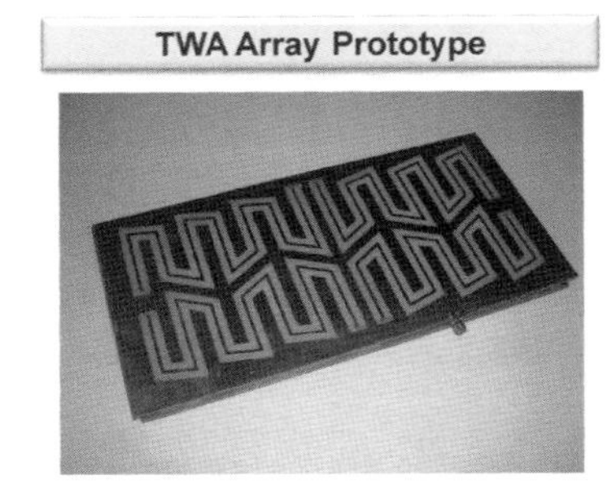

Measured Reflection Coefficient

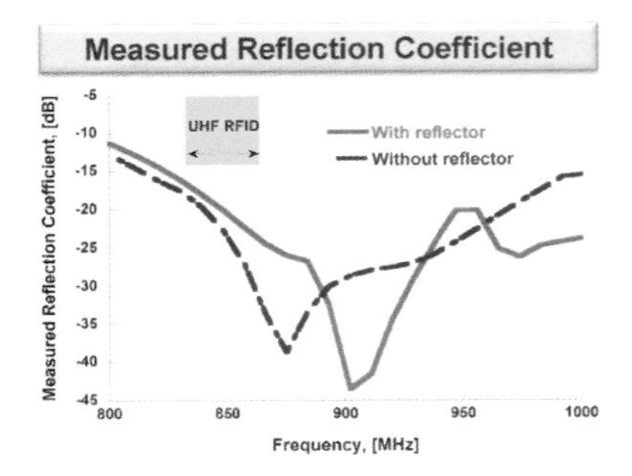

- S11 < -15dB in the UHF RFID Band (865-928 MHz)
- Wideband
- Low sensitivity to the presence of a metallic reflector

CST

Each of the four meandered CPW lines of the array is connected through a transition to a microstrip feeding line realized in the bottom FR-4 substrate. A detailed description of the microstrip to CPW transition can be found in the literature. Furthermore, since the reader antenna performance could be affected by the particular desk material, a 275 × 135 mm^2 reflector plane (not electrically connected to the antenna) has been placed very close to the bottom of the antenna substrate (about 10 mm of distance). The measured antenna reflection coefficient is shown, and it is below -14 dB in the entire UHF RFID band.

17

Measured tag detection

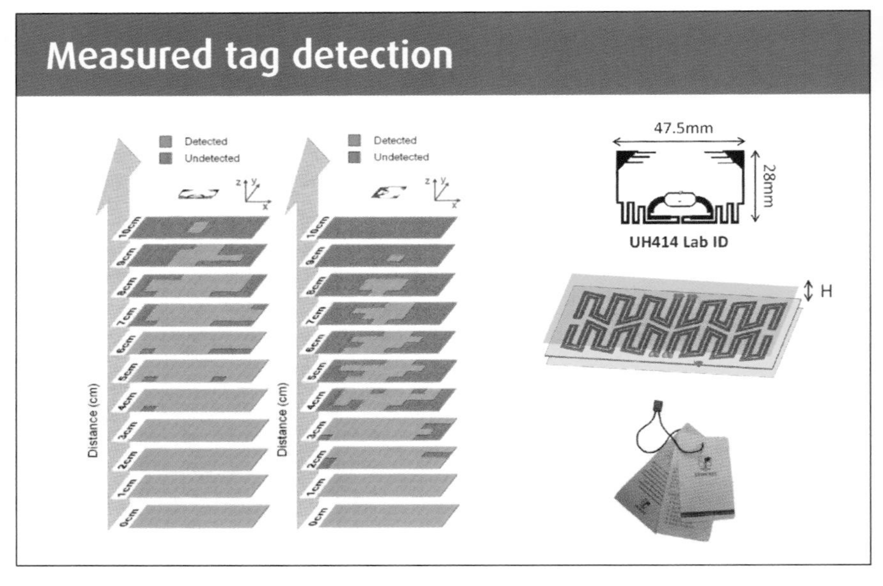

In order to evaluate the performance in a real scenario, the proposed antenna has been integrated into a commercial desktop reader and read range tests have been carried out. In this test, the UH414 (Lab-ID) tag has been chosen, and the antenna surface (275 × 135 mm^2) has been subdivided into 4×9 square cells. The detection tests have been repeated in each cell by varying the distance of the tag from the antenna surface, setting the input power to 23 dBm. The results are shown for two orthogonal orientations of the tag. Since the meandered lines cover almost the entire available area (275 × 135 mm^2), the tag is read in any location and orientation when it lies directly above the antenna surface. By increasing the tag distance from the antenna surface, the UH414 tag has been mainly detected in the central area of the TWAs array, as expected from the simulated field distributions. Moreover, the tag detection is only slightly dependent on the particular orientation. Finally, such tag detection tests show that the read range is limited up to 10 cm, so avoiding false positives readings in the desired detection volume.

18

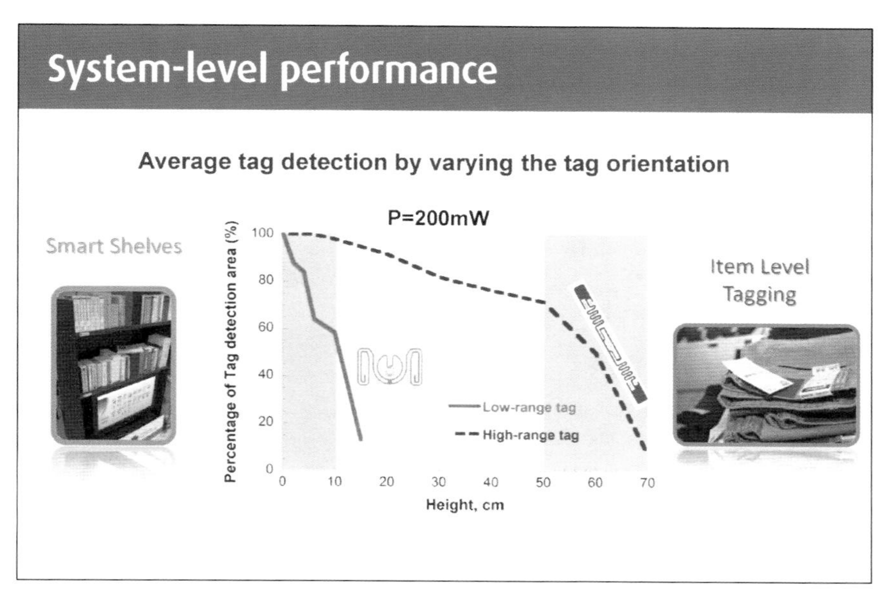

The performance of a near-field UHF RFID antenna must be also assessed in terms of read range. In particular, the average percentage of the whole area (275×135 mm²) in which the tags can be detected is obtained by varying the distance from the antenna surface. The reader input power is set to 23dBm (200mW). It should be noted that the tag readability strictly depends on the particular tag and its sensitivity to the direction of the incoming electromagnetic wave. With only 23dBm of input power, the short-range and long-range tags could be read on about 60% of the considered area up to 10cm and 55cm, respectively. It suggests that, depending on the considered tag, such an antenna could be easily exploited for different applications, such as smart shelves (short-range tag) or Item Level Tagging (long-range tag).

19

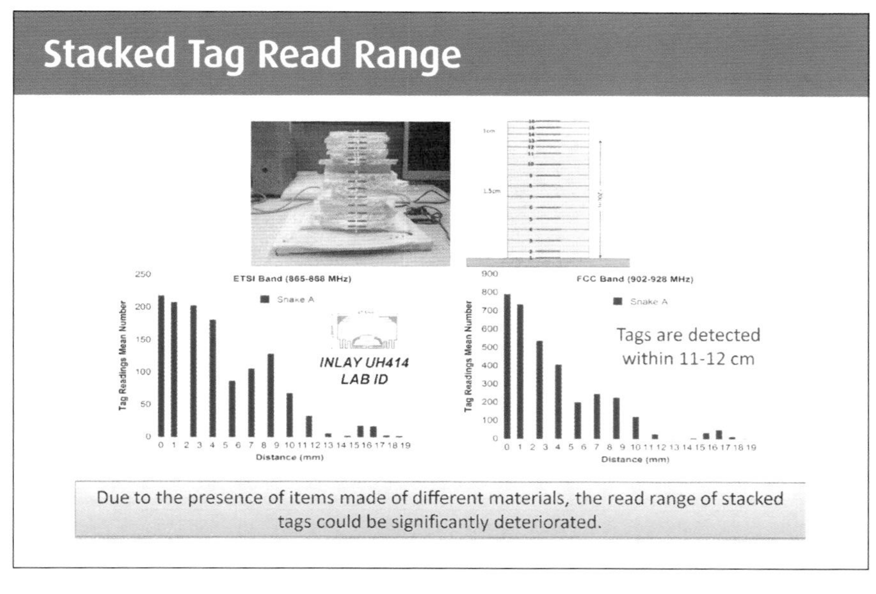

Another interesting test to be performed with a near-field antenna is the detection of UHF RFID tags arranged in a stacked configuration. In the example shown here, 20 LABID UH414 tags have been placed in a stacked configuration at a distance of 1 cm from each other, up to an overall height of 20 cm (tags are separated by a foam layer). The lower tag has been placed directly on the case surface. With a reader output power of 23dBm, tags can be detected up to 11-12cm. It is worth noting that in general the presence of close UHF RFID tags has an effect on the read range. Moreover, due to the presence of items made of different materials, the read range of stacked tags could be significantly deteriorated.

20

Modular antenna (I)

Combination of TL and Resonant Antennas
To extend the read range up to few decimeters
Travelling Wave Antenna (TWA) + low-gain resonating antenna

The TWA structure is effective to get a homogeneous and strong field right on the antenna surface (reactive near-field region)

The low-gain resonating antenna allows to extend tag readability up to a few decimeters from the antenna surface (radiative near-field region).

P Near-field Radiative Region
Q Near-field Reactive Region
ANT2
Series-feedings
Transmission Line Antenna for Near-field Coupling
Resonating Antenna for Broadside Radiation
Feeding Port
ANT1
UHF RFID Desktop Reader

A possible choice for Near-field UHF RFID antenna consists in the combination of two antenna modules used to separately match specifications in the two regions of interest, namely the radiative and reactive near-field regions. The first module is a TWA (namely, a transmission line designed to increase the field on the antenna surface with respect to a conventional low losses microstrip or CPW transmission line). A spiral or a meandered structure is preferred to uniformly distribute the EM energy among all the field components. The TWA module has to be located in the middle of the reader antenna surface, which represents a premium location since it is the region where a tagged item is more likely to be located. The region surrounding the TWA antenna can be used to place one or two resonating antennas that are supposed to cover the radiative near-field region with a circularly polarized field. Since most of the antenna surface has to be devoted to the TWA antenna, resonating antenna module has to be realized with miniaturized antennas. It is worth noting that the miniaturization helps to meet the requirement on low antenna gain.

21

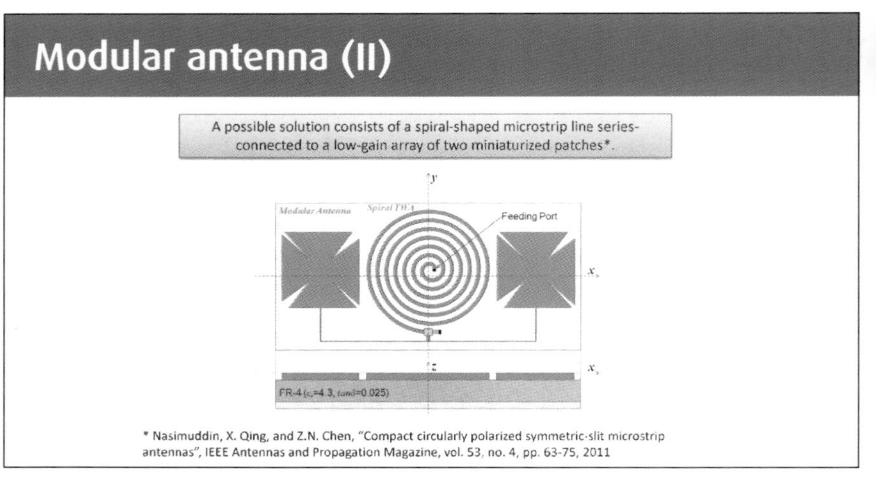

A possible implementation of the Modular Antenna concept is shown here. Specifically, a 50Ω coaxial cable, placed close to the antenna center, feeds a spiral-shaped microstrip TWA printed on a grounded 1.6 mm-thick FR4 dielectric substrate (ε_r=4.4, $\tan\delta$= 0.025). By means of a switch, the TWA can be series connected to either a planar array of two miniaturized circularly polarised (CP) square patches Modular Antenna configuration) or a matched load (Spiral TWA configuration). An ideal switch has been considered in the numerical simulations, without taking into account insertion or isolation losses. In detail, for the modular antenna configuration, the spiral TWA is directly connected to the patch array through a 3 dB power divider. On the other hand, for the spiral TWA configuration, the TWA is ended on a 50Ω resistor. It is worth noting that the replacement of that switch with a variable power divider would allow for a further degree of freedom to dynamically size and shape the reader detection volume, in addition to the power control that is already available in any commercial reader.

22

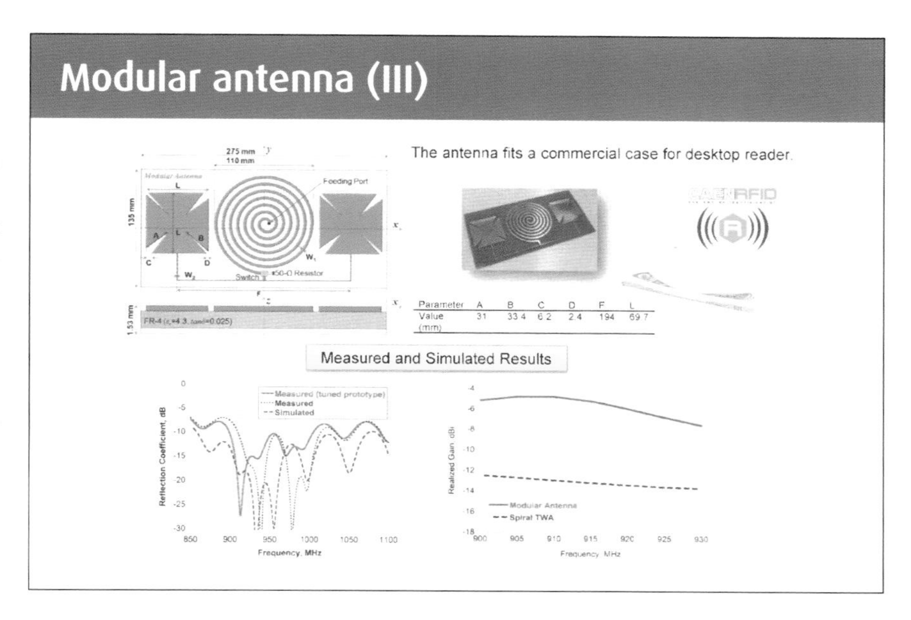

The whole antenna size is 275 mm × 135 mm, which can fit into commercial RFID reader cases. The spiral-shaped TWA overall length slightly affects the system performance since, in the modular antenna configuration, the spiral performs as a lossy transmission line feeding the resonating antenna. That is, if it is ended on a matched load, a stationary (non-uniform) current distribution as well as low-field minima are avoided on the antenna surface. The simulated and measured reflection coefficient in case of Modular Antenna are less than -10dB in the operative frequency band. The simulated gain of the Modular Antenna configuration is -5dB, while in case of Spiral TWA, it is definitely lower (less than -12dB), since the antenna should confine the e.m. field in proximity of the antenna surface.

23

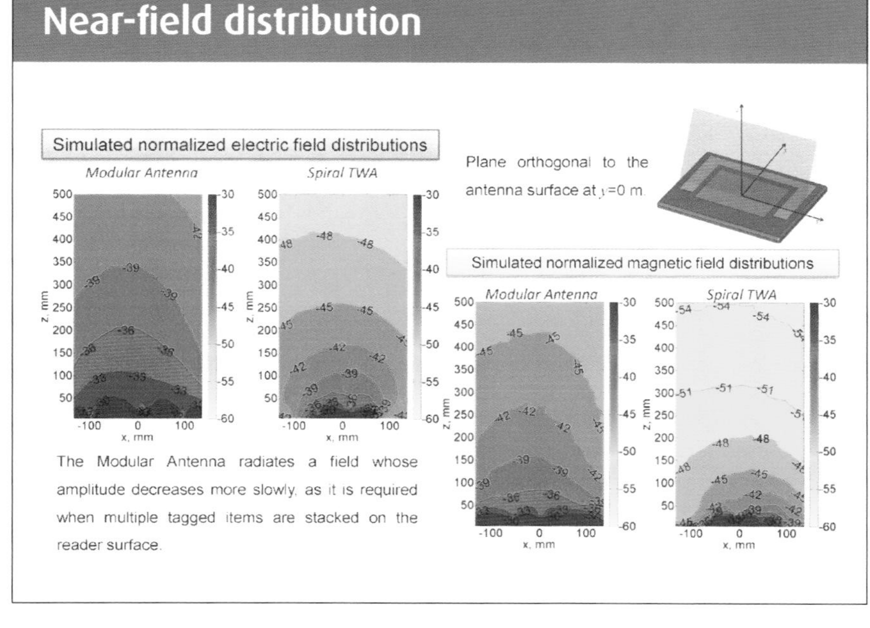

An important parameter to consider in the simulated analysis is the electric and magnetic field distributions on transverse planes. In particular, the Modular Antenna radiates a field whose amplitude decreases slowly, due to the presence of the miniaturized circularly polarized resonating patches. This is particularly useful when multiple tagged items are stacked on the reader surface. On the other hand, the field generated by the Spiral TWA configuration is characterized by a higher decay rate, so that the fields are strictly confined close to the antenna surface.

24

System-level antenna performance

System-level antenna performance is needed to validate the design and simulated analysis. As shown in these plots, a UHF RFID tag (namely Inlay UH414 LabID) can be read at any tag position and orientation on the reader surface. Also, a tag can be detected up to 10cm when considering the Spiral TWA Configuration, independently on the specific tag orientation. On the other hand, by using the Modular Antenna Configuration, the read range can be extended up to 60cm (radiative near-field region).

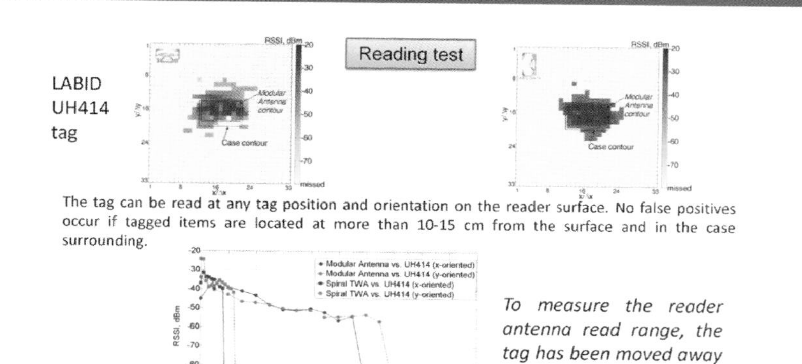

The read range is extended up to around 60 cm (radiative near field region) with respect to a conventional travelling wave antenna.

25

Stacked tag readability

The stronger electric and magnetic fields generated by the antenna operating in the Modular Antenna Configuration allow the detection of tags placed in a stacked configuration. In this example, 11 LabID UH414 tags have been placed in a stacked configuration at a distance of 2cm from each other, up to an overall height of 20cm. The tests demonstrate that multiple tags can be simultaneously read independently on their position and orientation. This is specifically due to the presence of the circularly polarized radiating patch directly fed through the spiral transmission line.

11 LABID UH414 tags have been placed in a stacked configuration at a distance of 2 cm from each other, up to an overall height of 20 cm.

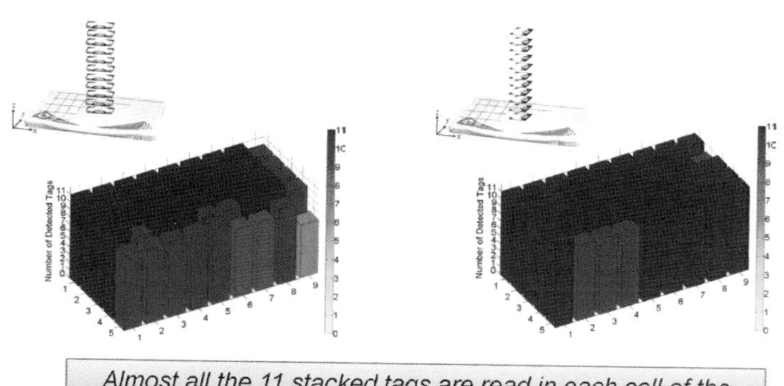

Almost all the 11 stacked tags are read in each cell of the antenna surface, for both orthogonal tag orientations.

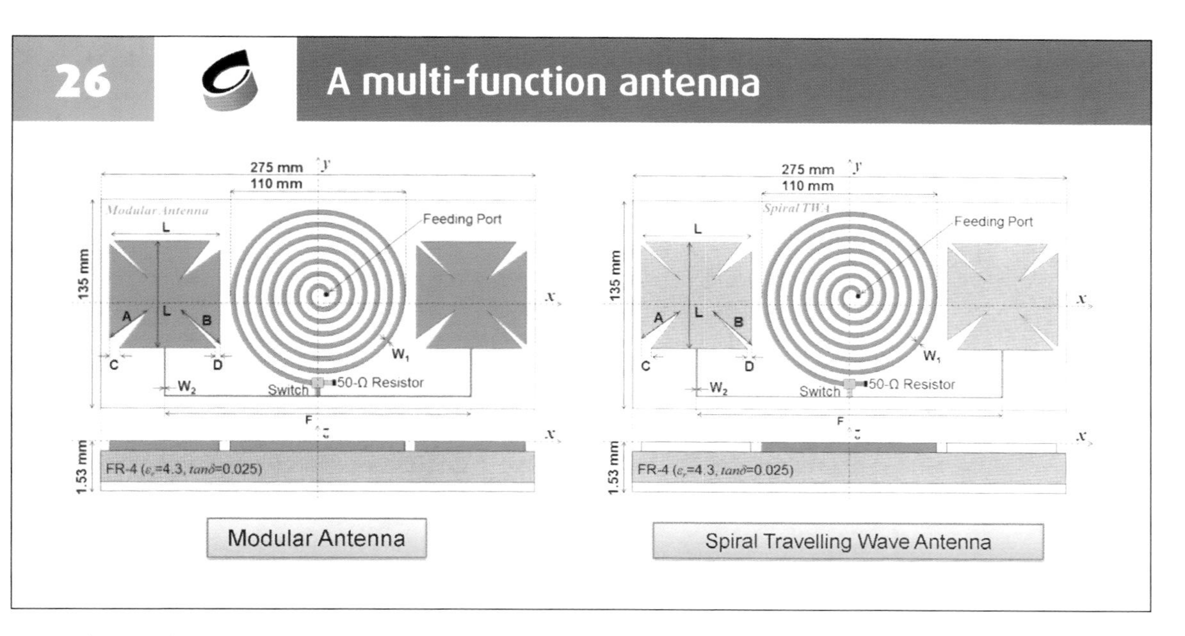

Besides reading capability, the antenna writing capability has to be also considered since a larger amount of radiated energy is here required at the tag side. In the proposed modular antenna, a traveling wave antenna configuration can be realized by adding a switch to end the spiral transmission line on a matched load. Thus, the electromagnetic field in the reader central area is maximized just on its surface.

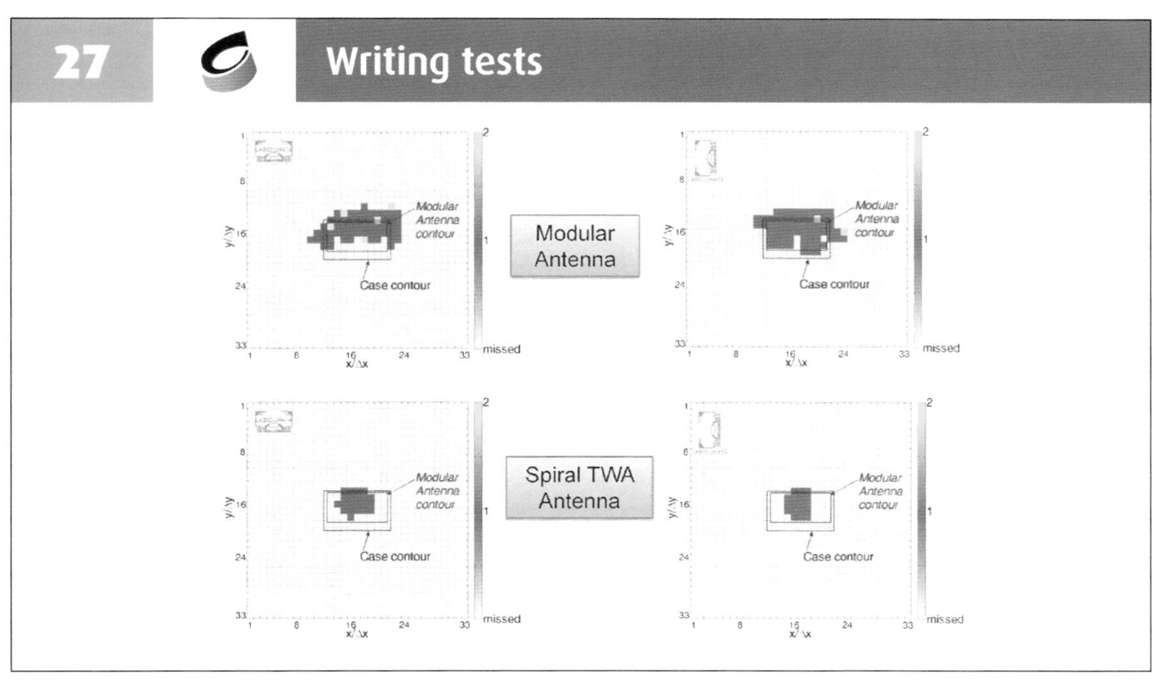

The Spiral TWA Configuration allows to get the area of successful writing concentrated at the antenna center, and only one attempt is required to initialize the tag for any tag orientation. This is the best configuration for tag writing operations, where the only tag that has to be read is most likely placed at the reader antenna center.

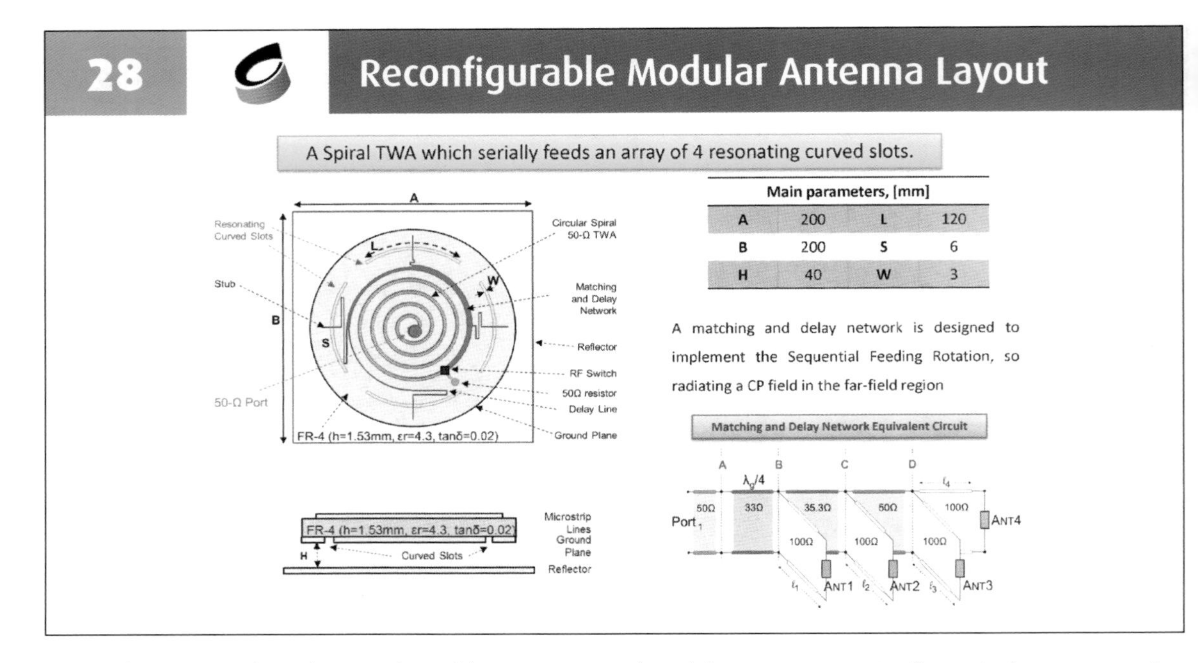

Main parameters, [mm]			
A	200	L	120
B	200	S	6
H	40	W	3

Another example of reconfigurable antenna for near-field UHF RFID applications is here shown. Specifically, a traveling wave antenna is combined with resonating antennas, which share the surface of the desktop reader antenna. When the TWA antenna ends on a matched load (*Spiral TWA Configuration*), strong and uniform electric and magnetic fields up to a few centimeters from the antenna surface (near-field reactive region) are generated. In the proposed layout, the TWA is represented by a spiral microstrip line in order to distribute the electromagnetic (EM) energy among all the field components, which is important to guarantee the detection of tags arbitrarily oriented with respect to the reader antenna. Alternatively, the spiral microstrip line can feed a resonating antenna or an array of resonating antennas (*Modular Antenna Configuration*), so covering the radiative near-field region up to a few tens of centimeters from the antenna surface, yet radiating a relatively low field in the antenna far-field region as required by antennas for desktop readers. The resonating antenna element is represented by an array of four curved slot antennas which share the TWA aperture (aperture-shared antenna configuration). The slots are 90-degree-rotated with respect to the antenna center, and they are fed by the microstrip transmission line through a matching/delay network. Such a network is responsible for feeding each radiating element with currents exhibiting the same amplitude but with a 90-degree phase difference, so implementing the sequential rotation feeding technique and achieving a circularly polarized radiated field.

29  Two States of the Reconfigurable Antenna

An absorptive RF switch is added to the end of the spiral microstrip line, and it is used to enable the proper antenna operating mode on the basis of the specific scenario. It is worth noting that by activating the proper radiating element, the field distribution generated by the *Reconfigurable Modular Antenna* changes, giving different system performance without increasing the reader output power level. In particular, the *Spiral TWA Configuration* is suitable especially for writing operations, where a higher field intensity is required. On the other hand, by activating the *Modular Antenna Configuration*, it is possible to improve the tag detection up to few decimeters from the antenna surface, even in the presence of stacks of tags.

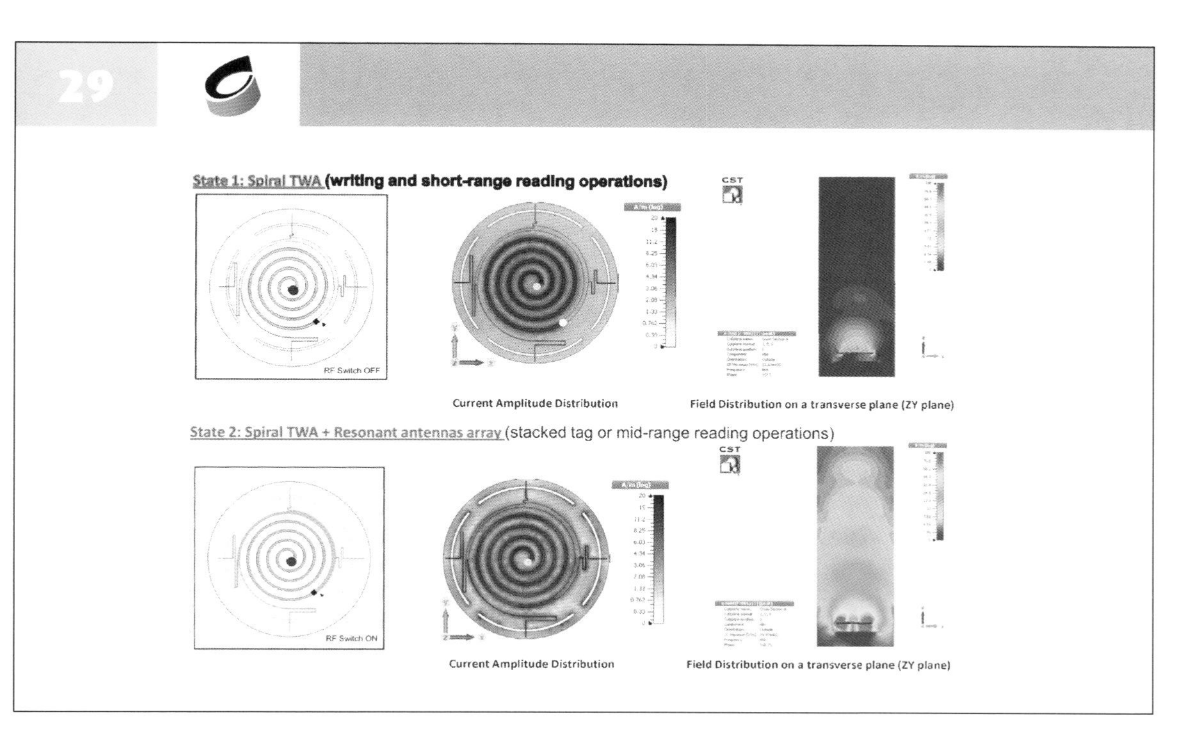

30

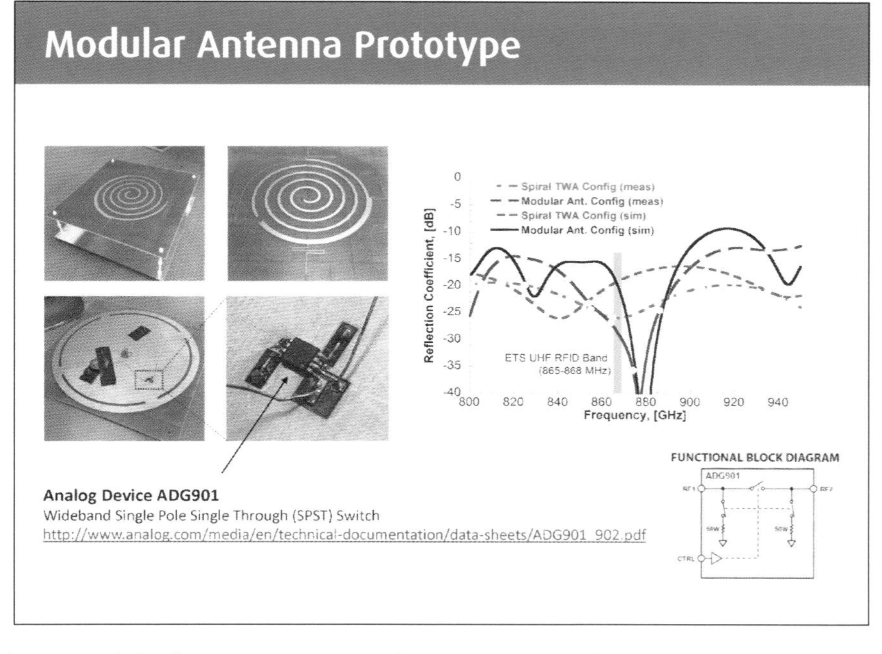

A prototype has been fabricated by using the fabrication facilities available at the University of Oviedo, Spain. A 3V battery is used to feed the RF switch (VDD) and to control the switch and select the antenna configuration. The RF switch has been integrated at the bottom of the FR-4 substrate.

The simulated and measured reflection coefficient is shown as a function of the frequency, for both the *Spiral TWA Configuration* and *Modular Antenna Configuration*. To limit the power reflected toward the reader RF front-end, in near-field UHF-RFID applications, the reflection coefficient is usually required to assume values lower than -14dB. Simulated and measured results show that such a requirement is satisfied by the proposed layout in a frequency band larger than the standard UHF RFID ETSI band (865-868 MHz), for both operating modes.

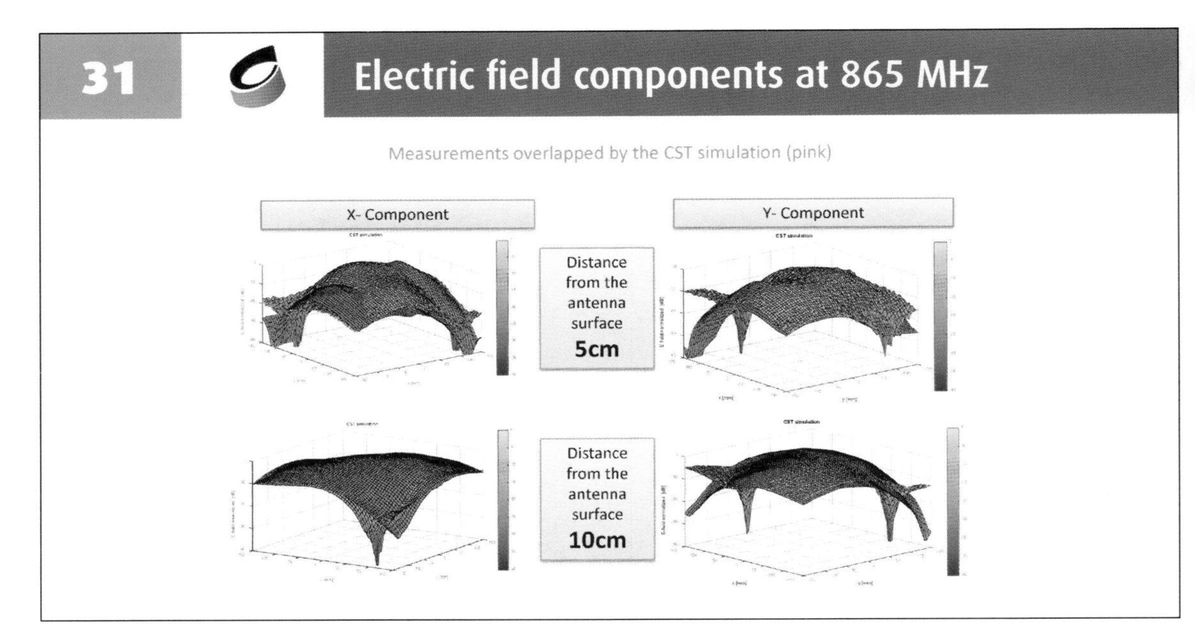

A near-field measurement system has been used to measure the electric field generated by the radiating element at a distance of 5cm and 10cm. As shown in the pictures reported here, the simulated and measured electric field normalized distribution are in a quiet good agreement, with a maximum field value in correspondence of the antenna central area.

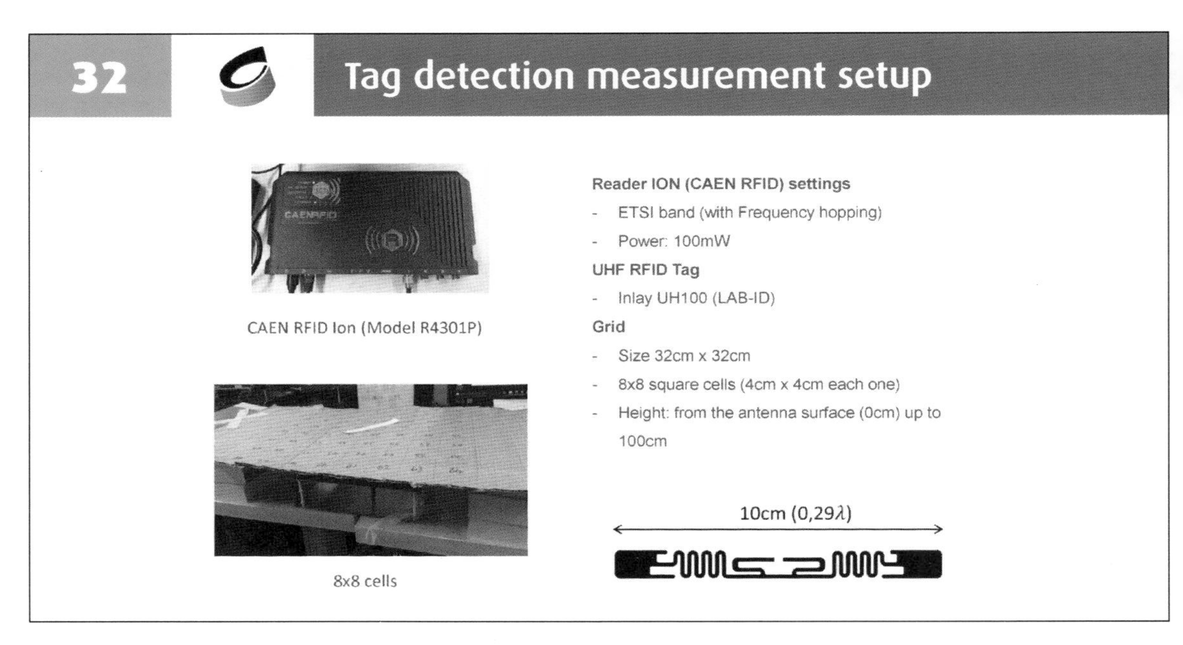

The proposed Reconfigurable Modular Antenna has been connected to a commercial UHF RFID reader (CAEN RFID Ion, Model R4301P) to carry out system level measurements in terms of reading range and tag detection. An 8×8 3-cm-square cells grid has been drawn on a cardboard sheet and aligned to the antenna center, at a fixed distance from the antenna surface. Then, a LAB-ID Inlay UH100 tag was placed in correspondence of each cell, with a specific orientation with respect to the radiating element.

33 

System-Level Performance

Spiral TWA Configuration

Modular Antenna Configuration

The UH100 (LAB-ID) tag has been placed in each grid cell at different distances from the antenna surface.

- ***Spiral TWA Configuration***, a reading range of **12cm** is achieved for both the tag orientations
- ***Modular Antenna Configuration*** a reading range of **70cm** is achieved for both the tag orientations

As a result, when the *Spiral TWA Configuration* is enabled, the generated field is confined close to the antenna surface, and the tag detection is limited up to around 10cm. By switching to the *Modular Antenna Configuration*, the reading range extends up to 70cm. It is worth noting that all the measurements have been performed with 20dBm reader output power. Thus, the different reading range is only due to the changed field shape generated by the two antenna configuration.

34 

Future trend

System limitations

- Shape and size of the antennas are normally fixed and may not be adapted to all the possible reader cases
- RFID system performance in terms of reading range is affected by the surrounding environment (*e.g.* various items made of different materials, presence of multiple tags in a stacked configuration)

RECONFIGURABILITY

is becoming an interesting antenna feature, which can make the UHF NF RFID system more adaptable to the specific application scenario

E or H
antennas capable to maximize the electric or the magnetic field would be effective to improve the entire RFID system performance

Shape the reading range
extend or reduce the detection volume size, and compensate the field attenuation due to the presence of items made of different materials

Choose field main direction
maximize a single field component, so that the relative position of the tag could be estimated

As a conclusion, antennas for near-field UHF RFID readers must be compact and low-profile, but at the same time, they must be easily integrated in reader cases with different size and shape. Thus, the radiating element is required to be as modular as possible to be easily adapted to the operative scenario. For this reason, reconfigurable antennas are being investigated, so that the performance of a UHF RFID can be maximized. In this context, future trends are represented by reconfigurable antennas capable of selectively maximizing the electric or magnetic field, or able to adaptively shape the reading range and detection volume size on the basis of the operative scenario or tags.

CHAPTER 07

Ultra-low-power Devices, and Application of New Materials to mm-wave Antennas and Circuits

Massimo Macucci

University of Pisa, Italy

1. The problem of power consumption 184
2. Three-terminal devices 184
3. Bipolar Junction Transistor 185
4. Field Effect Transistor 185
5. Power dissipated by devices 186
6. Scaling and power dissipation (I) 186
7. Scaling and power dissipation (II) 187
8. Amdahl's law 187
9. Nonideal scaling 188
10. MOS transistor transfer characteristic 188
11. Subthreshold slope (I) 189
12. Subthreshold slope (II) 189
13. Subthreshold slope (III) 190
14. Ways to overcome the subthreshold slope limitation 190
15. Tunnel FET concept 191
16. Tunnel FET operation (I) 191
17. Tunnel FET operation (II) 192
18. Tunnel FET operation (III) 192
19. Tunnel FET operation (IV) 193
20. Tunnel FET operation (V) 193
21. Tunnel FET operation (VI) 194
22. Tunnel FET implementation (I) 194
23. Tunnel FET implementation (II) 195
24. Adiabatic CMOS logic (I) 195
25. Adiabatic CMOS logic (II) 196
26. Adiabatic CMOS logic (III) 196
27. NAND SCRL 197
28. Adiabatic MOS 197
29. Limits of computation (I) 198
30. Limits of computation (II) 198
31. Limits of computation (III) 199
32. Maxwell's demon and the limits of computation (I) 199
33. Maxwell's demon and the limits of computation (II) 200
34. Maxwell's demon and the limits of computation (III) 200
35. Limits of computation 201
36. Example of low-power sensor nodes based on traditional CMOS technology 201
37. Test nodes 202
38. Example of Low-power strategy 202
39. Low-power strategy 203
40. Graphene basics (I) 203
41. Graphene basics (II) 204
42. Graphene basics (III) 204
43. Properties of graphene 205
44. Graphene Antennas 205
45. Plasma oscillations 206
46. Surface plasmon polaritons (I) 206
47. Surface plasmon polaritons (II) 207
48. Surface plasmon polaritons (III) 207
49. Graphene antennas (I) 208
50. Graphene antennas (II) 208
51. Graphene antennas (III) 209
52. Graphene antennas (IV) 209
53. Graphene antennas (V) 210
54. Graphene antennas (VI) 210
55. Graphene antennas (VII) 211
56. Graphene antennas (VIII) 211
57. Graphene phase shifters 212
58. Graphene RF transistor (I) 212
59. Graphene RF transistor (II) 213
60. Graphene RF transistor (III) 213

Power dissipation represents one of the most relevant issues in the development of new electronic systems, both because it has effectively halted the increase in clock frequency in processors and because it represents the main limitation for the implementation of autonomous IoT devices, such as smart dust. It is indeed relatively easy to miniaturize all the electronics needed, e.g., for a sensor, but it is not as easy to scale down in size a battery capable of keeping the device working for months or years.

On the other hand, novel materials and their properties are essential to improve the performance of existing devices or to introduce new device concepts. Some of the most important breakthroughs in the history of electronics and information technology can be traced back to the introduction of specific material systems.

In this chapter, we will discuss ultra-low-power operation of electronic systems suitable for IoT applications as well as the opportunities afforded by novel materials for RF circuits and millimeter-wave antennas. In particular, we will first focus on the limitations to information processing power resulting from power dissipation and on the fundamental limits for computation. We will then cover a few examples of devices, circuits, and systems designed to achieve ultra-low-power operation.

After a brief introduction on the main properties of graphene, we will discuss its application for millimeter wave antennas, specifically pointing out the advantage resulting from the reduced electrical length in the presence of surface plasmon polariton propagation. Also a few proposed applications of graphene to active devices (RF transistors) and electrically tunable delay lines will be introduced.

1  The problem of power consumption

- **Power consumption is one of the most important issues in IoT devices, since they often need to operate without connection to an external power supply or to operate for a very long time on a battery that cannot be replaced**
- **Energy that can be scavenged from the environment is very limited, therefore one of the main objectives is the reduction of the power consumption**
- **Strategies to reduce power consumption include having very short active times, smart circuit design, transferring most of the data processing to master nodes that have no strict energy limitations, but this may not be enough, and we need to act upon device technology**

In this lecture, we will first discuss the problem of power consumption in IoT applications and on the device, circuit, and architecture solutions that can be adopted to minimize it. We will also look into the thorny issue of the minimum energy requirements for computation, which set a fundamental limit to the performance of digital circuits. Then, we will move on to the discussion of the advantages that novel materials, such as graphene, can offer in the implementation of RF circuits.

2  Three-terminal devices

Almost all of the electronics we know has been based on three-terminal devices, vacuum tubes in the old days and, since the invention the transistor, bipolar and field effect transistors

To be useful for practical applications the three-terminal device needs to have at least three basic properties: a) provide power gain, b) be unilateral (i.e. the input quantities influence the output ones but the vice versa is not true; c) provide an output quantity that can drive another identical device

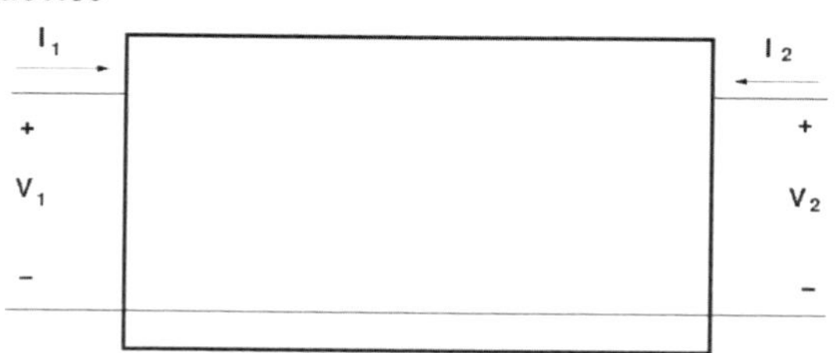

We first introduce the concept of three-terminal device, which has represented the fundamental building block of electronic circuits since the invention of the vacuum triode (and existed even previously in the form of the electromechanical relay). We list the basic properties for a useful three-terminal device that were formulated by Robert Keyes at IBM in the 1970s. The lack of some of these properties convinced Keyes that many new technologies that were being proposed at that time, such as logic based on tunnel diodes were indeed not competitive with classical silicon transistors.

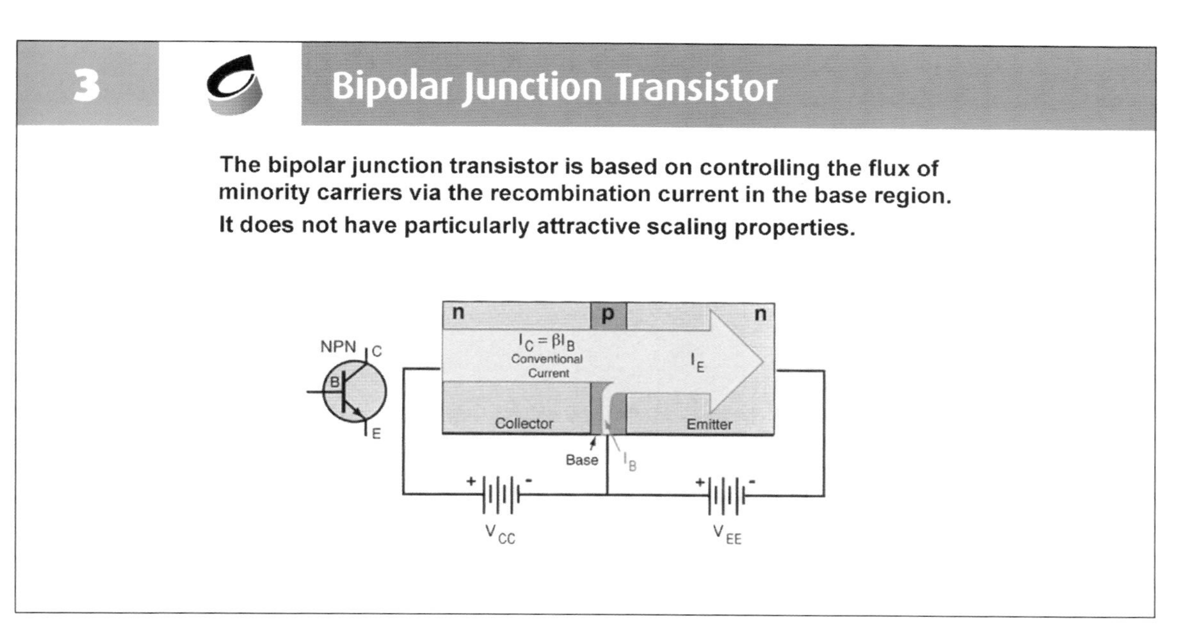

The first solid-state three-terminal device was the bipolar transistor, developed in 1948 as a result of the work by Bardeen, Brattain and Shockley. Its principle of operation consists in the injection of minority carriers into the base from the emitter and in the modulation of such a current (which for the most part reaches the collector as a result of the thickness of the base being much smaller than the minority carrier recombination length) via the much smaller recombination current flowing through the base electrode.

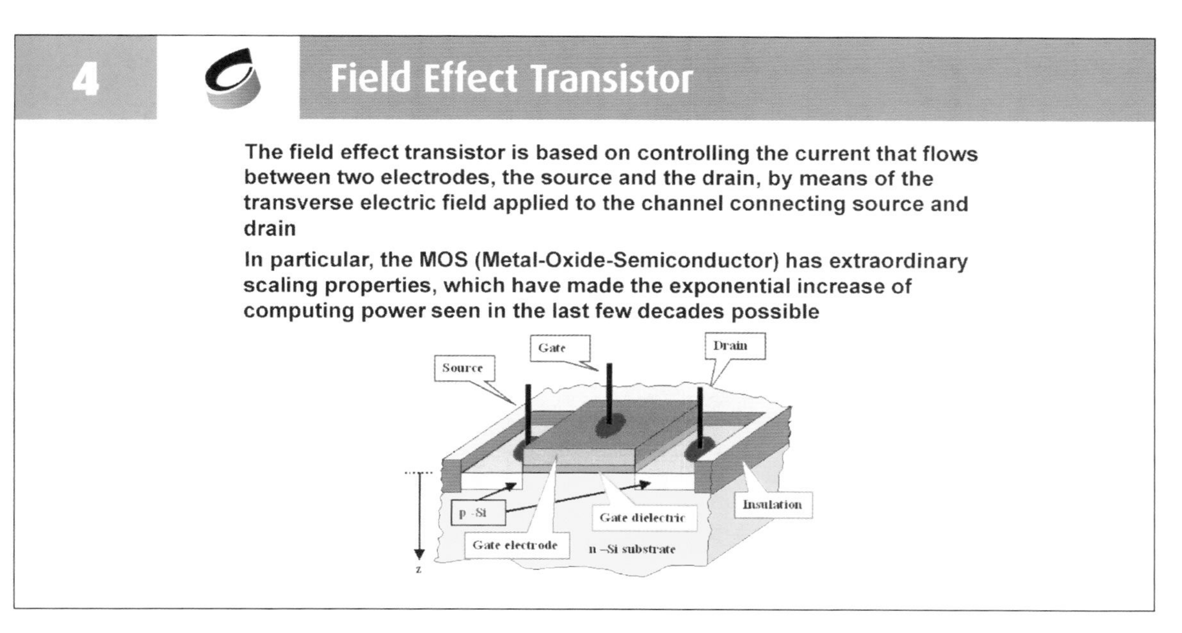

The principle of operation of the Field Effect Transistor (FET) was first conceived by Julius Lilienfeld in 1930, but at the time the technology was not mature to achieve a working device, which was then developed (on the basis of a different and improved design) in 1950 in the form of a Junction Field Effect Transistor (JFET) and in 1959 in the form of the Metal-Oxide-Semiconductor Field Effect Transistor (MOSFET), which is currently the workhorse of the electronic industry.

5  Power dissipated by devices

- **The problem of power dissipation is not restricted to devices for IoT applications, but it is of a more general nature, involving all high-performance circuits**

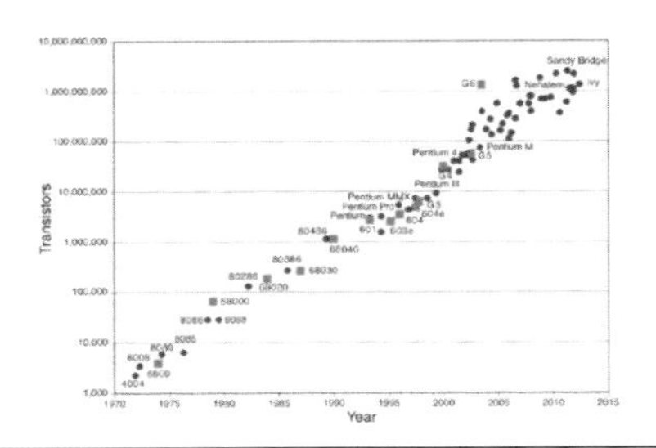

- **The well-known empirical scaling law by Moore has dictated the development of micro and nanoelectronic technologies in the last 5 decades, but power dissipation is the single most important obstacle**

The power dissipated by a classical CMOS gate is approximately proportional to the clock frequency, since it is mainly just dynamic power, i.e. power dissipated when a switching event occurs.

With the latest ultra-scaled devices, also static power dissipation plays a role, but the proportionality between speed and dynamic power dissipation still holds and represents one of the main limitations to achieving faster circuits.

6  Scaling and power dissipation (I)

- **According to Dennard scaling [R. Dennard et al., IEEE J. Solid State Circuits SC-9, 256 (1974)], the power dissipated per unit area was constant; with a clock frequency increasing at each new generation, this led to the extraordinary increase in computing power observed until the early 2000**

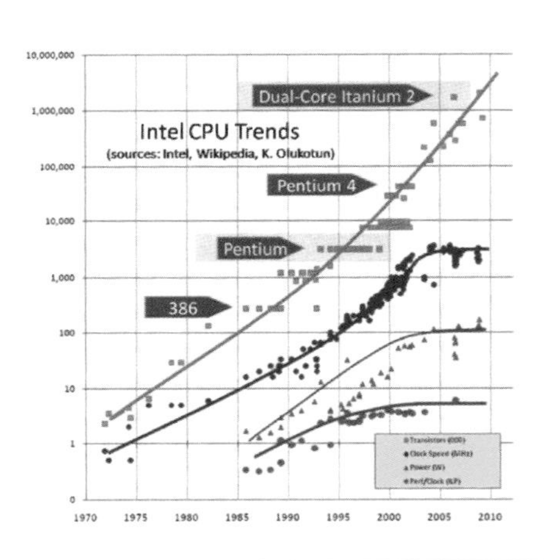

Unfortunately real-life scaling is not ideal Dennard scaling, and, while the number of transistors on a single chip has kept on increasing according to Moore's law (currently there are up to a few tens of billions of transistors on a single chip), power dissipation per unit area has increased significantly, mainly as a result of the impossibility of scaling down the supply voltage proportionally to the physical dimensions of the transistors (generalized scaling vs. constant field scaling).

7 Scaling and power dissipation (II)

- The reason for the abrupt stop in 2004 was the inability to limit the rise in power dissipation, resulting from non-ideal scaling
- The response of Intel and other microprocessor manufacturers was to increase the number of cores, but this is not the same as increasing clock speed

Processor Frequency Scaling Over Time

- Already in 1967 Amdahl formulated his famous expression for the speedup of a multiprocessor computer

In 2004, clock frequency stopped increasing: this was a major setback in the otherwise constant evolution of computing power that was experienced up to then.

The only alternative microprocessor manufacturers saw was the development of multi-core chips to make low-cost parallel computing possible.

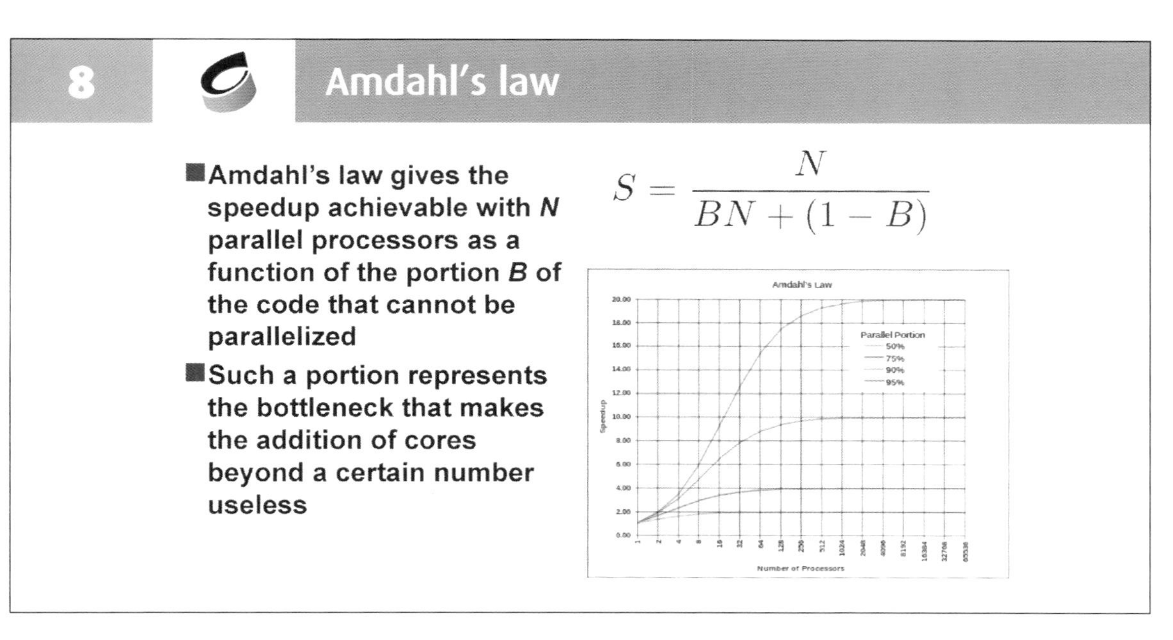

From the plot of the speedup vs. number of cores it is possible to see that, as the number of cores is increased above 1/B, the speedup starts growing more slowly and finally levels off at 1/B.

Thus, increasing the number of cores is definitely not as good as increasing the clock speed, in terms of running generic algorithms.

9  Nonideal scaling

- Excessive power dissipation is the consequence of nonideal scaling in which the electric field is not constant, but increases with decreasing feature sizes because of reduced scaling of the supply voltage
- There are a number of reasons why the supply voltage cannot be reduced, but one of the most important is keeping a large enough I_{on}/I_{off} ratio

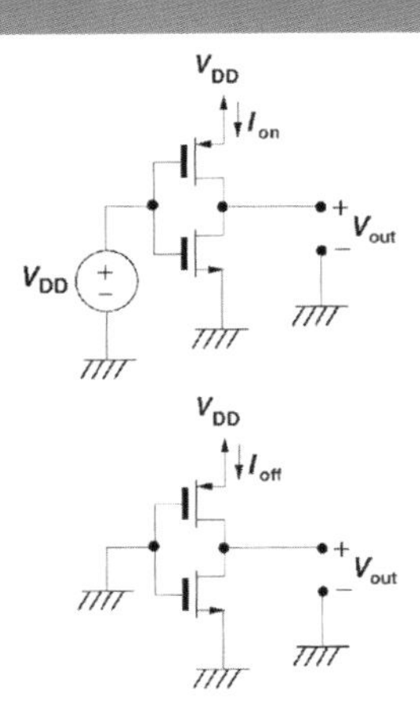

Generalized scaling, resulting from the impossibility of scaling down the power supply voltage proportionally to the geometrical sizes leads to increased power dissipation per unit area.

Besides the need to keep a large enough Ion/Ioff ratio, the power supply voltage cannot be reduced too much if a reasonable signal-to-noise ratio must be preserved.

Furthermore, standard values of the power supply voltage must be kept across a number of new device generations in order to warrant interoperability.

10  MOS transistor transfer characteristic

- For values of the gate voltage below V_T, the dependence of the drain current on gate voltage is exponential

$$I_D \propto \exp\frac{qV_{GS}}{kT}$$

- Therefore it appears as a line in a semilogarithmic plot, with a slope that is limited by physical reasons

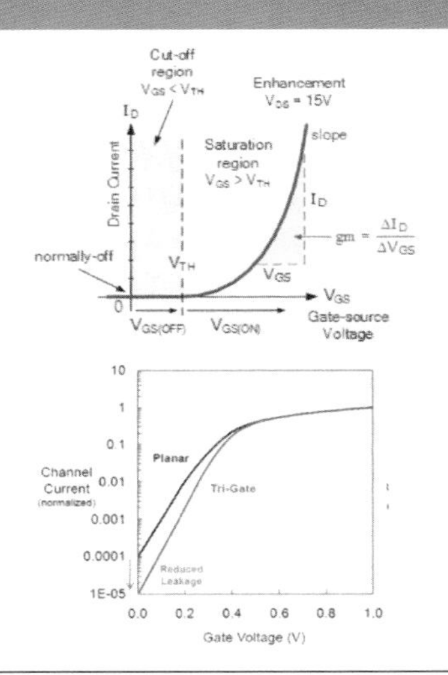

Let us focus on the subthreshold behavior of MOS transistors, in particular on the dependence of the drain current on the gate voltage, when this is below the threshold voltage.

Such a dependence is exponential because of the Fermi-Dirac distribution of the electron energies

As the difference between the top of the source-drain potential barrier and the Fermi level in the source is reduced, there is an exponentially larger number of electrons available for conduction.

11 Subthreshold slope (I)

The drain current depends actually on the surface potential ψ_s at the silicon-oxide interface

$$I_D \propto \exp\frac{q\psi_s}{kT} \quad \textbf{but} \quad V_{GS} = \frac{C_{dm} + C_{ox}}{C_{ox}}\psi_s = m\psi_s$$

and thus $I_D \propto \exp\frac{qV_{GS}}{kT}$

The subthreshold slope is defined as actually the reciprocal of the actual slope in the semilogarithmic representation:

$$S = \left[\frac{d\log_{10}\left(\frac{I_D}{I_0}\right)}{dV_{GS}}\right]^{-1}$$

The slope (or more precisely its reciprocal) of the drain current vs. gate voltage below threshold in a semilogarithmic scale is known as "subthreshold slope".

It can be derived considering the dependence of the current on the surface potential at the silicon-oxide interface.

The actual surface potential is a function of the applied gate voltage, as a result of the capacitive partition between the oxide capacitance and the depletion capacitance. The ratio of the gate voltage to the surface potential is usually indicated with *m*.

12 Subthreshold slope (II)

Switching base for the logarithm:

$$S = \left[\frac{d\frac{1}{\ln 10}\ln\left(\frac{I_D}{I_0}\right)}{dV_{GS}}\right]^{-1}$$

and substituting the expression for the current, we get:

$$S = \left[\frac{d\frac{1}{\ln 10}\ln\frac{\gamma\exp\frac{qV_{GS}}{mKT}}{I_0}}{dV_{GS}}\right]^{-1}$$

$$S = \left\{\frac{1}{\ln 10}\left[\frac{d\ln\frac{\gamma}{I_0}}{dV_{GS}} + \frac{d\left(\frac{qV_{GS}}{mKT}\right)}{dV_{GS}}\right]\right\}^{-1}$$

Performing some simple algebra and including the expression for the drain current as a function of gate voltage, we obtain an expression that can easily be further manipulated.

13 

Subthreshold slope (III)

Thus:

$$S = \left(\frac{1}{\ln 10}\frac{q}{mKT}\right)^{-1} = \ln 10\, m\,\frac{kT}{q} = m\,60\,\frac{\text{mV}}{\text{dec}}$$

which means that the maximum variation of the drain current is one decade for a variation of 60 mV of the gate voltage

This is a fundamental limit due to the principle of operation of the MOS transistor and the thermal distribution of carriers at 300 K

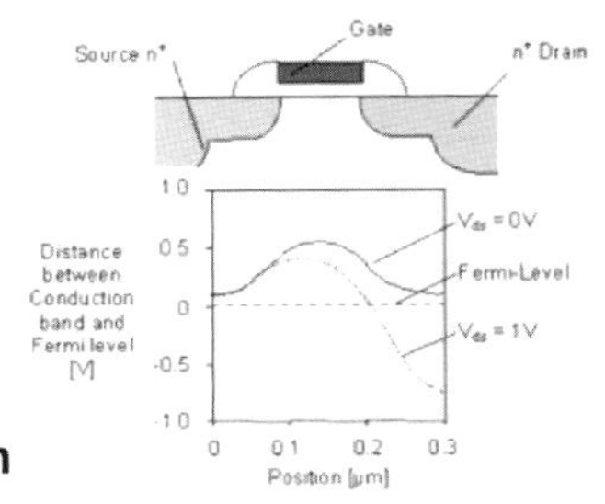

This expression tells us that the subthreshold slope cannot go beyond 60~mV, even for a surface potential identical to the gate potential (in the limit of depletion capacitance negligible with respect to the oxide capacitance).

Thus, to achieve a current variation of 4 decades (the minimum value considered acceptable for logic architectures) a gate voltage variation of at least 240 mV is necessary, which means that the power supply voltage must be larger than 240 mV.

14 

Ways to overcome the subthreshold slope limitation

- **One possibility would be that of lowering the temperature of operation below room temperature, but this is not a real option because it would require costly cooling equipment**
- **Another option would be that of injecting into the channel electrons with a sharp, narrow energy distribution by means of some filter**

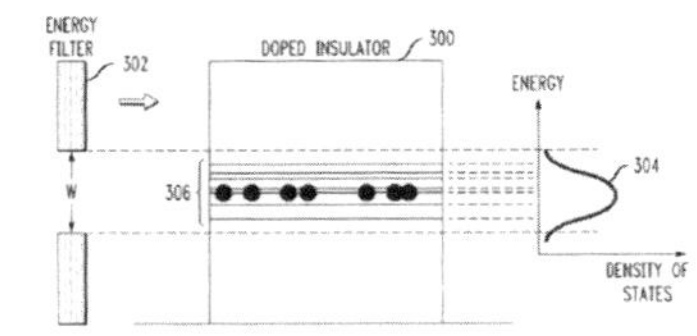

In the perspective of reducing the power supply voltage and thus the power dissipation, It would be extremely important to raise the subthreshold slope beyond the mentioned limit.

Operation below room temperature is impractical and there is no interest from the industry, due to the complexity and energy requirements of cooling systems.

A narrow peaked energy distribution can be achieved not only as a result of a low temperature but also through the action of a narrow-band energy filter, which can be obtained, for example, with a Fabry-Perot filter implemented with a semiconductor heterostructure. However, this approach is impractical because it would involve a rather complex fabrication process.

15  Tunnel FET concept

- **The tunnel FET is based on a different principle of operation with respect to the MOS transistor, a principle that circumvents the subthreshold slope limitation, not relying on a simple potential barrier**
- **In the tunnel FET (TFET) the current flows from source to channel via interband Zener tunneling, an effect that is usually deleterious in normal transistors**
- **Depending on the applied gate bias, the states in the valence band in the source that initially line up with states in the conduction band of the channel end up facing the bandgap and therefore no current can flow any longer, with an abrupt switching effect**

Another way of getting around the physical limitation for the subthreshold slope is that of implementing a transistor whose operation is based on some novel physical effect, instead of the traditional field effect.

An example is the tunnel FET, which is based on interband tunneling and is in principle implementable without too many technological difficulties.

Due to the novel operating principle, the power supply voltage can, at least in principle, be reduced well below 240 mV.

16 Tunnel FET operation (I)

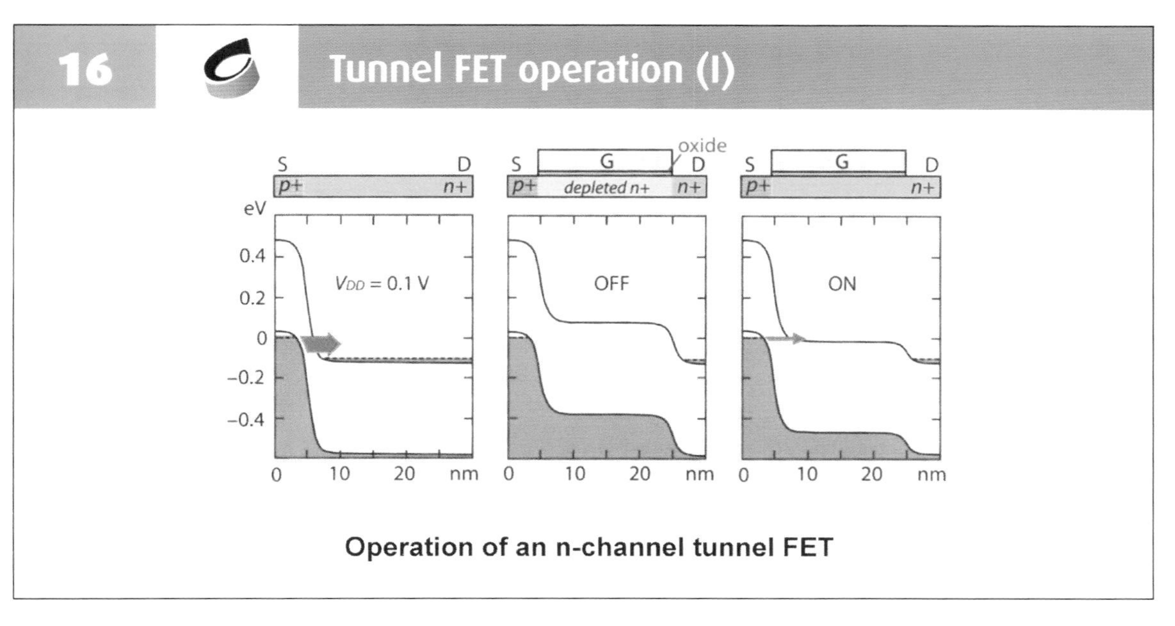

Operation of an n-channel tunnel FET

The tunnel FET consists, in its n-channel implementation, of a junction between a p+ region (source) and an n+ region (channel and drain).

The high doping leads to a very thin depletion region and to a band alignment such that filled states in the conduction band of the p+ region face empty states in the conduction band of the n+ region, thus leading to Zener (interband) tunneling.

17 Tunnel FET operation (II)

We can model the interband current assuming a triangular barrier and using the WKB approximation [A. C. Seebaugh and Q. Zhang, Proc. IEEE 98, 2095 (2010)]

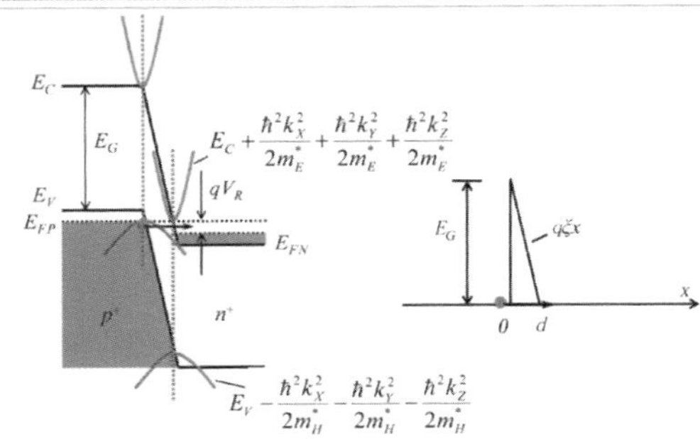

$$J^{3D} = \frac{\sqrt{2m_R^*}q^3\xi V_R}{4\pi^2\hbar^2 E_G^{1/2}} \exp\left(\frac{-4\sqrt{2m_R^*}E_G^{3/2}}{3q\xi\hbar}\right)$$

$$m_R^* = \left(\frac{1}{m_E^*} + \frac{1}{m_H^*}\right)^{-1}$$

It is possible to derive an analytical expression for the interband tunneling current.

A possible derivation exploits a few approximations and is based on considering a triangular shape for the tunnel barrier.

The tunnel current is then evaluated exploiting the WKB (Wentzel-Kramer-Brillouin) approximation.

Since the considered tunneling effect occurs between the valence band on one side and the conduction band on the other side, an effective tunneling mass is considered, which is the "parallel" of the effective masses for the electrons and holes.

The bands are assumed to be parabolic.

18 Tunnel FET operation (III)

We can thus write

$$I_D = aV_R\xi \exp\left(-\frac{b}{\xi}\right)$$

with

$$a = Aq^3\frac{\sqrt{2m_R^*/E_G}}{4\pi^2\hbar^2}$$

$$b = \frac{4\sqrt{m_R^*}E_G^{3/2}}{3q\hbar}$$

We can then compute the subthreshold slope for the TFET

$$S = \left[\frac{d\log\frac{I_D}{I_0}}{dV_{GS}}\right]^{-1}$$

$$= \left\{\frac{1}{\ln 10}\frac{d}{dV_{GS}}\left[\ln a + \ln V_R + ln\xi - \frac{b}{\xi} - \ln I_0\right]\right\}^{-1}$$

$$= \ln 10\left(\frac{1}{V_R}\frac{dV_R}{dV_{GS}} + \frac{1}{\xi}\frac{d\xi}{dV_{GS}} + \frac{b}{\xi^2}\frac{d\xi}{dV_{GS}}\right)^{-1}$$

$$= \ln 10\left\{\frac{1}{V_R}\frac{dV_R}{dV_{GS}} + \frac{\xi+b}{\xi^2}\frac{d\xi}{dV_{GS}}\right\}^{-1}$$

We obtain an expression for the tunnel current that is also a function of ξ, the electric field in the depletion region and V_R, i.e. the built-in potential.

From this, we can compute the subthreshold slope for the TFET following the same steps we have already considered for the MOS transistor.

19  Tunnel FET operation (IV)

$$S = \ln 10 \left\{ \frac{1}{V_R} \frac{dV_R}{dV_{GS}} + \frac{\xi + b}{\xi^2} \frac{d\xi}{dV_{GS}} \right\}^{-1}$$

The two terms appearing in this equation are not limited by *kT/q*

It is thus possible, at least in principle, to achieve a subthreshold slope less than 60 mV/dec

The first term can approach $1/V_R$, while the second term is maximum when the gate field aligns with the internal field of the junction

From the expression for the subthreshold slope for the TFET, we see that there is no fundamental limitation such as that observed for the Field Effect Transistors.

Particular geometries can be designed to achieve the best possible alignment of the gate field with the junction field, in order to maximize the second term in the expression of the subthreshold slope.

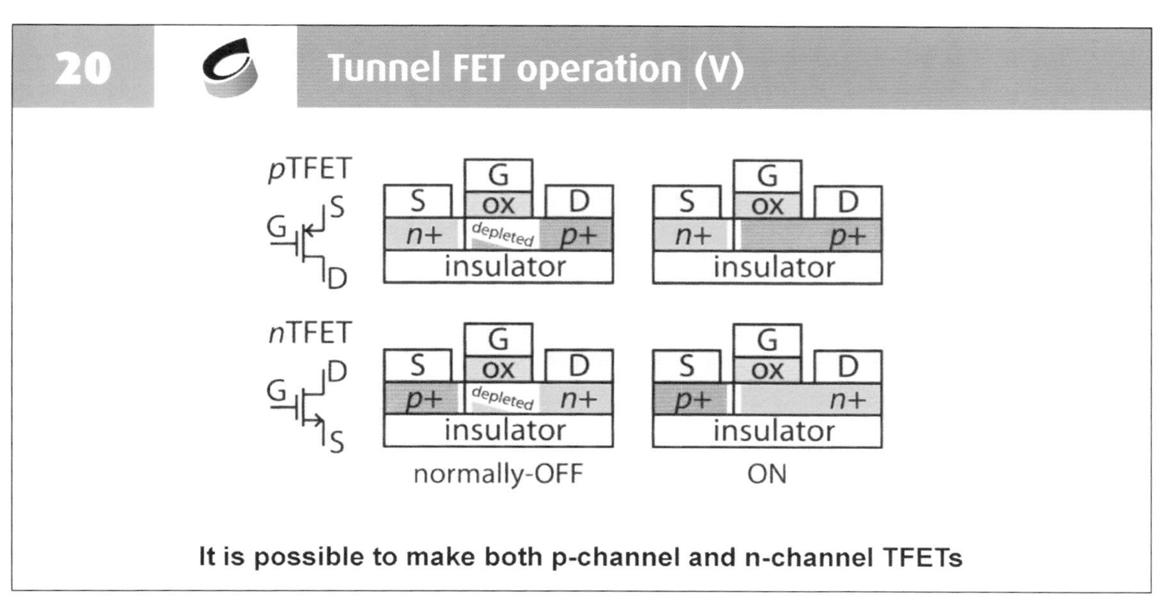

Tunnel FETs can have either an n channel or a p channel: the p-channel version has semiconductor regions with the opposite doping with respect to the n-channel version.

21  Tunnel FET operation (VI)

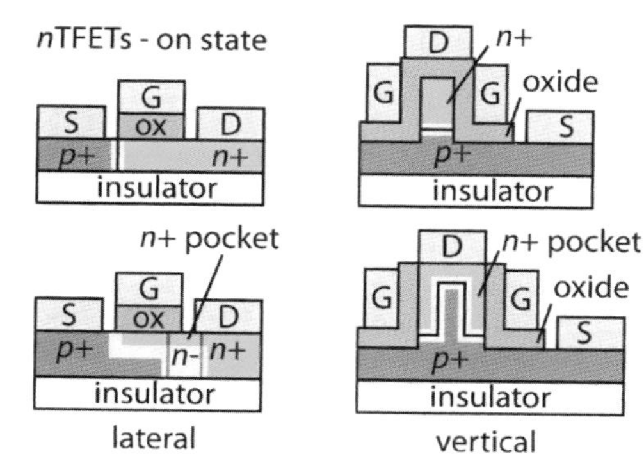

TFETs can have a single or a double gate; a highly doped pocket may be added in some cases to line up the internal electric field with that from the gate, to improve the subthreshold slope

Specific geometries can be introduced in the fabrication of tunnel FETs, in order to achieve the best possible subthreshold slope, on the basis of the expression that we have previously seen.

22  Tunnel FET implementation (I)

We can draw a comparison between the established CMOS technology and TFETs implemented with different material systems

Attribute	MOSFET	TFET			
V_{DD} (V)	1	0.4	0.3		0.2
L_G (nm)	29	35	15	20	40
Channel	Si [83]	Ge-Si [46]	CNT [51]	Broken gap [52]	GNR [50]
I_{ON} (μA/μm)	**1210**	**400**	**400**	**550**	**225**
C_G (fF/μm)	**0.76**	0.69	-	-	0.69
$\tau = C_G V_{DD}/I_{ON}$ (fs)	628	690	**100**	**120**	613
PDP = $C_G V_{DD}^2$ (fJ/μm)	0.76	0.11	0.012	0.020	0.028

This table lists the main parameters of experimentally fabricated TFETs in comparison with a MOSFET that represented the state of the art 9-10 years ago.

It is apparent that for comparable device sizes, the TFET allows a significantly smaller power supply voltage, which implies an important reduction in power dissipation (the power dissipation is proportional to the square of the power supply voltage and directly proportional to the capacitance which stays more or less constant). The on current is instead smaller for the TFET, which implies that it will be slower (in charging the load capacitance) with respect to traditional MOSFETs.

23  Tunnel FET implementation (II)

- **Several attempts have been made at the implementation of tunnel FETs, with a variety of material systems**
- **So far, the technology is not mature yet for large-scale application, but promising results start appearing**

Author	Ref.	Structure	*LG* (nm)	Gate dielectric	EOT (nm)	*S* (mV/dec)
Appenzeller	6	SG CNT	200	4 nm Al_2O_3	2	40
Lu	35	SG CNT	75	2 nm HfO_2	0.3	50
Mayer	37	SG Si	100	3 nm HfO_2	0.5	42
Jeon	38	SG Si	20000	HfO_2	0.9	46
Leonelli	39	MuG Si	160	2 nm HfO_2	1.3	46
Krishnamohan	40	DG Ge	1000	35 nm SiO_2	35	50
Kim	41	SG Ge/Si	5000	3 nm SiO_2	3	40

Overall, some interesting results have been achieved, but the TFET technology does not appear to be ready for industrial application, also because the improvement in the subthreshold slope achieved so far is not as impressive as one would expect. In the mean time, other approaches to the increase of the subthreshold slope, such as the introduction of ferroelectric dielectric materials, have been proposed and are being investigated.

24  Adiabatic CMOS logic (I)

- **In traditional CMOS logic the main source of energy dissipation is in charging and discharging capacitors (a somewhat different situation can occur for the latest technology nodes with decananometer devices)**
- **If we charge a capacitor connecting it to a constant voltage source trough a resistor of any kind, an energy of ½ CV_{DD}^2 will be dissipated and this happens also if the connection is performed through a nonlinear device, such as a MOS transistor**

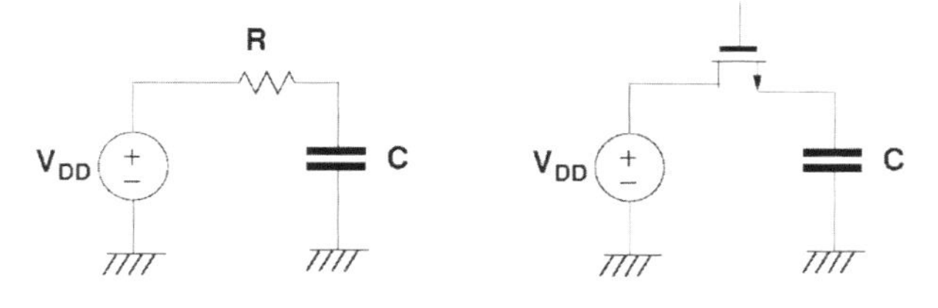

Ultra-low-power operation can be achieved not only with new devices but also with innovative architectures: one example is represented by the adiabatic CMOS logic, in which a properly devised approach to capacitor charging and discharging allows an important reduction in power dissipation per switching event.

25  Adiabatic CMOS logic (II)

- **This limit can however be overcome by trading off speed for reduced power consumption, i.e. charging the capacitor with a voltage source that varies following that of the capacitor itself**
- **This can be achieved with multiple voltage sources (ideally an infinite number) or with time-varying power supply voltages**
- **The power dissipation will be proportional to *RC/T***

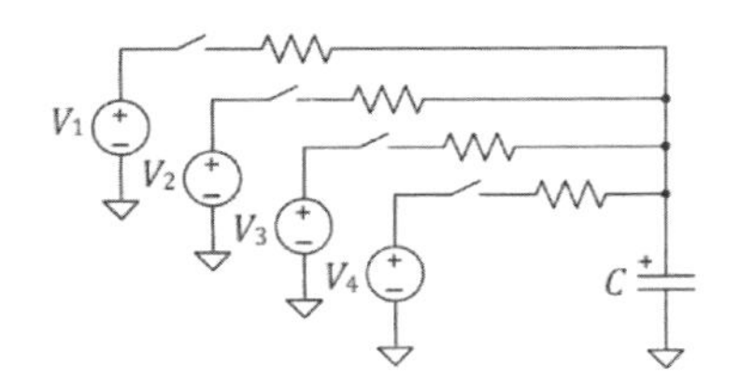

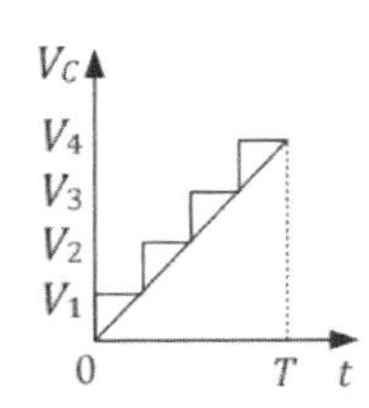

In order to reduce the amount of energy dissipated during capacitor charging, the voltage of the external source must track that of the capacitor: this is the adiabatic charging, i.e. a charging procedure that takes place in a quasi-equilibrium condition.

For an ideal adiabatic charging, the voltage drop across the resistor should be vanishingly small.

The slower the charging, the smaller the power dissipation will be. Adiabatic logic is thus suitable for applications in which power saving is essential, but speed is not a primary requirement.

26  Adiabatic CMOS logic (III)

- **There is an additional requirement, in order to have a working adiabatic logic: we need to be able to vary the output data smoothly from the previous to the new value, and the strategy to do this depends on the previous value, which disappears when the transition begins**
- **Therefore this is not a trivial issue, which requires the design of a reversible computing scheme**
- **An example is the SCRL (Split Level Charge Recovery Logic) by Younis and Knight ["Asymptotically zero energy computing using split level charge recovery logic", Technical Report AITR-1500 MIT AI Laboratory, June 1994]**
- **In place of the ordinary power supply rails, there are two trapezoidal clock phases:** ϕ **and** $\bar{\phi}$

A fully adiabatic logic also involves reversibility, i.e. the preservation of all the information needed to retrieve, at any stage, the initial data of the computation.

The reason for such a reversibility requirement will be clearer later when we will discuss the Bennet-Landauer principle on the limits of computation.

27  NAND SCRL

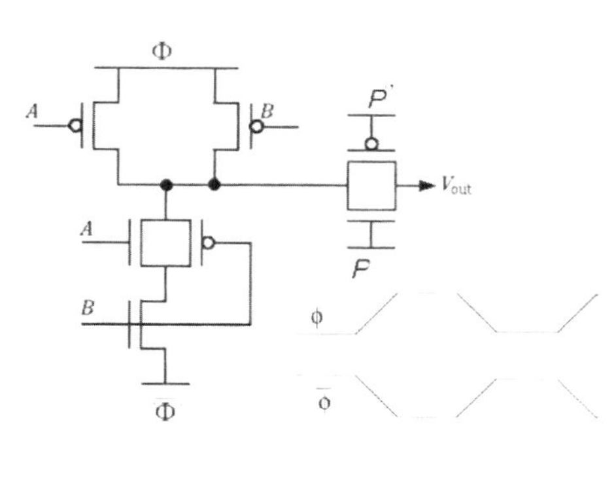

At the beginning all nodes are at $V_{DD}/2$ except for *P'* (at V_{DD}) and *P* (at ground potential), so that the transmission gate is off

Then *P'* and *P* are gradually switched, thus enabling the transmission gate while ϕ and $\bar{\phi}$ reach V_{DD} and ground, respectively

At this point the gate works as a normal NAND. Voltages then go back to the initial values and we are ready for a new cycle. The input value must not be changed until the rails are back at $V_{DD}/2$

Within the Split-level Charge Recovery Logic approach, it is possible to implement logic gates that can switch dissipating an arbitrarily small amount of energy, as long as slow enough operation is accepted.

The result of the operation is transferred to the following gate only when a well-defined logical value has been reached at the output, through gradual activation of the power supply rails.

28  Adiabatic MOS

- In general, properly designed adiabatic MOS circuits satisfy the 2 rules:
  - A transistor must not be turned on when a voltage is present between source and drain
  - A transistor must not be turned off when the drain current is not zero
- The dissipated energy per cycle decreases linearly with increasing cycle duration, therefore the dissipated power decreases quadratically
- In principle, if all computation is reversible, the dissipated power could become vanishingly small

The requirements of adiabatic operation can be summarized into two simple rules expressing the concept that during the switching process each transistor must not undergo sudden changes of the drain current or of the drain-source voltage drop.

The dissipated power decreases quadratically with decreasing clock frequency and, if the operation is made fully reversible, then it can asymptotically drop down to zero.

29 

Limits of computation (I)

- **An important question is: what is the minimum amount of energy that needs to be spent to perform a single logical operation?**
- **Since a relatively long time it has been known that in the case of reversible computation this limit can be assumed to be zero**
- **However a paper appeared in 2003 claiming that such limit existed [V. V. Zhirnov, R. K. Cavin, J. A. Hutchby, G. I. Bourianoff, Proceedings of the IEEE 91, 1934 (2003)]**

The conclusions we have reached for the operation of adiabatic logic raise a more general question on the actual limits of computation from the point of view, in particular, of the energy needed to complete an elementary logical operation.

30 

Limits of computation (II)

- **The main argument in the paper by Zhirnov et al. is that a barrier of at least a few kT is needed to make two states distinguishable in the presence of thermal fluctuations, and this barrier has to be lowered to perform a logical operation**

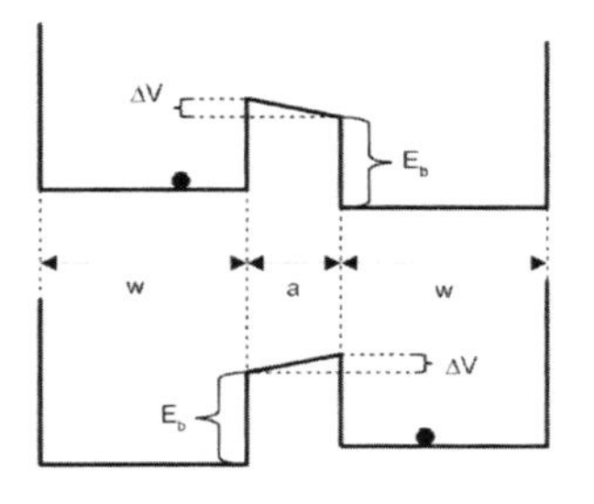

- **Zhirnov et al. claim that the minimum energy to perform an operation is equivalent to the height of the barrier, but does lowering and raising a barrier of height ΔE really cost an energy ΔE?**

Zhirnov *et al.* argue that completion of an elementary logical operation would require a minimum amount of energy, corresponding to kT ln 2.

Their argument, which is somewhat similar to an argument in a paper by Robert Keyes in the 1970s, has however been refuted by several authors [e.g. CS. Lent, M. Liu, and Y Lu, Nanotechnology 17, 4240 (2006)].

31

Limits of computation (III)

- An experiment performed by the Notre Dame group [G. P. Boechler et al., Appl. Phys. Lett. 97, 103502] tells otherwise: that there is no bottom limit for the power dissipation needed to perform an operation in reversible computing
- In particular they show that an argument raised by Zhyrnov to deny the possibility of zero-energy computation, i.e. the minimum energy cost for charging a capacitor, does not stand

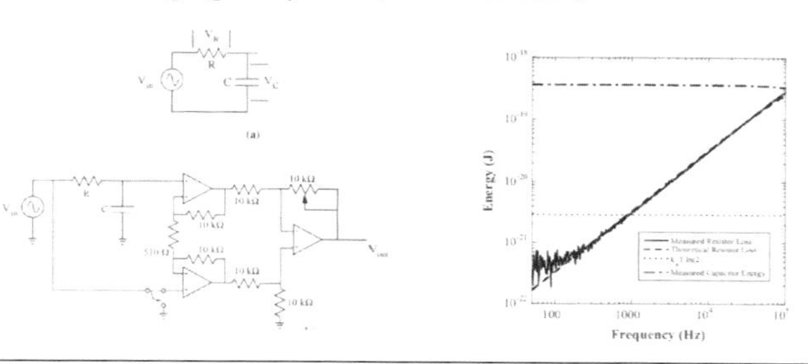

The argument by Zhirnov *et al.* was applied also to switching of Quantum Cellular Automata cells, in which simple calculations show that switching between two logical states with an energy dissipation less than kT ln2 should be possible. In particular, it was argued that it would be impossible to adiabatically charge a capacitor because thermal noise would anyway give rise to a finite voltage across the resistor connecting the voltage source to the capacitor. The experiment by Boechler *et al.* demonstrated instead that it is indeed possible to charge a capacitor with a vanishingly small dissipation on the part of the voltage source (the reason being that the additional dissipation on the resistor comes from the thermal bath and would be present anyway, even if we did not charge the capacitor).

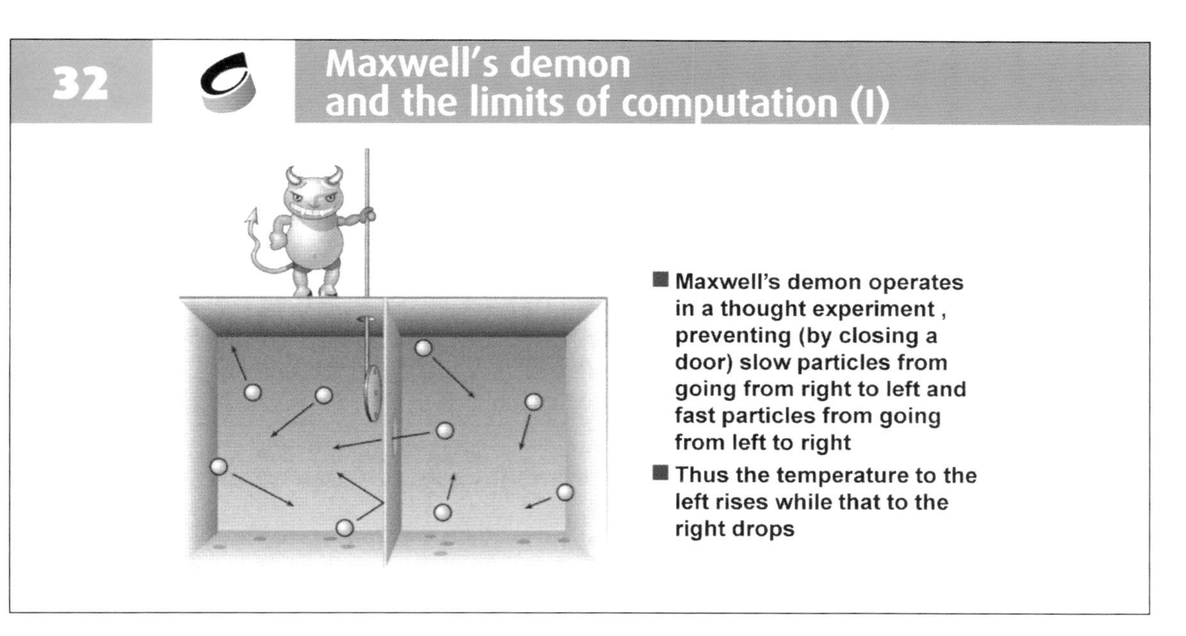

If slow particles are let through the door from left to right and the opposite happens for fast particles, with time there will be more fast particles in the left portion of the box and more slow particles on the right portion. Once the particle distribution on each side has thermalized, we will observe a lower temperature in the right region with respect to the left region.

If the demon can perform such a job dissipating a vanishingly small amount of energy, a temperature gradient could be created without supplying work from the outside and therefore the total entropy of the system would decrease.

33 

Maxwell's demon and the limits of computation (II)

- **If the demon could complete his task with vanishingly small energy, the second law of thermodynamics would be violated, because a temperature difference would appear starting from an isothermal system, without work being spent**
- **This apparent contradiction was finally solved by Bennett and Landauer [R. Landauer, IBM J. Res. Develop. 5, 183 (1961)], who reached the conclusion that the demon can operate with vanishingly small energy but will cause an increase in entropy, by kT ln2 whenever he erases a bit of information from the memory where he stores the results of his measurements**

Maxwell was not able to solve the paradox, and neither were able to solve it many others who worked on it, including prominent physicists such as Leo Szilard and Leon Brillouin. Only in 1960 Landauer and then Bennett provided an explanation based on the increase in entropy resulting from the erasure of the demon memory.

This has far-reaching consequences, because it implies that, as long as no information is erased (and thus computation is reversible) it can be performed with a vanishingly small energy dissipation. To reach the goal of reversible computation, it is also necessary to operate adiabatically and therefore asymptotically slow.

34 

Maxwell's demon and the limits of computation (III)

- **In this way, Bennet and Landauer established a direct and strong connection between thermodynamical entropy and information entropy**
- **Very recently a paper appeared that claims to have violated the "Landauer limit" by implementing an irreversible logic switch that uses up less than kT energy per operation [M. Lopez-Suarez, I. Neri, L. Gammaitoni, Nature Communications 7, 12068 (2016)]. This needs however some further verification.**

35  Limits of computation

- **Thus, there is no lower limit to the energy dissipated in performing a single logical operation, as long as we accept to perform it as slow as needed and we do it reversibly**
- **Indeed the time-energy uncertainty relation must be satisfied:** $\Delta E \Delta t \geq \hbar/2$

  Thus the minimum energy dissipated depends on the time within which the operation must be completed: $\Delta E \geq \hbar/(2\Delta t)$

  For example, for Δt = 10 ps the minimum energy dissipation is 5.28 x 10^{-24} J, corresponding, at room temperature, to 1.273 x 10^{-3} *kT*.

If computation is performed in a finite amount of time, an elementary logical operation has a cost in terms of energy that is determined by the time-energy uncertainty relation.

Therefore, this is the actual physical limit to be taken into account if we strive to reach the ultimate reduction in power dissipation in devices and circuits that process information.

36  Example of low-power sensor nodes based on traditional CMOS technology

The problem is the timely detection of a freight car derailment: the nodes must operate in an environment where no data or power bus is available

The solution is to use wireless communication and energy scavenging from vibration to power the system

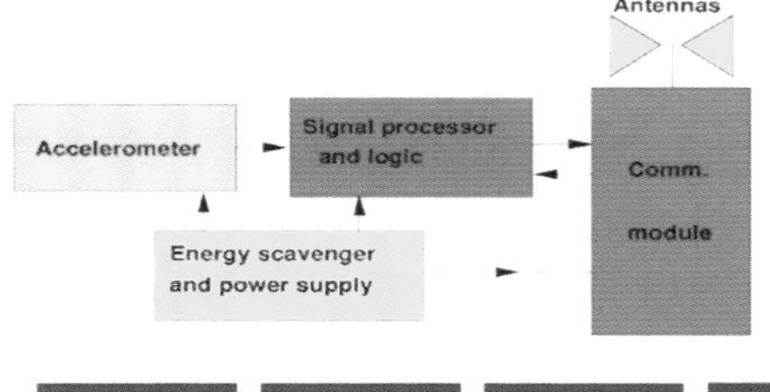

Nodes are located on each car of a train, at both ends, and relay the information to the locomotive

[M. Macucci et al., IEEE Trans. Instrum. Meas., DOI: 10.1109/TIM.2016.2556925]

We now examine an example of an ultra-low-power application based on conventional low-power components that are kept in sleep mode for most of the time.

The application is a wireless network of nodes that detect a derailment condition in a freight train and alert the driver. Without dedicated equipment, derailments, especially in the case of freight trains, may go undetected for several kilometers and may lead to extremely serious accidents, especially if the derailed cars are transporting hazardous materials (notable examples are the derailments and subsequent explosions of liquid propane gas tank cars in Crescent City, IL in 1970 and in Viareggio, Italy in 2009).

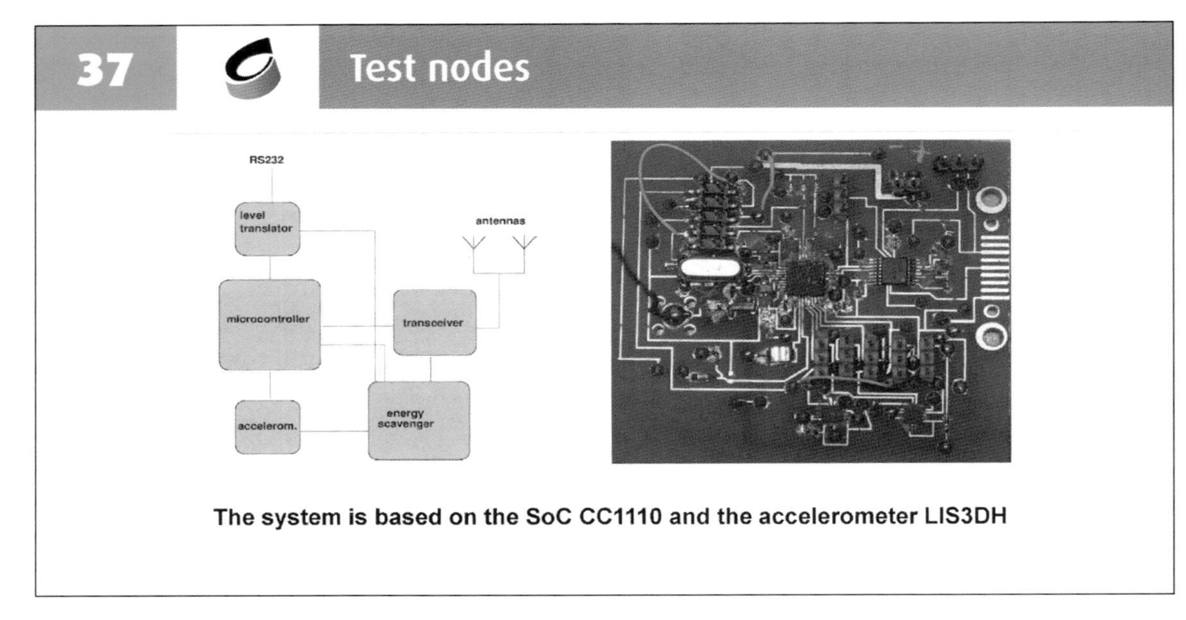

The nodes must be applied at each end of each car. A node consists of an integrated microcontroller-transceiver CC1110, a solid-state accelerometer LIS3DH and an energy scavenger exploiting the mechanical vibrations that are always present during the normal operation of the freight car, as soon as the speed is above 20-30 km/h.

Antennas are to be designed in such a way as to guarantee optimal communication with the two nearest nodes.

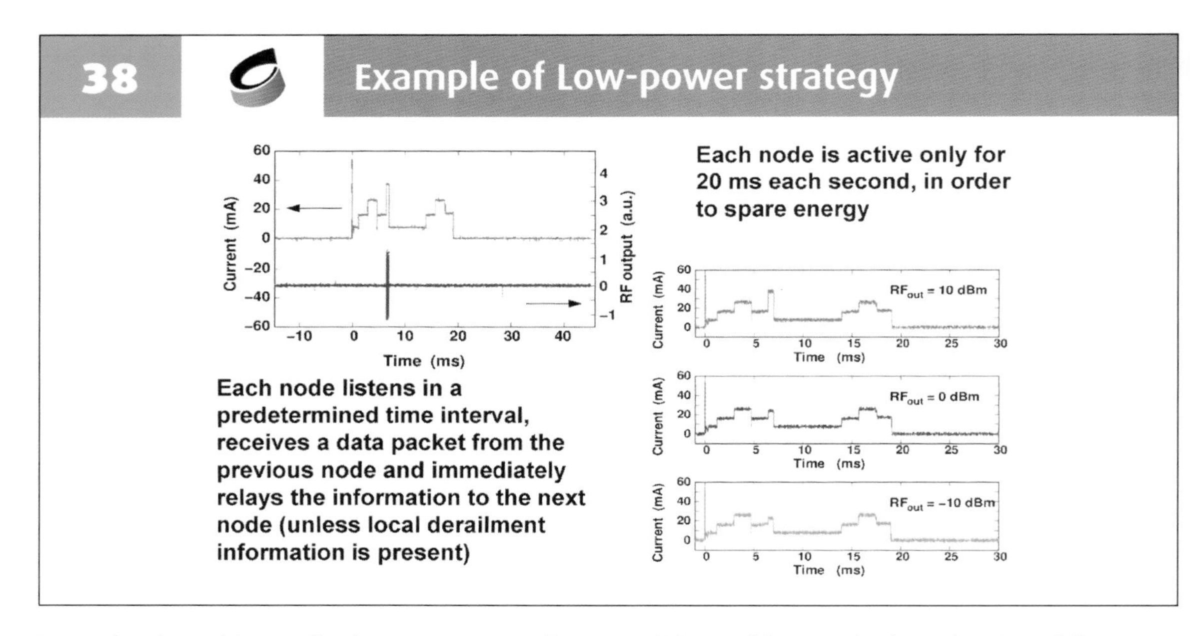

In order to achieve ultra-low-power operation, nodes are kept for most of the time in a sleep mode and turned on only to perform a measurement of the vertical acceleration, relay the information from the neighboring node, or send a possible derailment alert.

It is possible to notice from the plots of the power supply current as a function of time that the energy spent during the transmission burst is negligible compared to that needed for the initialization phase after exiting the sleep mode and for the operation of the receiver.

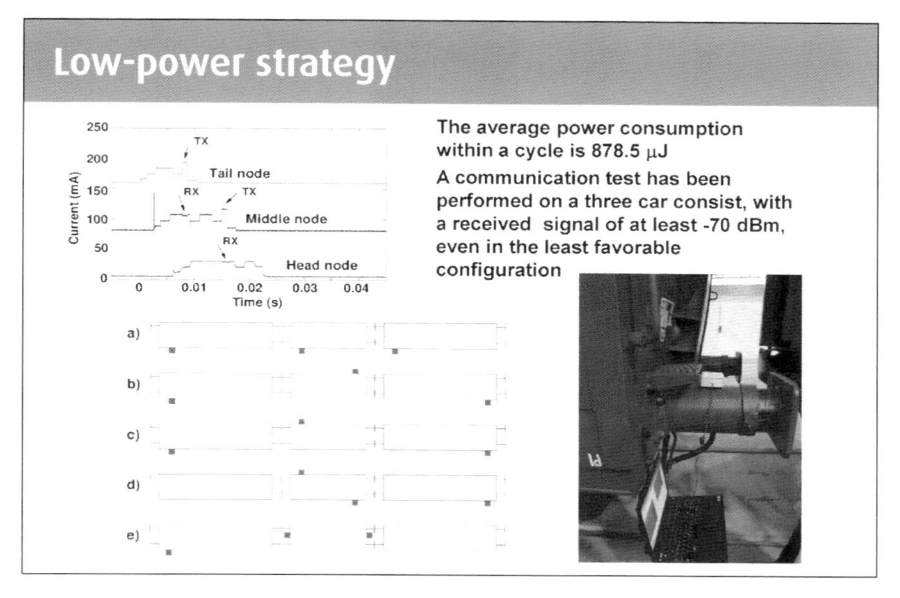

A wireless network is set up automatically based on the relative intensities of the received signals. Once each node has been attributed a progressive number, the one with the highest number (tail node) starts sending periodic messages that are relayed all the way to the head node (which in operational conditions would be in the locomotive)

Intermediate nodes synchronize with the periodic transmissions and then exit from the sleep mode only in a window of a few milliseconds around the expected time of arrival of the message from the previous node.

Static tests performed on a three-car consist demonstrated reliable operation for any choice of positioning of the nodes.

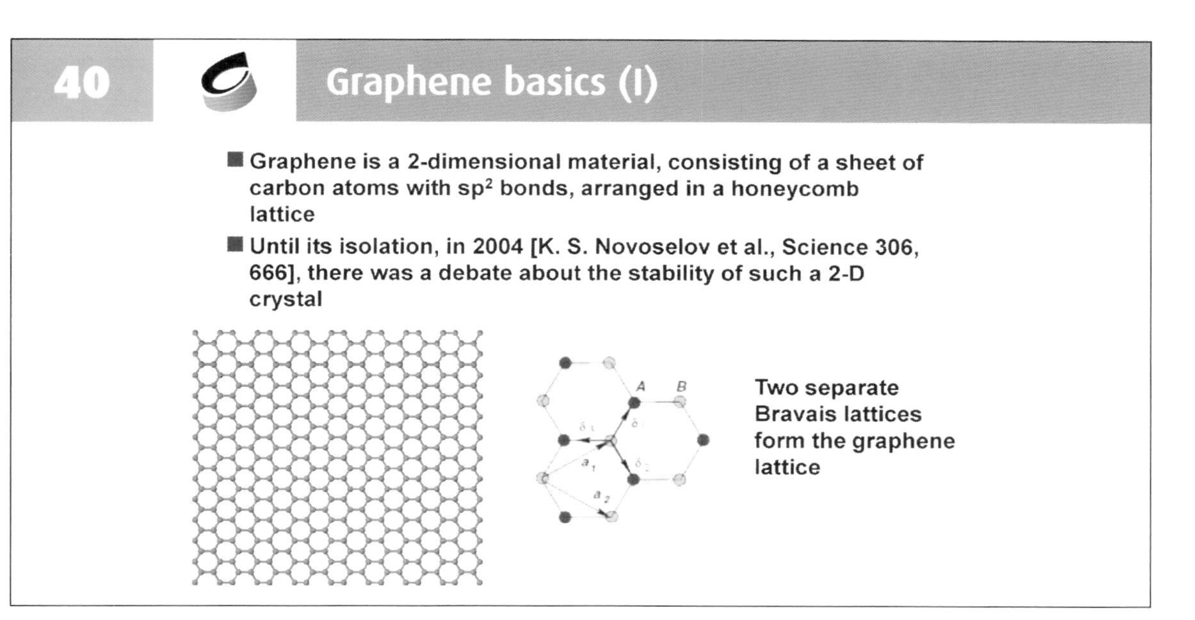

In the last decade, graphene, a two-dimensional crystal of carbon atoms, has received a lot of attention, because of its very peculiar electronic and mechanical properties. Because of the particular symmetries, graphene is a zero-gap semiconductor, whose conductivity can be modulated by varying the chemical potential and thereby altering the electron and hole concentrations

However, the absence of a gap, as we will discuss later, severely limits its potential for applications in the field of digital electronics.

41  Graphene basics (II)

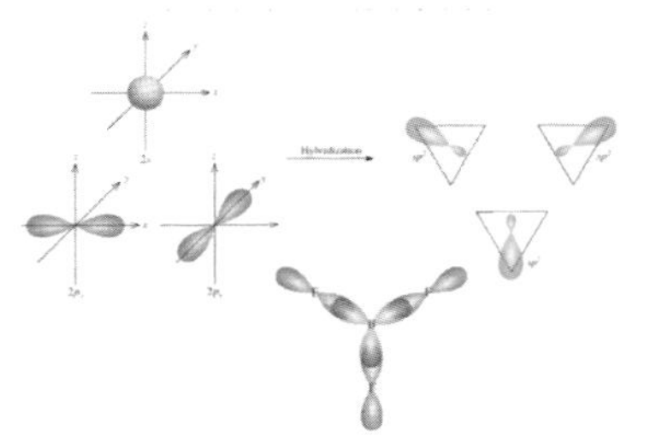

sp2 bonds are the result of the hybridization of the 2s orbital and two 2p orbitals: this leads to three equally spaced bonds in the plane of the 2p orbitals

In graphene the third 2p orbital (in the direction orthogonal to the sheet) provides a delocalized electron that contributes to conduction

The peculiar electronic structure affords special properties, such as high carrier mobility, optical transparency, mechanical flexibility and high thermal conductivity

The carbon atoms forming the graphene lattice have sp2 hybridization, which leads to three coplanar bonds 120° apart.

Electrons belonging to such orbitals (σ orbitals), which form the covalent bonds holding the crystal together, are localized within the bonds and are not available for conduction.

Electrons from the third 2p orbital (ϖ orbital), which is orthogonal to the plane of the crystal, are delocalized and support electrical conduction.

42  Graphene basics (III)

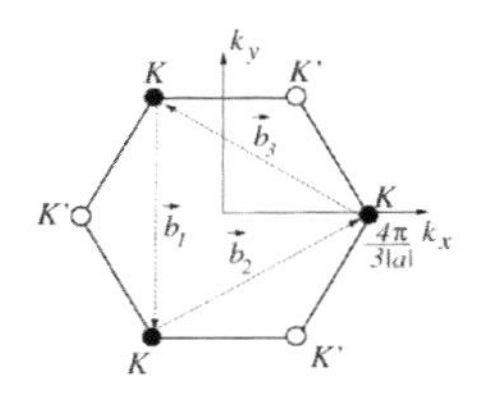

Graphene has a hexagonal Brillouin zone, with K (green) and K' (red) points, corresponding to the two inequivalent sublattices

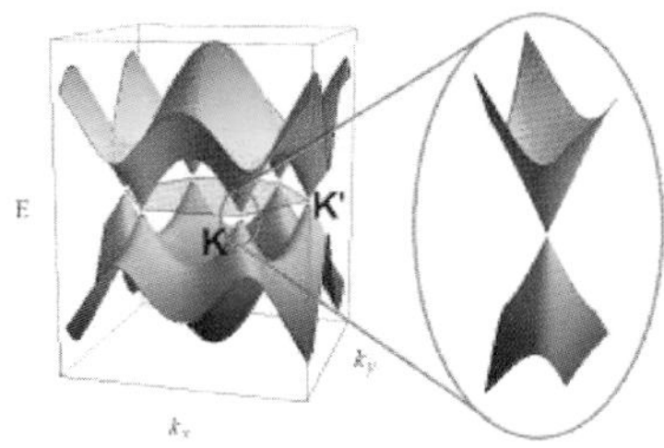

The bandstructure is very peculiar, with cones touching each other at the K and K' points

Thus graphene is a zero-gap semiconductor (semiconductor because the density of states in the K and K' points is zero)

The bandstructure of graphene was studied a long time ago [P.R. Wallace, "The Band Theory of Graphite," *Physical Review* 71, 622 (1947)], since graphite is just a collection of graphene sheets held together by van der Waals forces and some electron delocalization [J.-C. Charlier, "Graphite Interplanar Bonding: Electronic Delocalization and van der Waals forces," *Europhys. Lett.* 28, 403 (1994)].

43  Properties of graphene

- For general electronic applications graphene has important advantages, but also severe disadvantages:
- Pros: graphene is characterized by high mobility (which allows near-ballistic transport) and one-atom-thickness (and thus good gate control on the channel).
- Cons: graphene has a zero energy gap (which severely limits the I_{on}/I_{off} ratio of the transistor) and an ambipolar electrical behavior (which makes the development of a complementary logic architecture quite difficult).

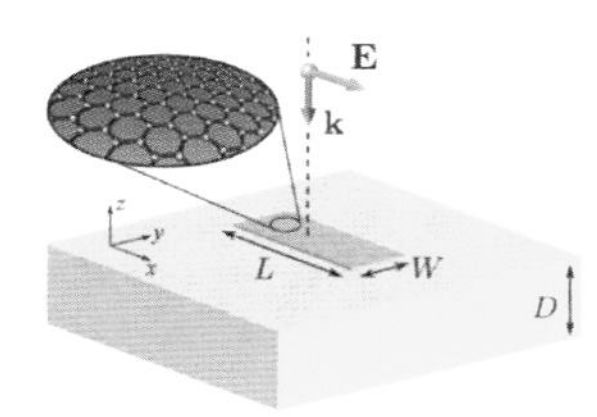

There are however particular applications in which the advantages of graphene largely surpass its drawbacks: one such application is as a material for antennas in the upper THz band [I. Llatser et al., Photonics and Nanostructures – Fundamentals and Applications 10, 353 (2010)]

As mentioned before, the absence of an energy gap makes graphene unsuitable as a general replacement for currently used semiconductors (silicon in particular) in digital electronics.

There are however some applications in which graphene could offer significant advantages, due to its high carrier mobility (high-frequency devices, interconnects), relatively low plasma frequency (antennas for THz communications), transparency at optical wavelengths (contacts for photovoltaic devices), flexibility (wearable electronics), and high surface-to-volume ratio (gas sensors).

44  Graphene Antennas

- The frequency range where graphene antennas have important advantages is that above 100 GHz, where the presence of surface plasmon polaritons makes resonance possible with dimensions much smaller than the free-space wavelength
- This is very interesting in terms of enabling wireless sensors with a size in the tens of micrometer range, thus open the perspective to the implementation of real "smart dust"
- To explain the operation of graphene antennas, and in particular why they can be resonant at frequencies that are much lower that those corresponding to free-space wavelengths of the order of their physical size, we need to introduce a few concepts:
  - The plasma frequency
  - Surface plasmon polaritons

There is currently a very significant interest in the possible applications of THz frequencies (millimeter waves), both in short-range high-bandwidth applications and in imaging systems that can detect objects hidden under clothes without resorting to ionizing (and therefore hazardous) radiations such as X-rays.

45 Plasma oscillations

- Plasma oscillations take place in an electron gas when the electron density undergoes a perturbation
- Then, as a result of the restoring force of the positive ions, an oscillation appears
- The frequency at which such oscillation occurs is ω_p

$$\omega_p = \sqrt{\frac{nq^2}{m^* \epsilon}}$$

- In metals the plasma frequency is extremely high, at frequencies that are in the range of visible light
- In graphene, instead, the plasma frequency is in the THz band

Plasma oscillations are the result of the restoring force that appears when the electron density in a conductor is altered with respect to the equilibrium configuration.

46 Surface plasmon polaritons (I)

Surface plasmon polaritons are the result of a combination of charge motion in a conductor and of electromagnetic waves

They propagate along the conductor surface (conductor-dielectric interface) and are evanescent in the direction perpendicular to the surface

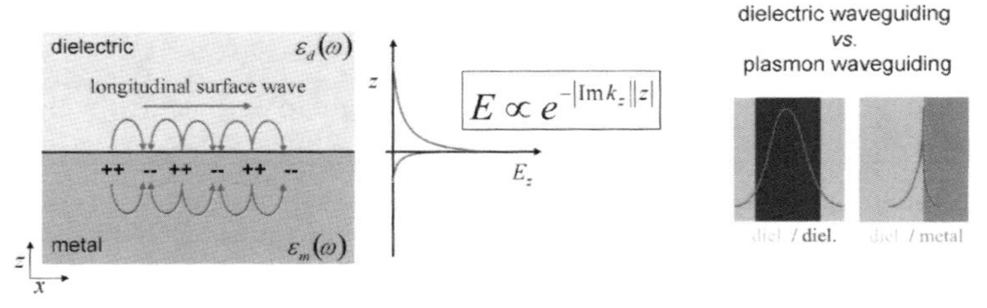

Surface plasmon polaritons result from a particular solution of the Maxwell equations at the interface between a conductor and a dielectric.

They are a particular form of waveguiding, in which there is confinement in the proximity of the surface and propagation occurs along the surface.

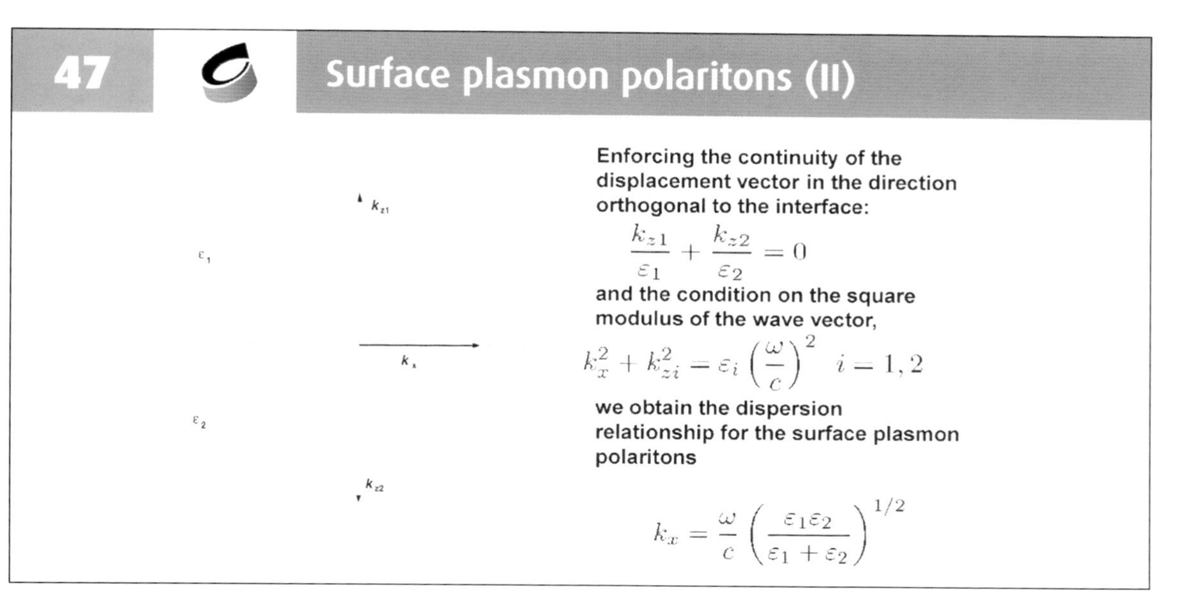

The dispersion relationship of surface plasmon polaritons in the direction parallel to the interface can be derived by combining the continuity of the electric displacement vector across the conductor-dielectric surface with the relationship between the modulus of the wave vector and the angular frequency in the dielectric and in the conductor.

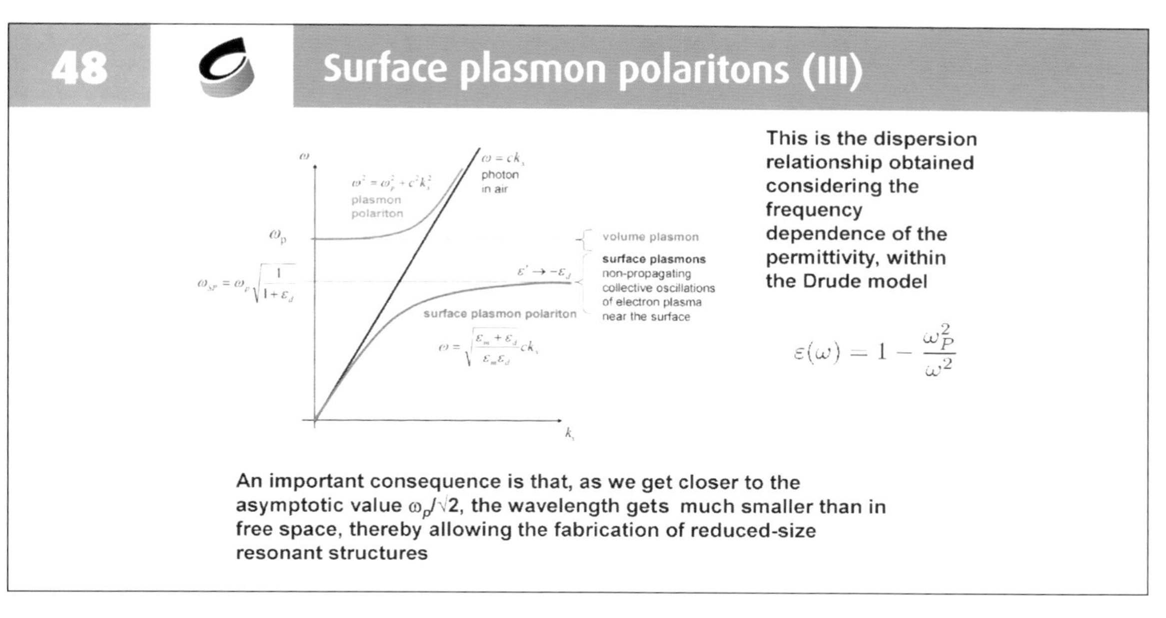

By inserting the frequency dependence of the permittivity in the conductor from the Drude model (which is a function of the plasma frequency) into the previously obtained dispersion relationship, we obtain an explicit expression of the angular frequency as a function of the wave vector parallel to the dielectric-conductor interface.

We see that there are two different branches of this dispersion relationship, separated by a gap of prohibited frequencies. Below the gap and close to its bottom boundary [corresponding to $\omega_p\ (1+\varepsilon_d)^{-1/2}$] we observe that for a given value of the angular frequency we have a much larger wave vector than what we would have in free space propagation, and, therefore, a much smaller wavelength.

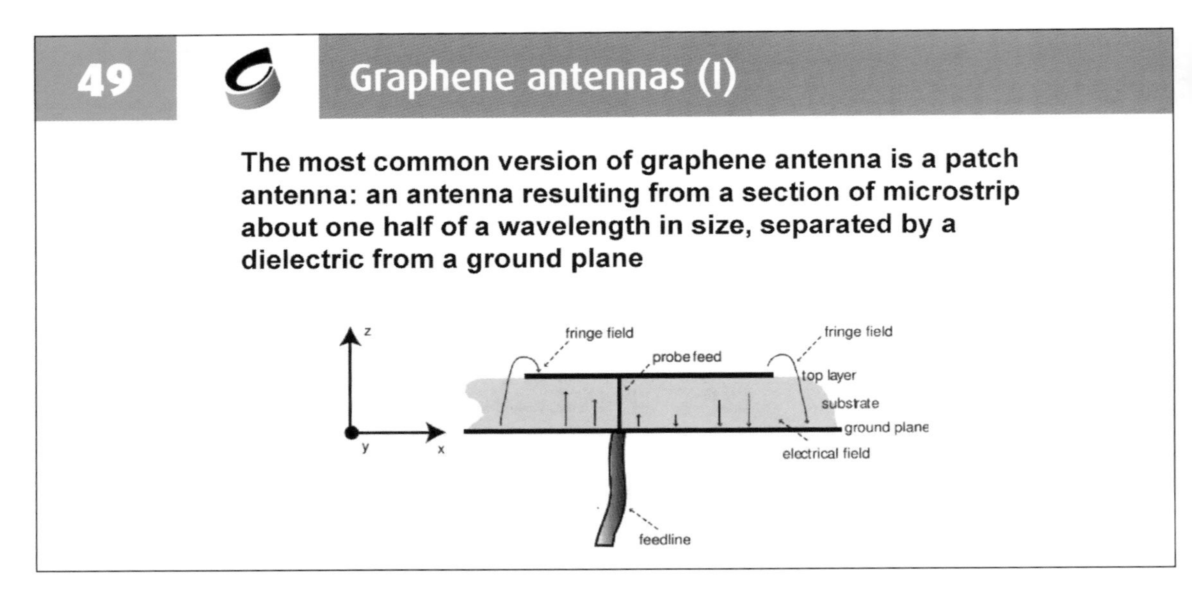

As a result of the shorter wavelength of the surface Plasmon polaritons, an antenna made of graphene can be made resonant with a size that is significantly smaller than that which would be necessary with a metal antenna.

This is particularly interesting for applications in extremely small devices, such as autonomous sensors forming the so-called smart-dust.

For frequencies below 100 GHz, graphene antennas, in the absence of surface plasmon polaritons, not only have the advantage of flexibility, but also have significant drawbacks in terms of efficiency, because of the higher resistivity of graphene compared to the metals typically used for antennas, such as aluminum or copper.

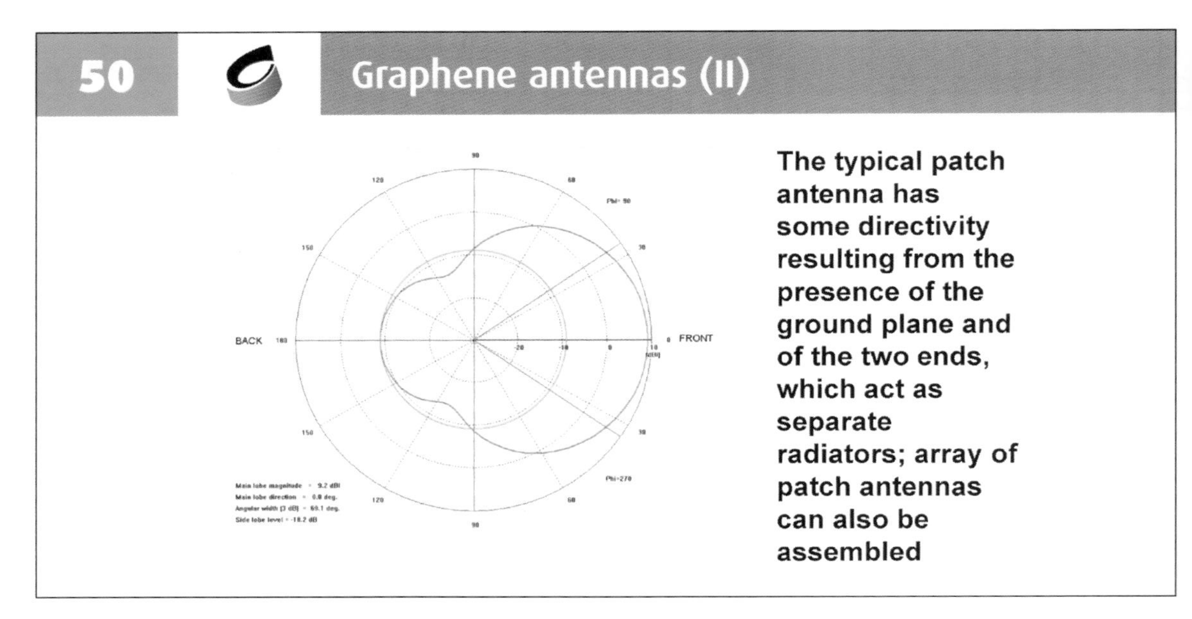

Arrays of patch antennas, even complex ones, can be easily obtained from a graphene sheet, using a lithographic process.

The radiation diagram of the array will correspond to the product of the radiation diagram of the same array made with isotropic radiators and the radiation diagram of a single patch antenna.

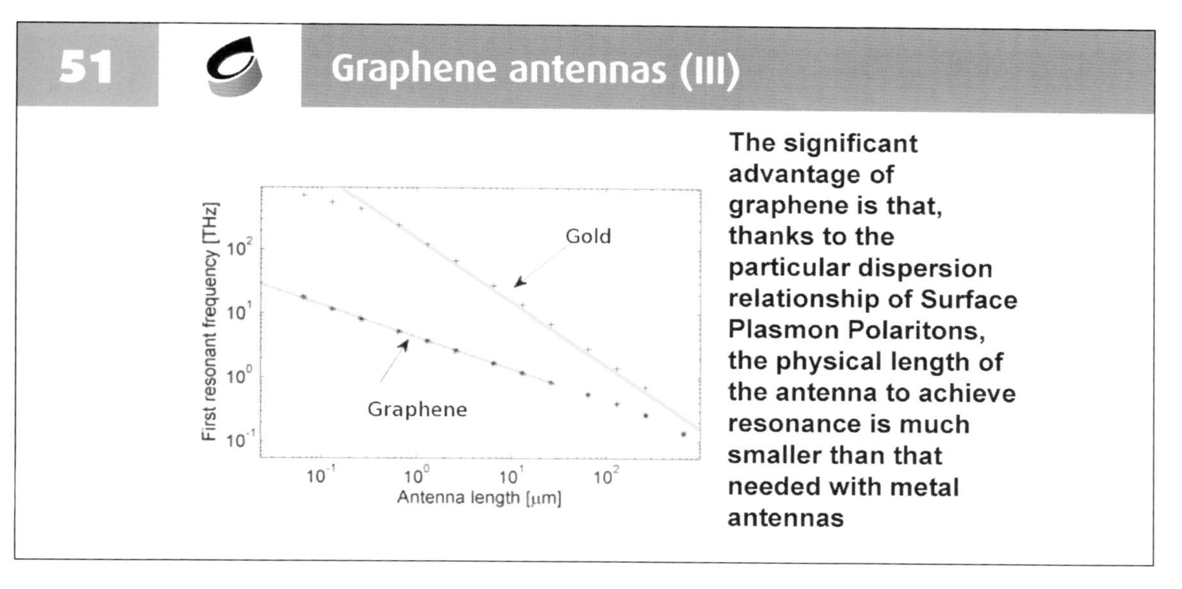

As the frequency is increased, the dimensional advantage of graphene antennas becomes more relevant, up to the beginning of the gap for surface plasmon polaritons.

It is important to notice that the advantage of a graphene antenna is in terms of obtaining a properly resonanting radiator (and therefore allowing good power transfer into it and from it) with a smaller footprint, which may be essential for some miniaturized devices (such as smart dust). However, with the physical dimensions being smaller, the gain will be lower with respect to that of a full-size antenna.

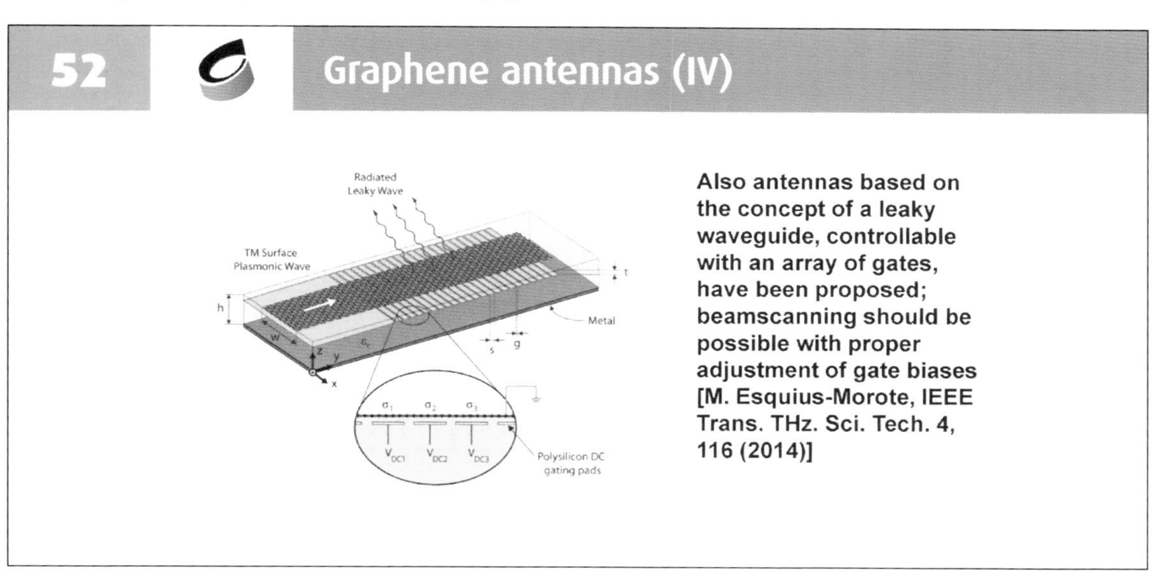

The concept of a leaky wave guide, i.e. a wave guide that radiates part of the electromagnetic energy flowing through it (a classical leaky wave guide can be implemented by cutting slots in its conducting walls), can be obtained with a graphene stripe on which plasmonic modes are propagating. Leakage can be obtained by means of backgates, whose bias will locally alter the graphene conductivity.

By adjusting the voltages applied to the gates, it would be possible to obtain radiation from different positions along the wave guide and therefore to control the radiation pattern.

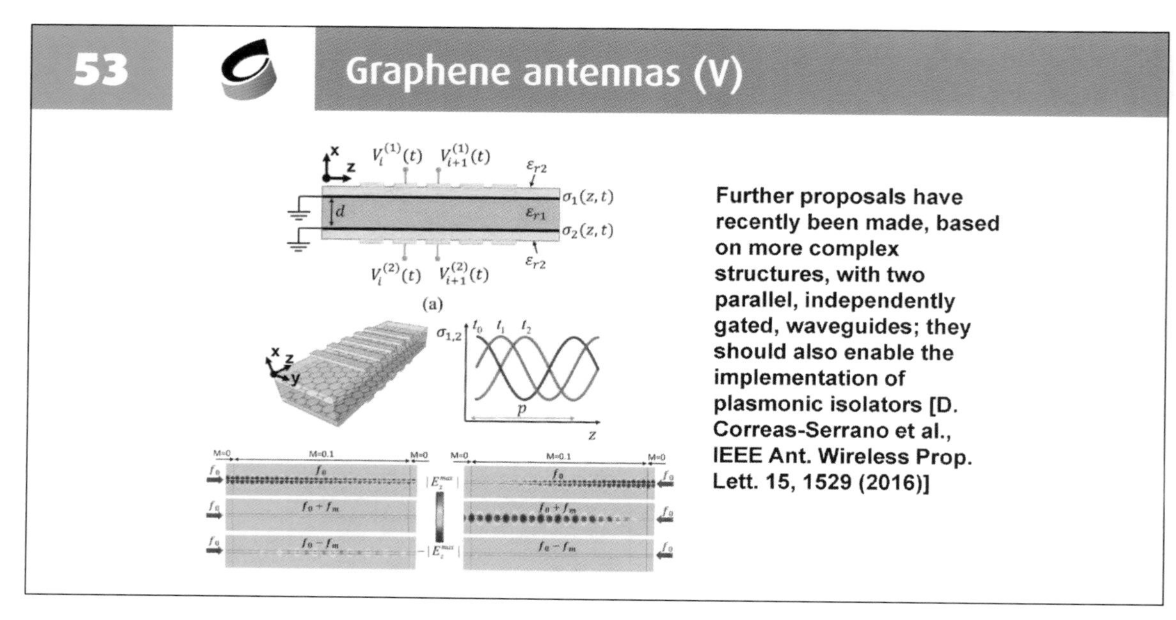

Authors have proposed combinations of leaky wave guides with a spatiotemporal modulation for the implementation of non-reciprocal devices, such as radio frequency insulators.

However, achieving the desired performance may require a non-negligible length of the device, which also implies an increase in the ohmic losses. A careful tradeoff should be sought.

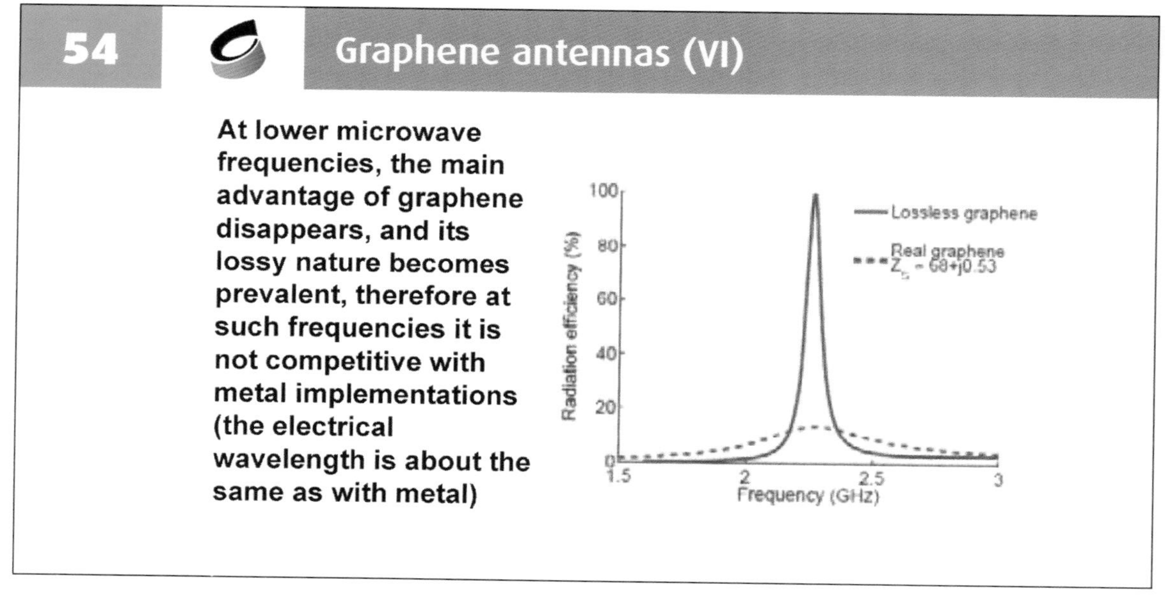

As already mentioned, the efficiency obtainable with graphene antennas is relatively low, as a consequence of the significant ohmic losses of this material.

This makes usage of graphene for antennas at frequencies below 100 GHz suboptimal for most applications.

55 Graphene antennas (VII)

Another proposed application of graphene in antenna systems is in reflectarrays, which are formed by a collection of sub-wavelength patches, for each of which the phase of the reflection can be individually controlled [E. Carrasco et al., Appl. Phys. Lett. 102, 104103 (2013)]

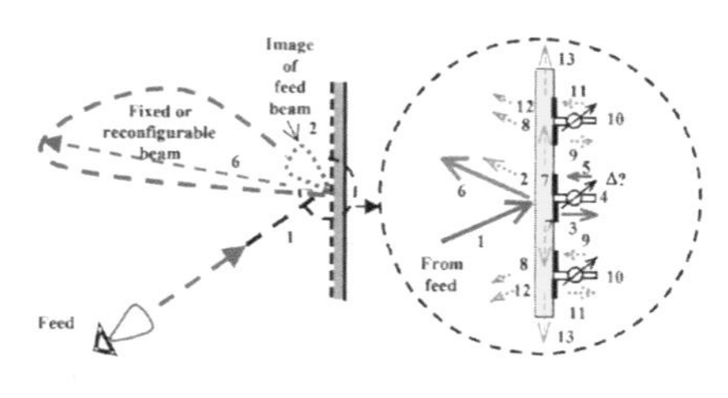

The shape of the beam depends on the intensity and phase of the reflection from each patch: the advantage with graphene is that beam reconfiguration can be performed continuously, just by changing the voltages applied to the patch backgates.

Reflectarrays are phased arrays in which, instead of having an array of radiators fed with individually controlled phases, there is a single radiator illuminating a reflector made up of small patches whose reflection phase can be controlled, by varying its electrical properties (e.g. a gated graphene patch).

56 Graphene antennas (VIII)

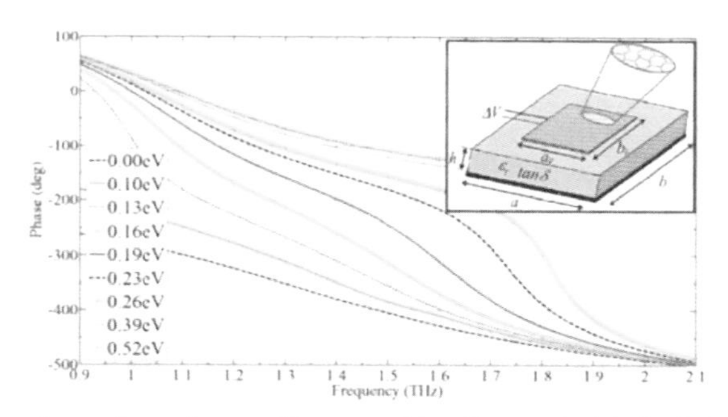

The advantage of graphene with respect to other materials is that the phase with which the impinging RF signal is reflected can be controlled by varying the chemical potential in the graphene sheet, via a DC-biased gate [J. Perruisseau-Carrier, Loughborough Antennas and Propagation Conference, 2012 UK]; it is to be noted, however, that the value of the chemical potential influences not only the phase but also the modulus of the reflection coefficient

In graphene reflectarrays, the reflection phase of graphene patches should be adjusted by tuning the backgate voltages, i.e. varying the chemical potential This, however, has been shown to affect the modulus of the reflection, too.

57 Graphene phase shifters

Graphene-based phase shifters have been proposed for the implementation of THz phased arrays [P.-Y. Chen et al., IEEE Trans. Ant. Prop. 61, 1528 (2013)]; the critical issue is the amount of isolation that can be obtained in the switches (based on varying the chemical potential in the graphene sheet)

Phase shifters are a critical element in the implementation of phased arrays. There are proposals to make phase shifters based on graphene transmission lines in which a selection between each pair of parallel paths is made by depletion of carriers, the graphene regions through which the non-selected path would be accessed. The two parallel paths have different lengths; therefore, they are associated with different phase shifts.

Overall, virtually all proposals for graphene antennas (except for just a few tests of flexible wearable antennas at relatively low frequencies) have been only theoretical: no experimental prototype has been implemented yet.

58 Graphene RF transistor (I)

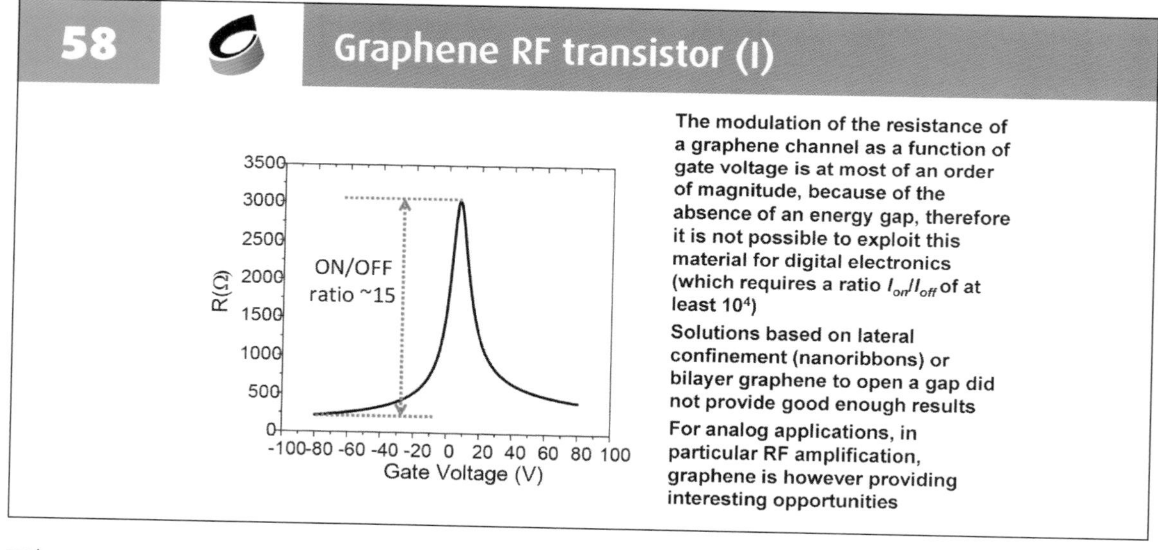

The modulation of the resistance of a graphene channel as a function of gate voltage is at most of an order of magnitude, because of the absence of an energy gap, therefore it is not possible to exploit this material for digital electronics (which requires a ratio I_{on}/I_{off} of at least 10^4)

Solutions based on lateral confinement (nanoribbons) or bilayer graphene to open a gap did not provide good enough results

For analog applications, in particular RF amplification, graphene is however providing interesting opportunities

There are many approaches that have been suggested to open up a gap in graphene, but none of them has been demonstrated to be suitable for practical applications. One possible way to create a gap is lateral confinement in the form of armchair nanoribbons; however, to achieve useful gaps, the transverse size should be of the order of a few nanometers and, due to the fact that the gap almost drops to zero for ribbons with 3*n*-1 dimer lines (rows of atoms) for any integer *n*, edge roughness would be likely to prevent operation of the device. Another possibility to obtain a gap, if bilayer graphene is used, is that of applying an orthogonal electric field, but the gap can reach at most about 250 mV and fabrication of devices would not be straightforward.

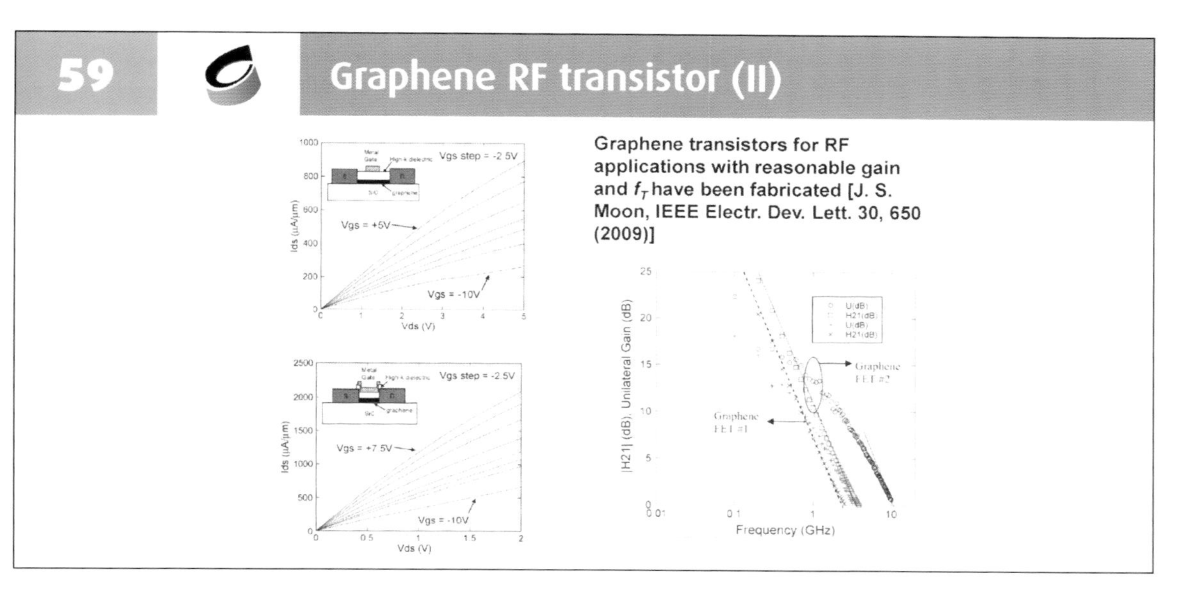

A more promising field of application of graphene is that of transistors for analog RF applications, where the very high mobility of this material can play an important role.

Some interesting results have been obtained and research is ongoing.

Currently, the main difficulty is represented by the relatively low values of the transconductance that have been achieved.

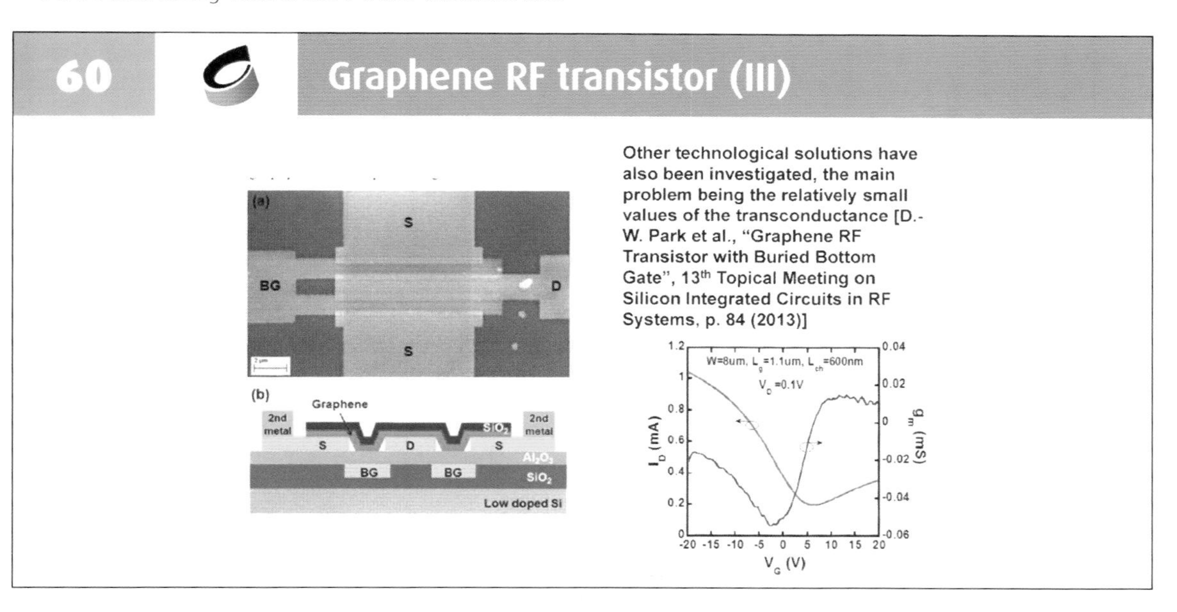

Besides "classical" topologies with a top gate separated from the graphene sheet by a dielectric layer, other solutions have been explored, using, for example, a buried bottom gate.

Another important issue is the compatibility with silicon technology: it would be extremely interesting to be able to fabricate graphene devices for some special functions on a silicon chip with CMOS devices. However, it does not seem feasible to transfer graphene flakes (as done in experiments) onto the chip within a large-scale production environment. On the other hand, direct in situ growth of graphene requires, with currently existing techniques, a temperature (of the order of 1000 °C) which is not compatible with the backend of the CMOS process.

CHAPTER 08

A Functionally Safe SW Defined Autonomous and Connected IoT

Riccardo Mariani

Intel Fellow
Chief Functional Safety Technologist

INTRODUCTION 218
1. Outline 218
2. The third phase of IoT 218
3. The essence of an Autonomous IoT platform 219
4. The levels of autonomy (example of autonomous cars) 219
5. The levels of autonomy (example of autonomous cars) 220
6. AutonomousIoT is E2E 220
7. Autonomous IoT is powered by data 221
8. Example of IoT platform (for industry 4.0) 221
9. The challenges of Autonomous IoT 222
10. Functional Safety 222

FOCUS ON ISO 26262 223
11. ISO 26262 2nd edition 223
12. ISO 26262 2nd edition roadmap 223
13. Safety architectures trend – H/W 224
14. ISO 26262 2nd edition hot topics – fail operational 224
15. ISO 26262 2nd edition hot topics –FTTI 225
16. ISO 26262 2nd edition hot topics – dependent failures 225
17. ISO 26262 2nd edition hot topics – sensors / 1 226
18. ISO 26262 2nd edition hot topics – sensors / 2 226
19. ISO 26262 2nd edition hot topics – sensors / 3 227
20. ISO 26262 2nd edition hot topics – fault injection 227
21. ISO 26262 2ndedition hot topics – tools 228
22. Safety architectures trend – S/W 228
23. ISO 26262 2nd edition hot topics – S/W 229
24. ISO 26262 2nd edition hot topics – SOTIF / 1 229
25. ISO 26262 2nd edition hot topics – SOTIF / 2 230
26. ISO 26262 2nd edition hot topics – SOTIF / 3 230
27. ISO 26262 2nd edition hot topics – SOTIF / 4 231
28. ISO 26262 2nd edition hot topics – SOTIF / 5 231
29. ISO 26262 2nd edition hot topics – SOTIF / 6 232
30. ISO 26262 2nd edition hot topics – SOTIF / 7 232

SAFETY AND SECURITY 233
31. Security 233
32. Safety & Security 233

A ZOOM ON H/W RANDOM FAILURES 234
33. Safety & Security in ISO 26262 2nd edition 234
34. HW random failures in a nutshell / 1 234
35. HW random failures in a nutshell / 2 235
36. FuSa architectural and absolute metrics 235
37. Challenges to mitigate random failures 236
38. Challenges to mitigate failure rates 236
39. Challenges to mitigate failure rates – hard faults 237
40. Challenges to mitigate failure rates – radiation effects 237
41. Challenges to identify failure modes 238
42. Challenges to identify failure modes - ISO 26262 view 238
43. Challenges to identify failure modes - examples 239
44. Challenges to measure safe faults 239
45. Challenges to measure safe faults - examples 240
46. Challenges to measure safe faults – vulnerability factors 240
47. Challenges to measure safe faults – AVF example 241
48. Challenges to measure safe faults 241
49. Challenges to detect random failures 242
50. Challenges to detect random failures - options 242
51. Challenges to detect random failures – ECC pitfalls 243
52. Challenges to detect random failures – in field tests 243
53. Challenges to detect random failures – SW Tests 244
54. Challenges to detect random failures – "safe island" 244
55. Challenges to detect random failures – LCLS 245
56. Challenges to detect random failures – other redund 245
57. Fault tolerant systems in automotive – examples / 1 246
58. Fault tolerant systems in automotive – examples / 2 246

CONCLUSION 247
59. Key takeaways 247
Bibliography 248

"A robot may not injure a human being or, through inaction, allow a human being to come to harm." (Isaac Asimov, *Runaround. Eventually Runaround*, 1942): the new world of software-defined autonomous things brings both technical challenges and liability concerns.

Autonomous things are composed of electronic platforms with many sensing inputs and also with many complex processing elements: today an autonomous driving platform involves tens of processor cores and millions of S/W code lines. As a consequence, H/W and S/W may go wrong and this may cause hazards if no countermeasures are taken. On top of H/W and S/W failures, they operate in a very complex environment (with many variants) as also in a multi-agent scenario. Last but not least, the increase of connectivity opens possibility for security attacks.

As a consequence, today's engineers of autonomous things work in a context in which they need to consider several potential issues. The following pages give an overview of the instruments (standards, techniques, methodologies, tools, and models) that engineers can use to plan for countermeasures.

1 Outline

- Introduction
- Focus on ISO 26262
- Safety & Security
- A zoom on H/W random failures
- Conclusions

Summer School IEEE-UNIPI Enabling Technologies for IoT - Riccardo Mariani - Intel

INTRODUCTION

2 The third phase of IoT

Summer School IEEE-UNIPI Enabling Technologies for IoT - Riccardo Mariani - Intel

3

The essence of an Autonomous IoT platform

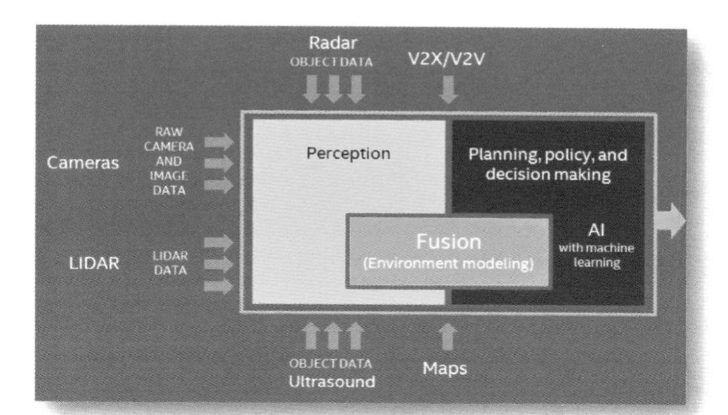

Intelligent eyes

- Vision.

Intelligent & powerful brain

- Perception and fusion.
- Modeling and planning.
- Decision making.
- Machine learning.

We need high-performance computing, flexibility, and programmability.

The levels of autonomy (example of autonomous cars)

SAE Level	Name	Narrative Definition	Execution of Steering and Acceleration/ Deceleration	Monitoring of Driving Environment	Fallback Performance of Dynamic Driving Task	System Capability (Driving Modes)
***Human driver* monitors the driving environment**						
0	No Automation	The full-time performance by the human driver of all aspects of the dynamic driving task, even when enhanced by warning or intervention systems	Human Driver	Human Driver	Human Driver	N/A
1	Driver Assistance	The driving mode-specific execution by a driver assistance system of either steering or acceleration/deceleration using information about the driving environment and with the expectation that the human driver perform all remaining aspects of the dynamic driving task	Human Driver and System	Human Driver	Human Driver	Some Driving Modes
2	Partial Automation	The driving mode-specific execution by one or more driver assistance systems of both steering and acceleration/deceleration using information about the driving environment and with the expectation that the human driver perform all remaining aspects of the dynamic driving task	System	Human Driver	Human Driver	Some Driving Modes

Source - SAE (J3016) Autonomy Level

5 

The levels of autonomy (example of autonomous cars)

SAE Level	Name	Narrative Definition	Execution of Steering and Acceleration/ Deceleration	Monitoring of Driving Environment	Fallback Performance of Dynamic Driving Task	System Capability (Driving Modes)
Automated driving system ("system") monitors the driving environment						
3	Conditional Automation	The driving mode-specific performance by an automated driving system of all aspects of the dynamic driving task with the expectation that the human driver will respond appropriately to a request to intervene	System	**System**	Human Driver	Some Driving Modes
4	High Automation	The driving mode-specific performance by an automated driving system of the dynamic driving task, even if a human driver does not respond appropriately to a request to intervene	System	System	**System**	Some Driving Modes
5	Full Automation	The full-time performance by an automated driving system of all aspects of the dynamic driving task under all roadway and environmental conditions that can be managed by a human driver	System	System	System	**All Driving Modes**

Source - SAE (J3016) Autonomy Levels

Summer School IEEE-UNIPI Enabling Technologies for IoT - Riccardo Mariani - Intel 

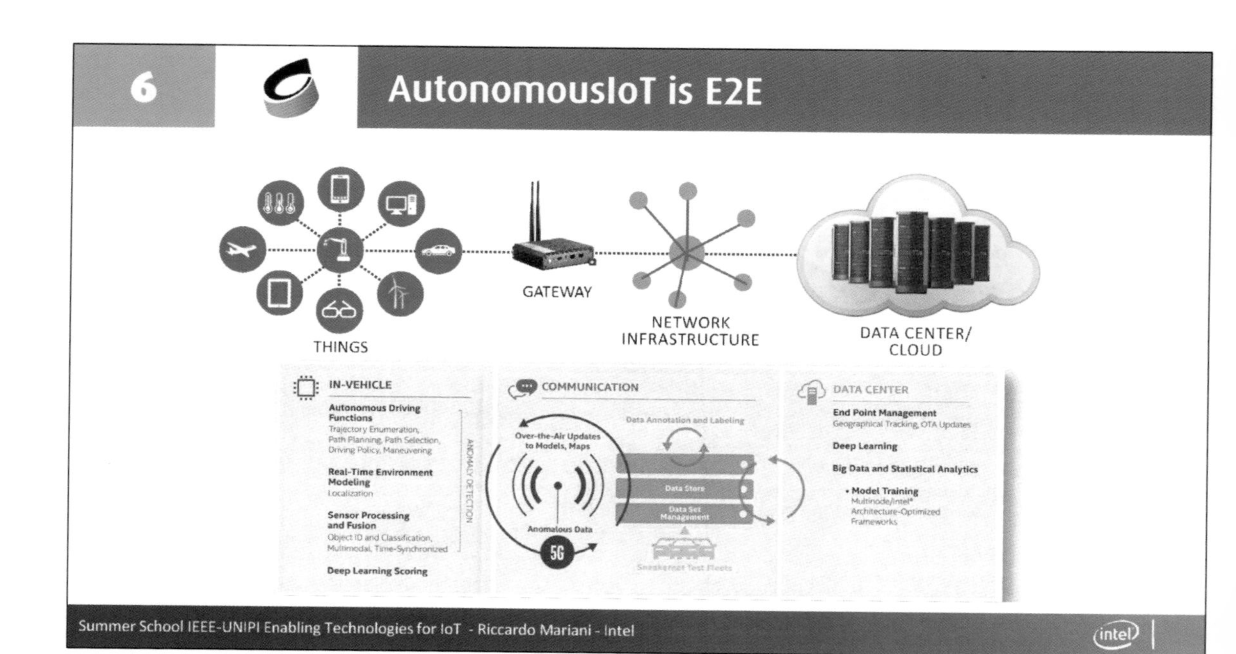

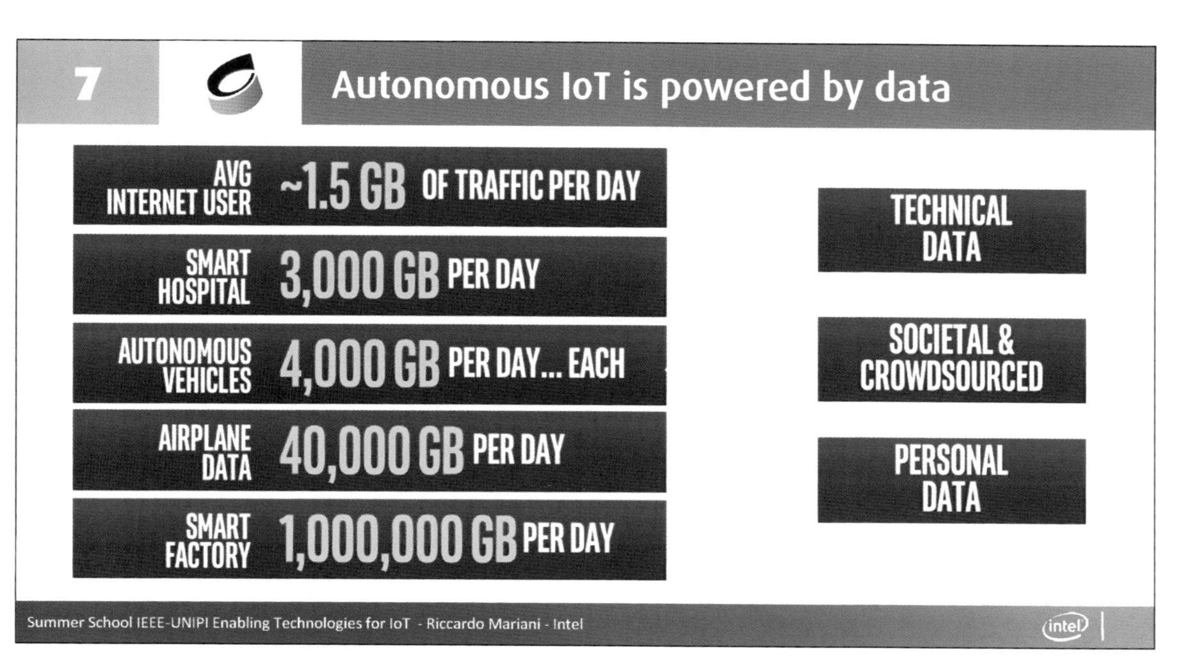
7
Autonomous IoT is powered by data
AVG INTERNET USER ~1.5 GB OF TRAFFIC PER DAY
SMART HOSPITAL 3,000 GB PER DAY
AUTONOMOUS VEHICLES 4,000 GB PER DAY... EACH
AIRPLANE DATA 40,000 GB PER DAY
SMART FACTORY 1,000,000 GB PER DAY
TECHNICAL DATA
SOCIETAL & CROWDSOURCED
PERSONAL DATA
Summer School IEEE-UNIPI Enabling Technologies for IoT - Riccardo Mariani - Intel
intel

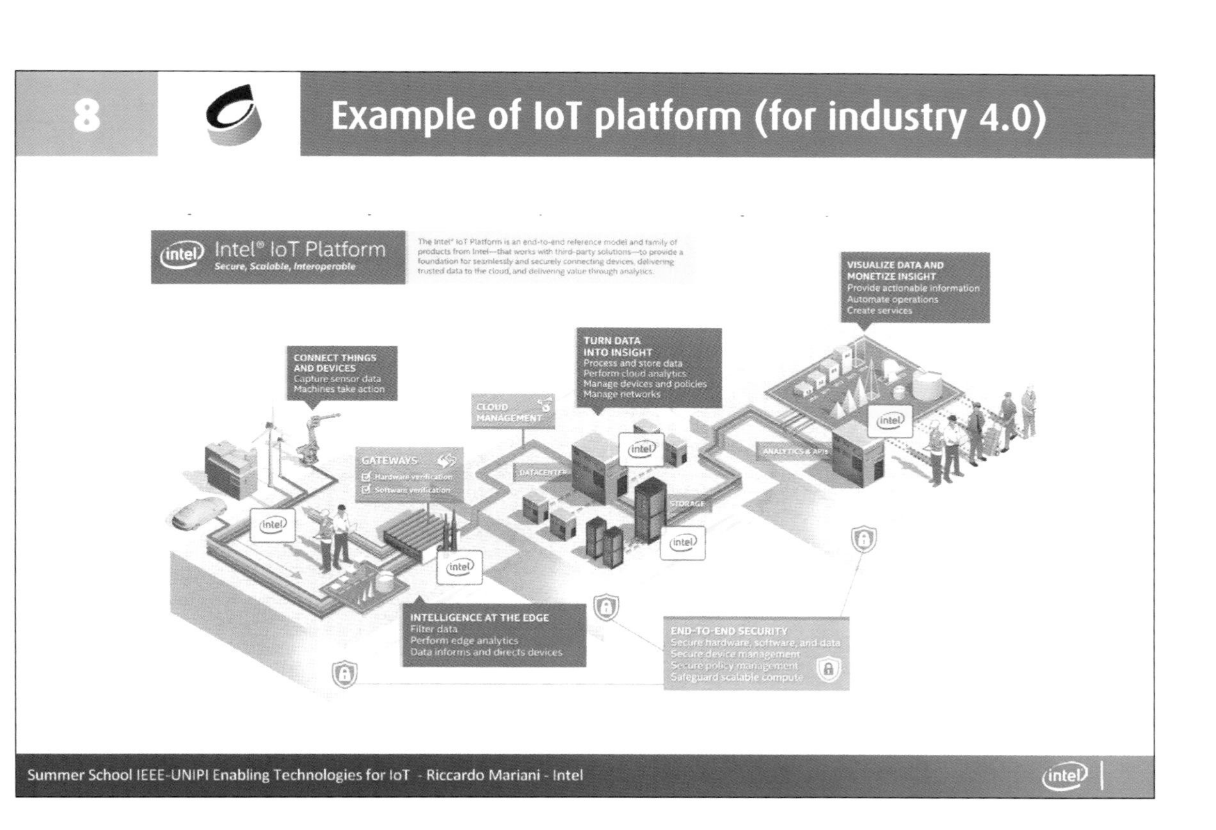
8
Example of IoT platform (for industry 4.0)
intel Intel® IoT Platform
Secure, Scalable, Interoperable
The Intel® IoT Platform is an end-to-end reference model and family of products from Intel—that works with third-party solutions—to provide a foundation for seamlessly and securely connecting devices, delivering trusted data to the cloud, and delivering value through analytics.
VISUALIZE DATA AND MONETIZE INSIGHT
Provide actionable information
Automate operations
Create services
TURN DATA INTO INSIGHT
Process and store data
Perform cloud analytics
Manage devices and policies
Manage networks
CONNECT THINGS AND DEVICES
Capture sensor data
Machines take action
CLOUD MANAGEMENT
GATEWAYS
Hardware verification
Software verification
DATACENTER
STORAGE
ANALYTICS & APIs
INTELLIGENCE AT THE EDGE
Filter data
Perform edge analytics
Data informs and directs devices
END-TO-END SECURITY
Secure hardware, software, and data
Secure device management
Secure policy management
Safeguard scalable compute
Summer School IEEE-UNIPI Enabling Technologies for IoT - Riccardo Mariani - Intel
intel

9 

The challenges of Autonomous IoT

H/W and S/W complexity is expected to grow at least by a factor of 20 in the next few years, so higher risk of **failures**....

Connectivity brings **security** threats....

...and autonomous systems are expected to detect & **control** failures !

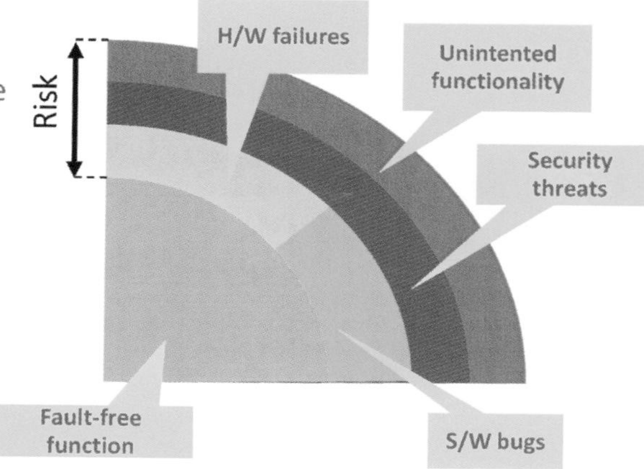

Summer School IEEE-UNIPI Enabling Technologies for IoT - Riccardo Mariani - Intel 

10 

Functional Safety

The absence of unreasonable risk due to hazards caused by malfunctioning behavior of E/E systems

Systematic failures
(Bugs in S/W, H/W design and Tools)

Random H/W failures
(permanent faults, transient faults occurring while using the system)

Ruled by International Standards
setting the "state of art" (for liability)

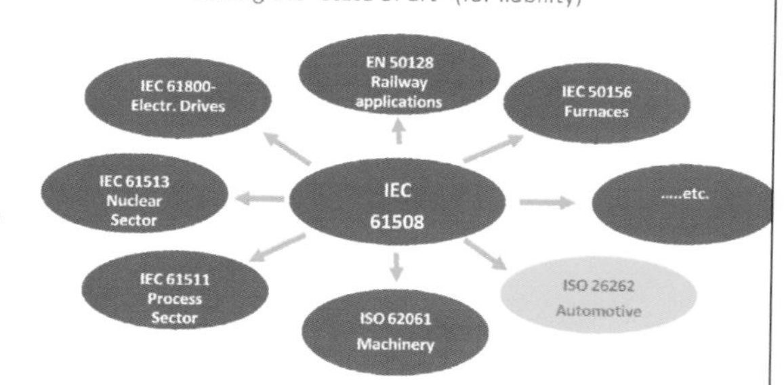

Summer School IEEE-UNIPI Enabling Technologies for IoT - Riccardo Mariani - Intel 

FOCUS ON ISO 26262

11 ISO 26262 2nd edition

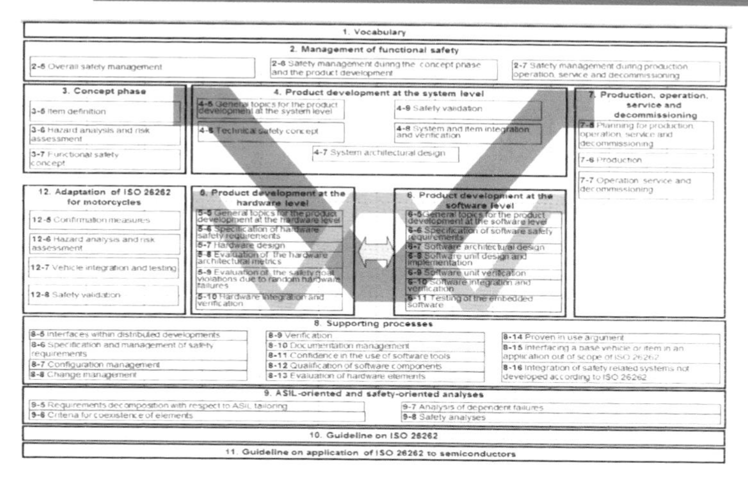

Source – ISO/DIS ISO 26262:2018

12 ISO 26262 2nd edition roadmap

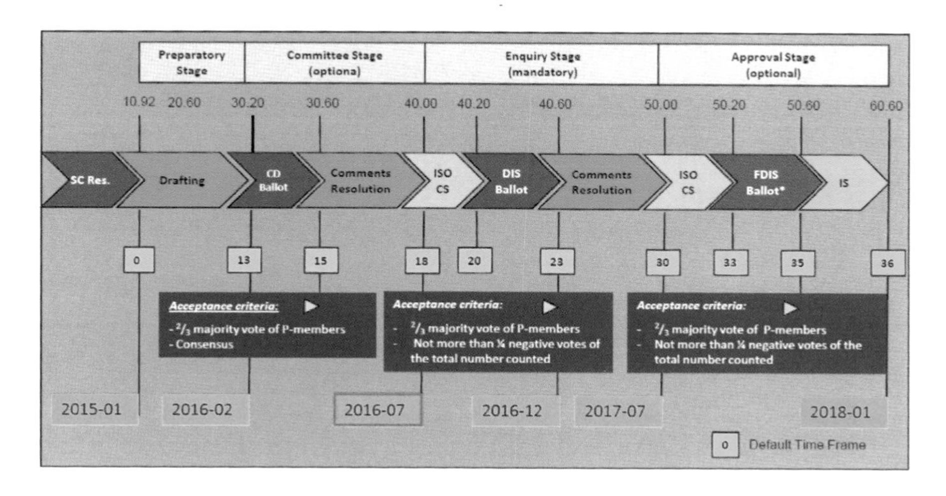

Source – ISO TC22/SC32/WG8

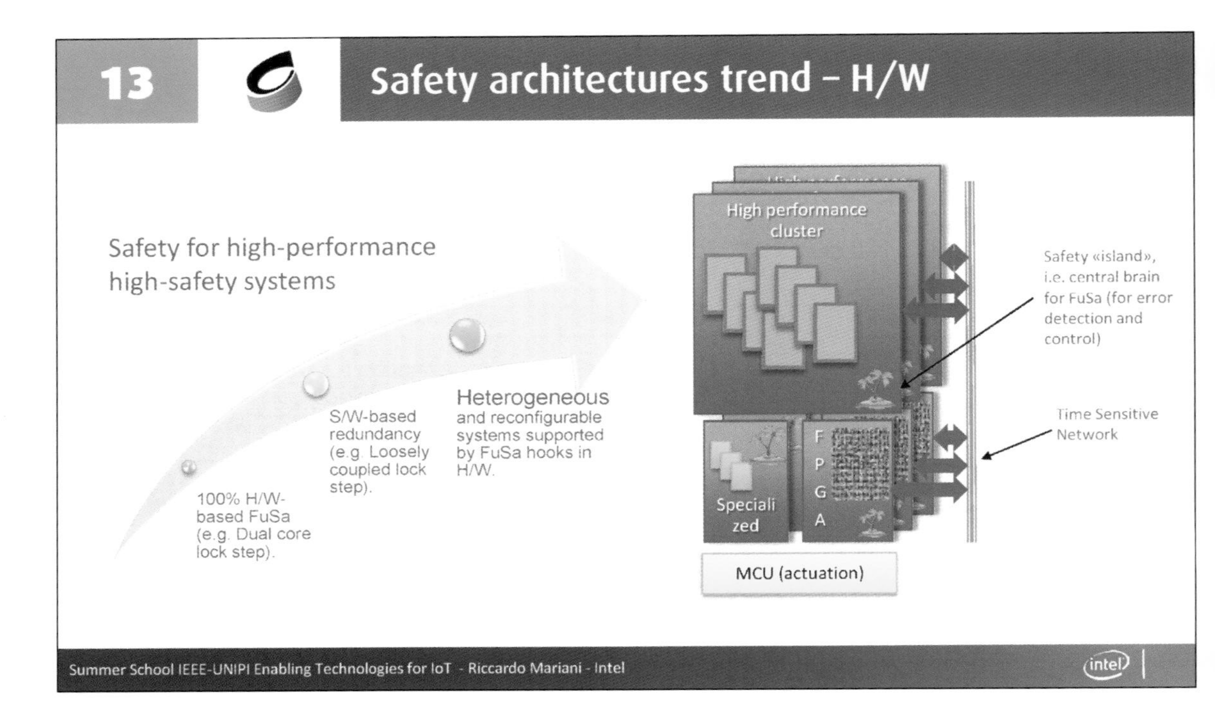

14

ISO 26262 2nd edition hot topics – fail operational

«Guidance for system development with safety-related availability requirements», i.e. fail operational systems including the possibility of degrading to a lower ASIL in case of a fault...

ISO 26262-10, clause 12.2.5.5

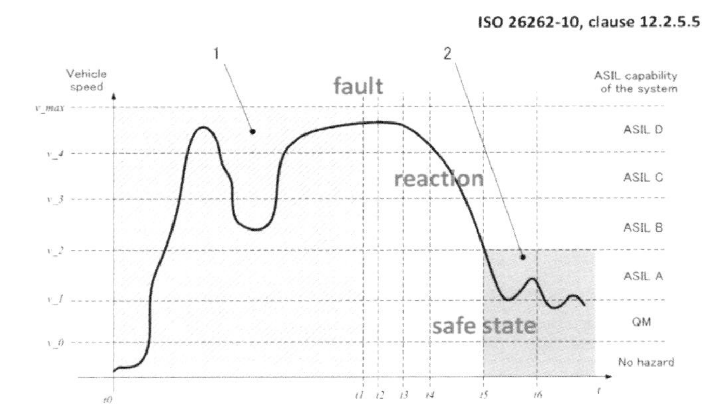

Summer School IEEE-UNIPI Enabling Technologies for IoT - Riccardo Mariani - Intel

intel

15

ISO 26262 2nd edition hot topics –FTTI

ISO 26262-10, clause 4.4.2.2

Time to detection and time to reaction are important characteristics of the safety architecture

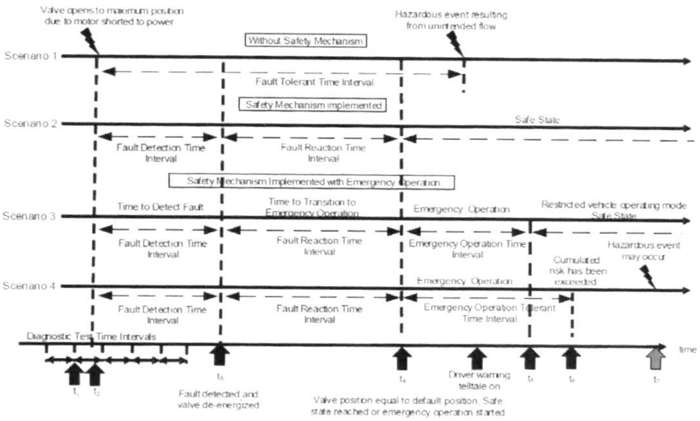

Valve opens to maximum position due to motor shorted to power

Hazardous event resulting from unintended flow

Without Safety Mechanism

Scenario 1

Fault Tolerant Time Interval

Safety Mechanism implemented

Scenario 2

Safe State

Fault Detection Time Interval

Fault Reaction Time Interval

Safety Mechanism Implemented with Emergency Operation

Time to Detect Fault

Time to Transition to Emergency Operation

Emergency Operation

Restricted vehicle operating mode Safe State

Scenario 3

Fault Detection Time Interval

Fault Reaction Time Interval

Emergency Operation Time Interval

Hazardous event may occur

Cumulated risk has been exceeded

Scenario 4

Emergency Operation

Fault Detection Time Interval

Fault Reaction Time Interval

Emergency Operation Tolerant Time Interval

Diagnostic Test Time Intervals

time

Fault detected and valve de-energized

Driver warning telltale on

Valve position equal to default position. Safe state reached or emergency operation started

16

ISO 26262 2nd edition hot topics – dependent failures

ISO 26262-11 has a chapter dedicated to dependent failures of semiconductor devices.

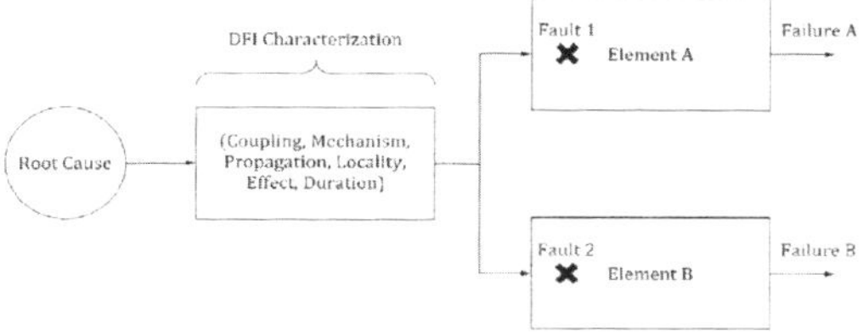

DFI Characterization

Root Cause

(Coupling, Mechanism, Propagation, Locality, Effect, Duration)

Fault 1 Element A Failure A

Fault 2 Element B Failure B

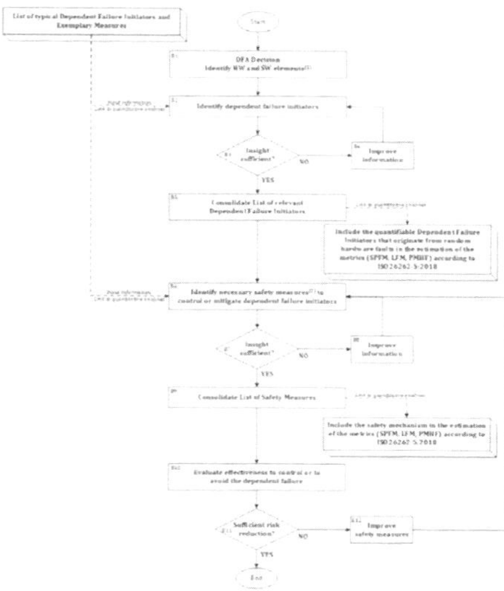

ISO 26262-11, clause 4.7.6

17 

ISO 26262 2nd edition hot topics – sensors / 1

ISO 26262-11 has a chapter dedicated to sensors and transducers, including MEMS.

ISO 26262-11, clause 5.5.2

Technical Specification	Failure mode	Description
Offset	Offset outside of specified range	Transducer output is offset from the ideal value in the absence of stimulus (input energy)
	Offset error over temperature	Offset error over temperature is beyond specified limits
	Offset drift	Offset value changes over time
Dynamic Range	Out of range	Transducer output is outside of prescribed operational range
Sensitivity (Gain)	Sensitivity too high/low	Sensitivity deviates beyond specified limits
	Stuck at	Sensitivity is zero due to mechanical and electrical failure (e. g. particle short, stiction)
	Nonparametric sensitivity	Sensitivity deviates from a mathematical relationship within its specified range including discontinuities or clipping of output response
	Noise, poor repeatability	Variable threshold required to overcome dynamic noise floor
	Sensitivity error over temperature	Sensitivity deviates beyond specified limits over temperature
NOTE 1 Possible effects at system level includes: Inaccurate switching threshold, Changes in switching threshold over temperature, Changes in switching threshold over time, Loss of function, Inaccurate switching threshold, Phase shift (leading, lagging), Changes in duty cycle, Variation of output switching threshold, Changes in switching threshold over temperature, Phase shift over temperature, Changes in duty cycle over temperature		

18 

ISO 26262 2nd edition hot topics – sensors / 2

Dependent failures initiators for sensors and transducers

ISO 26262-11, clause 5.5.3.2

Table 56 — Dependent failures initiators for sensors/transducers

DFI classes defined in 4.7.5	Examples
Dependent failures initiators due to random hardware faults of shared resources	Common calibration and/or configuration resources (e.g. eFUSE to control the CMOS based image sensor)
Dependent failures initiators due to random physical root causes	Temporal Noise or Fixed Pattern Noise
Systematic dependent failures initiators due to environmental conditions	Extended exposure to excessive heat, humidity, or strong sunlight Electrostatic discharge
Systematic dependent failures initiators due to development faults	Wrong design of image sensor
Systematic dependent failures initiators due to manufacturing faults	Sensor manufacturing defects
Systematic dependent failures initiators due to installation faults	Magnetic sensor target wheel mounted off axis (runout) Incorrect positioning of mirror in image sensor

19 ISO 26262 2nd edition hot topics – sensors / 3

Safety mechanisms and safety analyses

ISO 26262-11, clause 5.5.4

Table 57 —Example of Safety Mechanisms for Sensors/Transduce

Safety mechanism/measure	See overview of techniques	Notes
Sealed Proof mass Filter	5.5.4.1	MEMS specific implementation.
Redundant Diaphragms	5.5.4.2	MEMS specific on-chip calibrated reference.
Offset cancellation	5.5.4.3	Allows for offset optimization.
Transducer specific self-test	5.5.4.4	Various methods to test signal path integrity.
Automatic Gain Control	5.5.4.5	Accounts for low levels of environmental stimulu dynamic range.
Sensitivity adjustment	5.5.4.6	Allows for sensitivity centering.
MEMS specific non E/E safety measures	5.5.4.7	Measures that assess physical properties of MEM

ISO 26262-11, Annex D

Table D.5 — Example of quantitative analysis in the case of fine granularity – mission parts

Part	Sub part	Safety-related Component or No Safety related Component	Failure Mode	Potential Effect of Failure Mode in Absence of Safety Mechanism (SM) on IC level[a]	Fault Model[b]	Failure distribution	Failure rate (FIT)	Amount of Safe Faults	Safety mechanism(s) preventing the violation of the safety requirement	Failure mode coverage with respect to violation of safety requirement	Residual or Single Point Fault failure rate / FIT	Safety mechanism(s) to prevent latent faults	Failure mode coverage with respect to Latent failures	Latent Multiple Point Fault failure rate / FIT
	Regulator core	SR	Output voltage higher than a predefined high threshold of the prescribed range (i.e. Over voltage - OV)	Regulated voltage higher than VA_OV	P	14 %	1.49E-03	0 %	SM2	99.9 %	1.49E-06	SM2	100 %	0.00E+00
		SR	Output voltage lower than a predefined low threshold of the prescribed range (i.e. Under voltage - UV)	Regulated voltage lower than VA_UV	P	14 %	1.49E-03	0 %	SM1	99.9 %	1.49E-06	SM1	100 %	0.00E+00
		SR	Output voltage affected by spikes	Regulated voltage out of the expected range (VA_UV-VA_OV)	P	14 %	1.49E-03	0 %	SM1 SM2	99.9 %	1.49E-06	SM1 SM2	100 %	0.00E+00
		SR	Output voltage oscillation within the prescribed range	Regulated voltage within the expected range but with low accuracy	P	14 %	1.49E-03	0 %	SM3	97.0 %	4.46E-05	SM3	100 %	0.00E+00
			Output voltage		P									

20 ISO 26262 2nd edition hot topics – fault injection

Fault injection at different abstraction level...

ISO 26262-11, clause 4.8

4.8.2 Characteristics or variables of fault injection

With respect to fault injection, the following information can help the verification planning:

— the description and rationale about fault models, and related level of abstraction;

— type of safety mechanism including required confidence level;

— observation points and diagnostic points;

— fault site, fault list; and

— workload used during fault injection.

4.8.3 Fault injection results

Results of fault injection can be used to verify the safety concept and the underlying assumptions as listed in 4.8.1 (e.g. the effectiveness of the safety mechanism, the diagnostic coverage and amount of safe faults).

NOTE 1 Evidence of fault injection is maintained in the case of inspections during functional safety audits.

NOTE 2 An exact correspondence between the fault simulated and the fault identified in the safety analysis (e.g. for open faults) may not always exist. In such a case refinement of the safety analysis can be based on the results of other representative faults (e.g. N-detect testing as reported in 5.1.10.2).

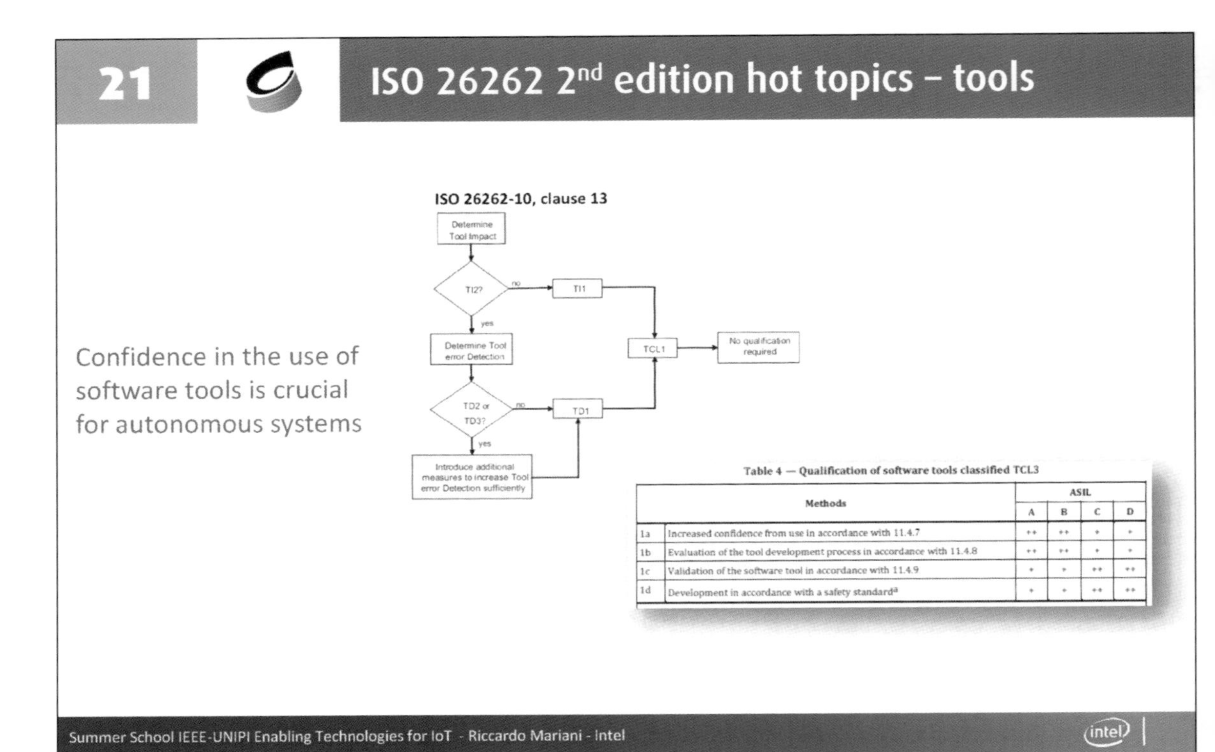

Table 4 — Qualification of software tools classified TCL3

	Methods	ASIL			
		A	B	C	D
1a	Increased confidence from use in accordance with 11.4.7	++	++	+	+
1b	Evaluation of the tool development process in accordance with 11.4.8	++	++	+	+
1c	Validation of the software tool in accordance with 11.4.9	+	+	++	++
1d	Development in accordance with a safety standard[a]	+	+	++	++

22  Safety architectures trend – S/W

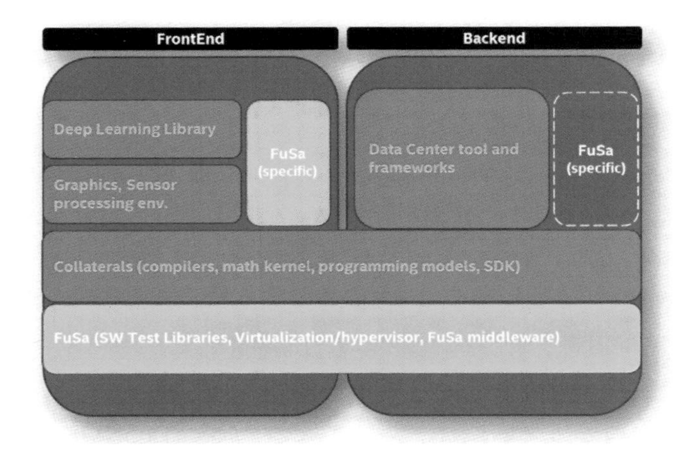

- A full stack of S/W components from front-end to back-end...... + FuSa.
- Need to support multiple and heterogeneous programming models.
- Complex OS compliant with Functional Safety standards.

Summer School IEEE-UNIPI Enabling Technologies for IoT - Riccardo Mariani - Intel

intel

23 

ISO 26262 2nd edition hot topics – S/W

Trending toward a combination of different techniques to achieve Functional Safety of S/W components, e.g.:

- Failure mode and effects analysis (FMEA) at least at S/W architecture level.
- ASIL/SIL decomposition and/or sand-boxing (e.g. for legacy F/W).
- S/W statistical analysis.
- Etc...

ISO 26262 2nd edition:
New examples on ASIL decomposition

Note: QM+QM cannot lead to ASIL

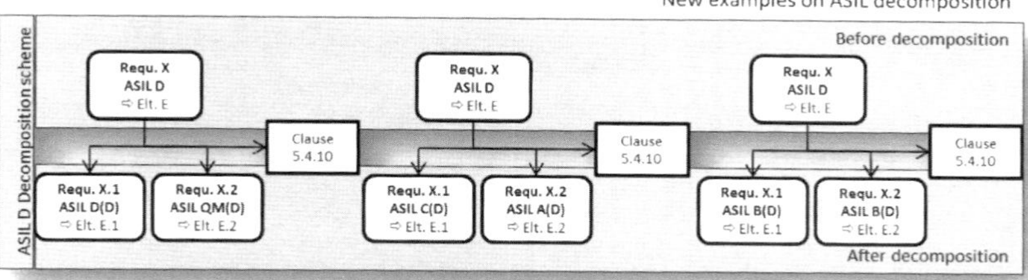

24 

ISO 26262 2nd edition hot topics – SOTIF / 1

Autonomous systems that rely on sensing, can miss their goal and cause safety violations in absence of H/W or S/W failure...

ISO WD PAS 21448 1
Safety of Intended Functionality (SOTIF)

Causal factor of hazard with example		Scope
System	E/E System failures	ISO 26262
	Unintended behaviour without fault or failure (including E/E System performance limit)	SOTIF guidance
	Foreseeable user misuses	SOTIF Annex
External factor	Security violation	Mentioned as necessary to ensure a safe behaviour, but not addressed in this document (See ISO21434:XXXX or SAE J3061)
	Impact from active Infrastructure and/or vehicle to vehicle communication.	Can be necessary for a safe behaviour but not fully addressed in this document (ISO 20077 can be considered)
	Impact from car surroundings (other users, "passive" infrastructure, environment: weather, EMC...)	Included in SOTIF scope

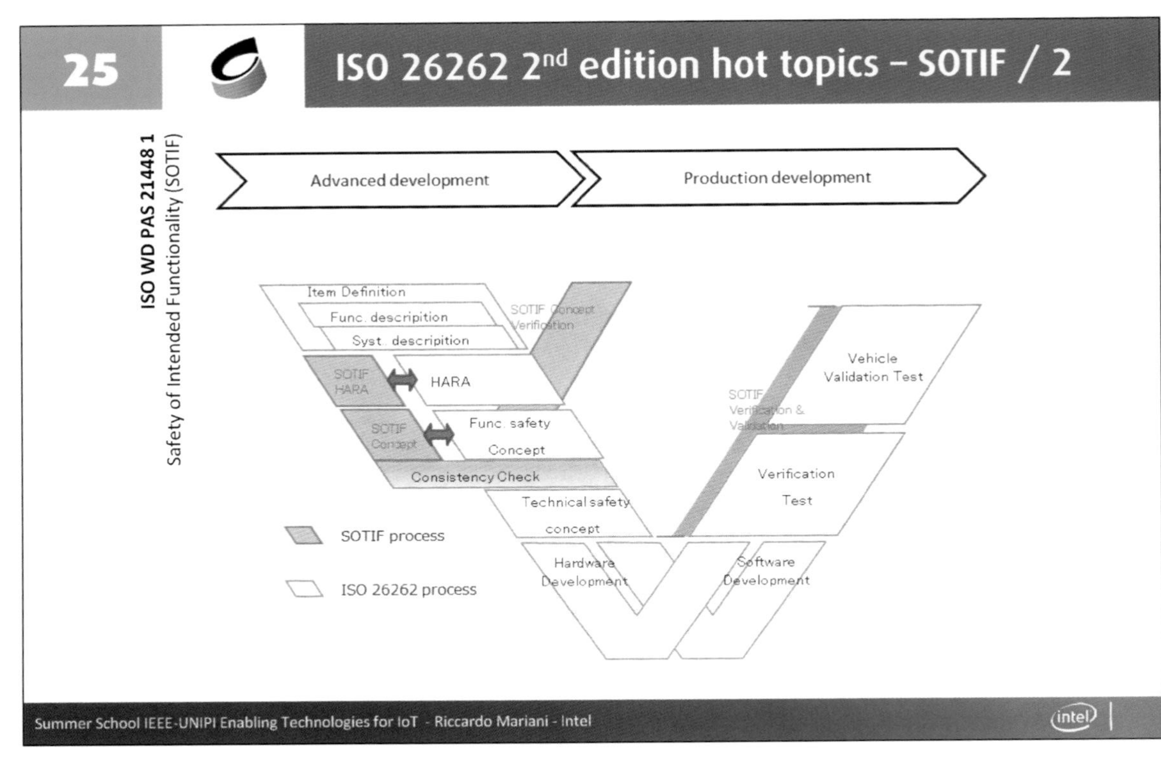
25
ISO 26262 2nd edition hot topics – SOTIF / 2
ISO WD PAS 21448 1
Safety of Intended Functionality (SOTIF)
Advanced development
Production development
Item Definition
Func. description
Syst. description
SOTIF Concept Verification
SOTIF HARA
HARA
SOTIF Concept
Func. safety Concept
Consistency Check
Technical safety concept
SOTIF Verification & Validation
Vehicle Validation Test
Verification Test
Hardware Development
Software Development
SOTIF process
ISO 26262 process
Summer School IEEE-UNIPI Enabling Technologies for IoT - Riccardo Mariani - Intel
intel

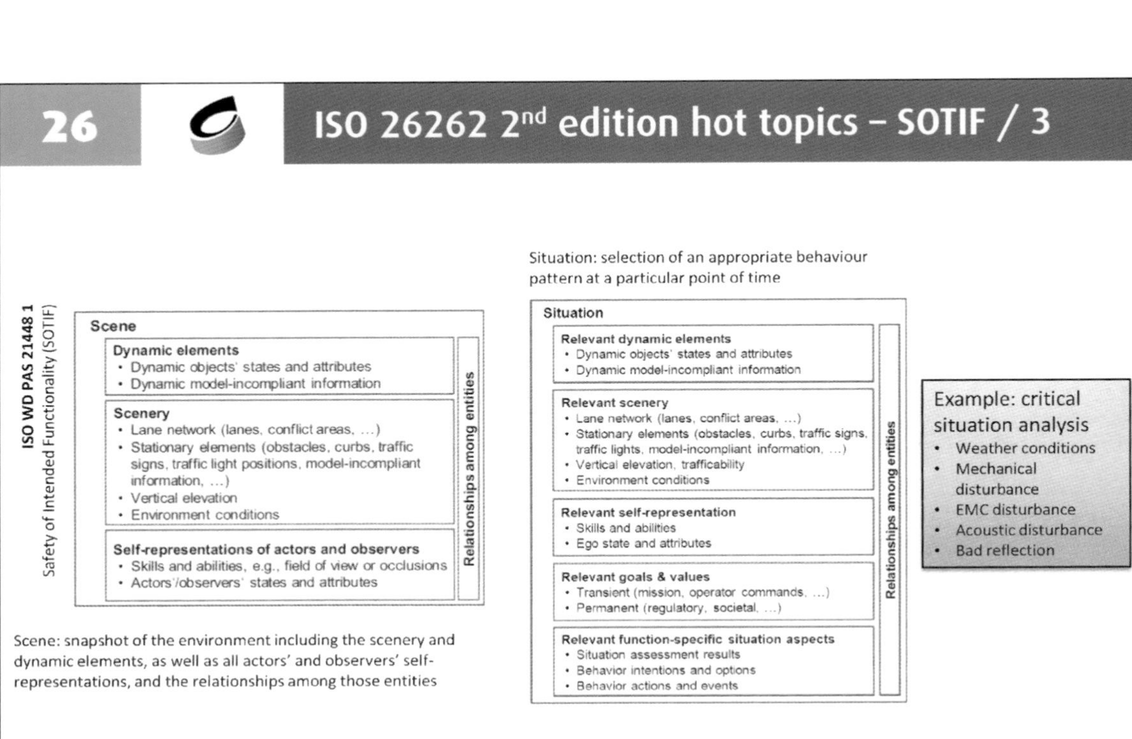
26
ISO 26262 2nd edition hot topics – SOTIF / 3
ISO WD PAS 21448 1
Safety of Intended Functionality (SOTIF)
Scene
Dynamic elements
• Dynamic objects' states and attributes
• Dynamic model-incompliant information
Scenery
• Lane network (lanes, conflict areas, ...)
• Stationary elements (obstacles, curbs, traffic signs, traffic light positions, model-incompliant information, ...)
• Vertical elevation
• Environment conditions
Self-representations of actors and observers
• Skills and abilities, e.g., field of view or occlusions
• Actors'/observers' states and attributes
Relationships among entities
Scene: snapshot of the environment including the scenery and dynamic elements, as well as all actors' and observers' self-representations, and the relationships among those entities
Situation: selection of an appropriate behaviour pattern at a particular point of time
Situation
Relevant dynamic elements
• Dynamic objects' states and attributes
• Dynamic model-incompliant information
Relevant scenery
• Lane network (lanes, conflict areas, ...)
• Stationary elements (obstacles, curbs, traffic signs, traffic lights, model-incompliant information, ...)
• Vertical elevation, trafficability
• Environment conditions
Relevant self-representation
• Skills and abilities
• Ego state and attributes
Relevant goals & values
• Transient (mission, operator commands, ...)
• Permanent (regulatory, societal, ...)
Relevant function-specific situation aspects
• Situation assessment results
• Behavior intentions and options
• Behavior actions and events
Relationships among entities
Example: critical situation analysis
• Weather conditions
• Mechanical disturbance
• EMC disturbance
• Acoustic disturbance
• Bad reflection
Summer School IEEE-UNIPI Enabling Technologies for IoT - Riccardo Mariani - Intel
intel

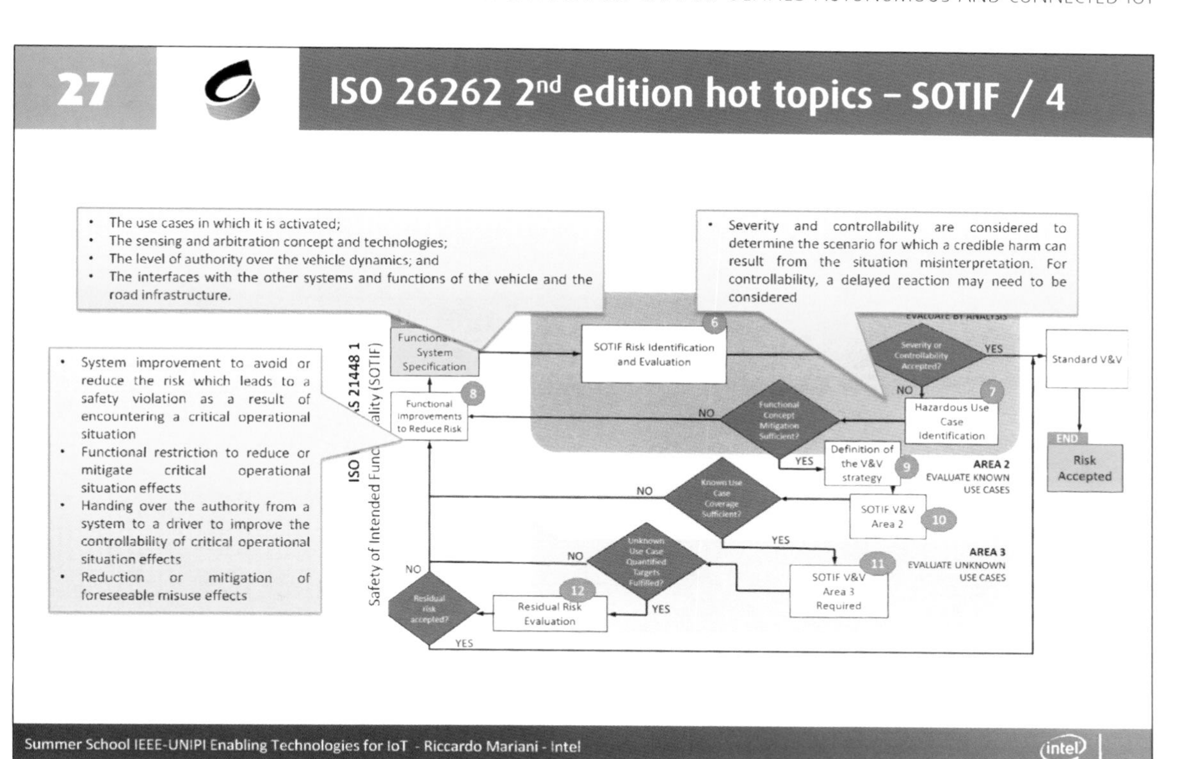
27
ISO 26262 2nd edition hot topics – SOTIF / 4
• The use cases in which it is activated;
• The sensing and arbitration concept and technologies;
• The level of authority over the vehicle dynamics; and
• The interfaces with the other systems and functions of the vehicle and the road infrastructure.
• Severity and controllability are considered to determine the scenario for which a credible harm can result from the situation misinterpretation. For controllability, a delayed reaction may need to be considered
• System improvement to avoid or reduce the risk which leads to a safety violation as a result of encountering a critical operational situation
• Functional restriction to reduce or mitigate critical operational situation effects
• Handing over the authority from a system to a driver to improve the controllability of critical operational situation effects
• Reduction or mitigation of foreseeable misuse effects
Safety of Intended Functionality (SOTIF)
SOTIF Risk Identification and Evaluation
Functional improvements to Reduce Risk
Standard V&V
Hazardous Use Case Identification
Risk Accepted
END
Definition of the V&V strategy
AREA 2
EVALUATE KNOWN USE CASES
SOTIF V&V Area 2
AREA 3
EVALUATE UNKNOWN USE CASES
SOTIF V&V Area 3 Required
Residual Risk Evaluation
YES
NO
Summer School IEEE-UNIPI Enabling Technologies for IoT - Riccardo Mariani - Intel
intel

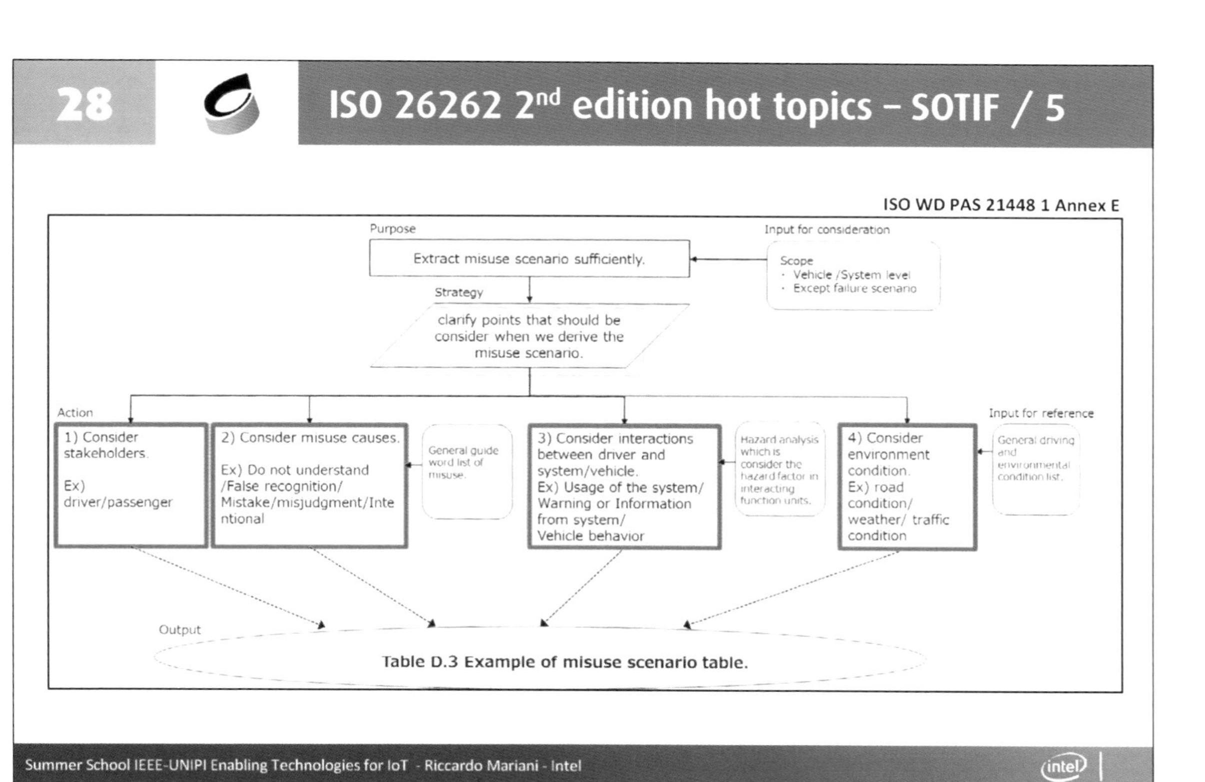
28
ISO 26262 2nd edition hot topics – SOTIF / 5
ISO WD PAS 21448 1 Annex E
Purpose
Extract misuse scenario sufficiently.
Input for consideration
Scope
· Vehicle /System level
· Except failure scenario
Strategy
clarify points that should be consider when we derive the misuse scenario.
Action
1) Consider stakeholders.
Ex) driver/passenger
2) Consider misuse causes.
Ex) Do not understand /False recognition/ Mistake/misjudgment/Intentional
General guide word list of misuse.
3) Consider interactions between driver and system/vehicle.
Ex) Usage of the system/ Warning or Information from system/ Vehicle behavior
Hazard analysis which is consider the hazard factor in interacting function units.
4) Consider environment condition.
Ex) road condition/ weather/ traffic condition
Input for reference
General driving and environmental condition list.
Output
Table D.3 Example of misuse scenario table.
Summer School IEEE-UNIPI Enabling Technologies for IoT - Riccardo Mariani - Intel
intel

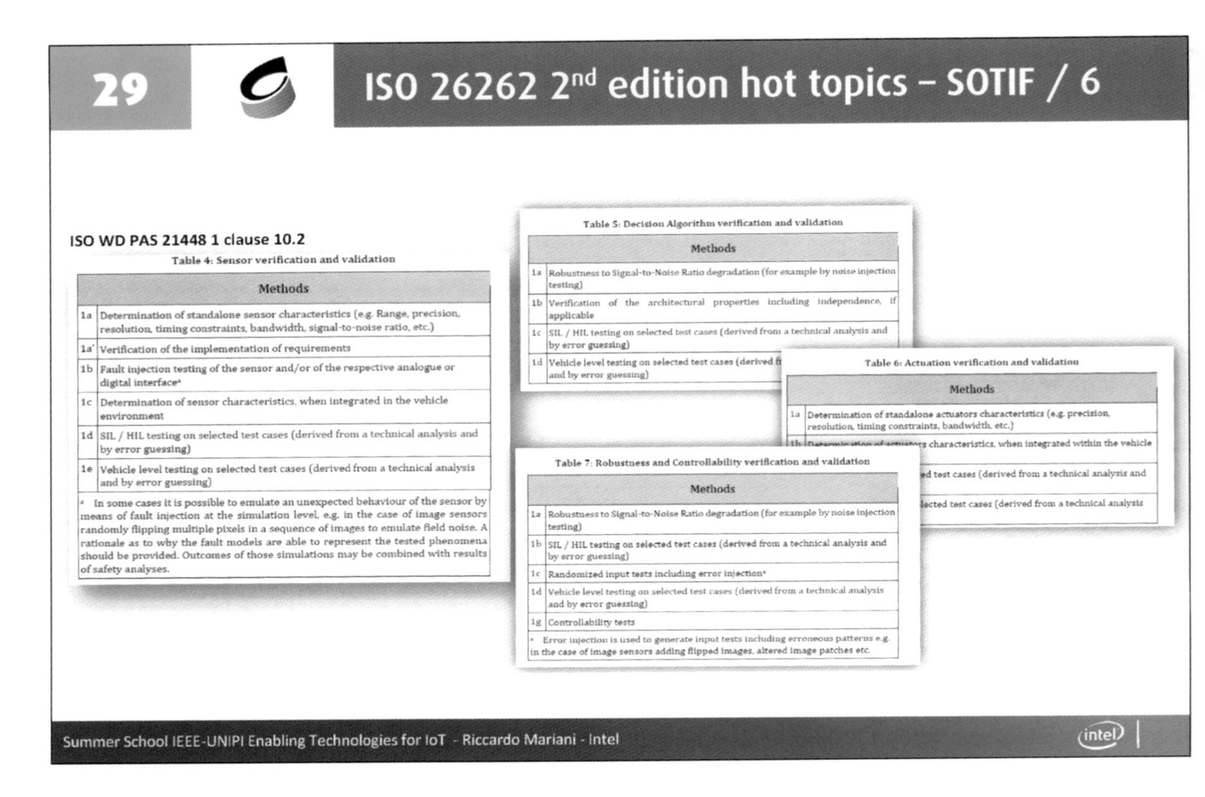
29
ISO 26262 2nd edition hot topics – SOTIF / 6
ISO WD PAS 21448 1 clause 10.2
Table 4: Sensor verification and validation
Methods
1a Determination of standalone sensor characteristics (e.g. Range, precision, resolution, timing constraints, bandwidth, signal-to-noise ratio, etc.)
1a' Verification of the implementation of requirements
1b Fault injection testing of the sensor and/or of the respective analogue or digital interface[a]
1c Determination of sensor characteristics, when integrated in the vehicle environment
1d SIL / HIL testing on selected test cases (derived from a technical analysis and by error guessing)
1e Vehicle level testing on selected test cases (derived from a technical analysis and by error guessing)
[a] In some cases it is possible to emulate an unexpected behaviour of the sensor by means of fault injection at the simulation level, e.g. in the case of image sensors randomly flipping multiple pixels in a sequence of images to emulate field noise. A rationale as to why the fault models are able to represent the tested phenomena should be provided. Outcomes of those simulations may be combined with results of safety analyses.
Table 5: Decision Algorithm verification and validation
Methods
1a Robustness to Signal-to-Noise Ratio degradation (for example by noise injection testing)
1b Verification of the architectural properties including independence, if applicable
1c SIL / HIL testing on selected test cases (derived from a technical analysis and by error guessing)
1d Vehicle level testing on selected test cases (derived
and by error guessing)
Table 6: Actuation verification and validation
Methods
1a Determination of standalone actuators characteristics (e.g. precision, resolution, timing constraints, bandwidth, etc.)
characteristics, when integrated within the vehicle
ed test cases (derived from a technical analysis and
lected test cases (derived from a technical analysis
Table 7: Robustness and Controllability verification and validation
Methods
1a Robustness to Signal-to-Noise Ratio degradation (for example by noise injection testing)
1b SIL / HIL testing on selected test cases (derived from a technical analysis and by error guessing)
1c Randomized input tests including error injection[a]
1d Vehicle level testing on selected test cases (derived from a technical analysis and by error guessing)
1g Controllability tests
[a] Error injection is used to generate input tests including erroneous patterns e.g. in the case of image sensors adding flipped images, altered image patches etc.
Summer School IEEE-UNIPI Enabling Technologies for IoT - Riccardo Mariani - Intel
intel

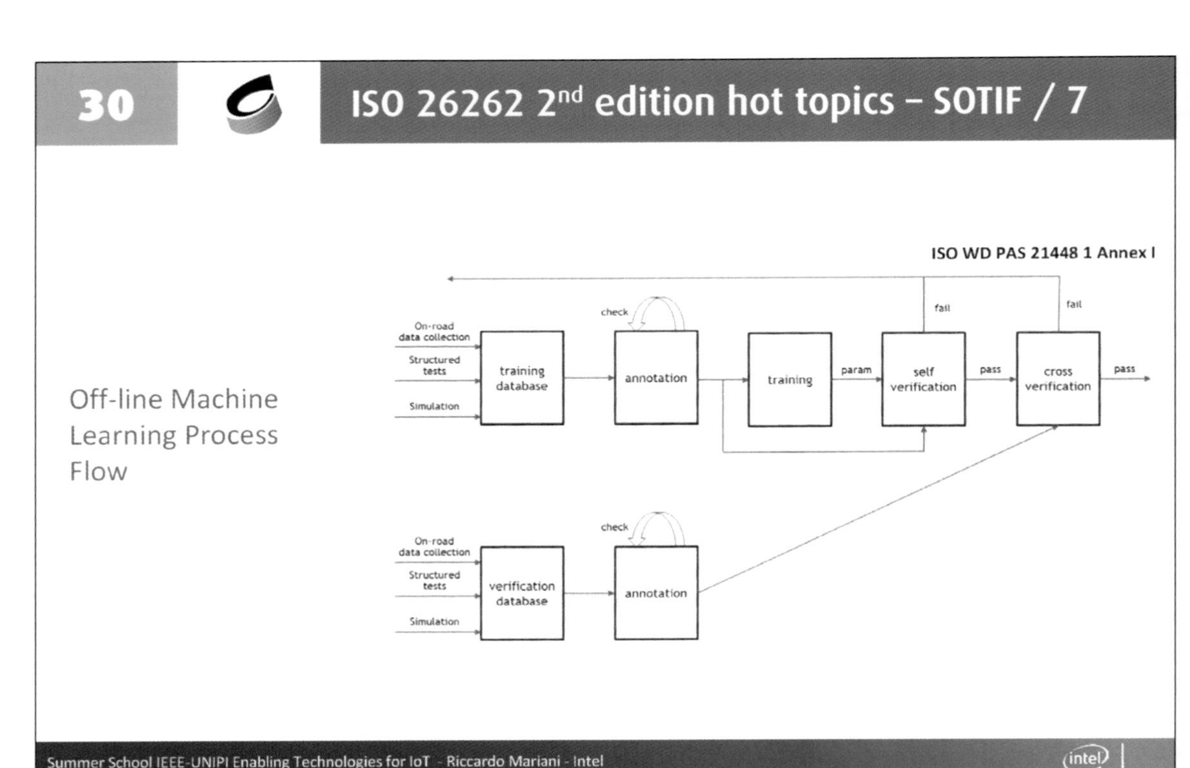
30
ISO 26262 2nd edition hot topics – SOTIF / 7
ISO WD PAS 21448 1 Annex I
Off-line Machine Learning Process Flow
On-road data collection
Structured tests
Simulation
training database
check
annotation
training
param
self verification
pass
cross verification
pass
fail
fail
On-road data collection
Structured tests
Simulation
verification database
check
annotation
Summer School IEEE-UNIPI Enabling Technologies for IoT - Riccardo Mariani - Intel
intel

SAFETY AND SECURITY

31  **Security**

The prevention of risks related to malicious intrusions, through computer and/or communication networks

SAE J3101 - Requirements for Hardware-Protected Security for Ground Vehicle Applications

- Secure Boot.
- Secure Storage.
- Secure Execution Environment.
- OTA, authentication.
- ...

SAE J3061 - Cybersecurity Guidebook for Cyber-Physical Vehicle Systems

- Attacks enumeration.
- Threat analysis.
- Reduction of attacks surface.
- Security testing.
- ...

Summer School IEEE-UNIPI Enabling Technologies for IoT - Riccardo Mariani - Intel 

32  **Safety & Security**

Security threats affecting FuSa	Security measures helping FuSa
FuSa measures as vulnerable points for security attacks	FuSa measures helping security

Hazard Analysis And Risk Assessment (HARA)

Hazards and Safety Goals from HARA

Safety-Related Threats from TARA

Threat Analysis and Risk Assessment (TARA)

Functional safety concept

Fu..Sa. And Security Concepts defined

Cybersecurity concept

Safety concept review

Fu..Sa. And Security Concepts consolidation

Cybersecurity concept review

Source: Guidance on Potential Safety-Cybersecurity Interface Points (Informative Only)

Summer School IEEE-UNIPI Enabling Technologies for IoT - Riccardo Mariani - Intel 

33  Safety & Security in ISO 26262 2nd edition

ISO 26262-4 :2018
Cybersecurity related changes

6.4.1 Specification of the technical safety requirements

a) the system configuration and calibration requirements

NOTE 1 The ability to reconfigure a system for alternative applications is a strategy to reuse existing systems.

EXAMPLE 3 Calibration data (see ISO 26262-6:2018, Annex C) is frequently used to customise electronic engine control units for alternate vehicles.

NOTE 2 The cybersecurity concept (cybersecurity strategy and requirements), if applicable.

6.4.2 Safety mechanisms

a) the safety mechanisms that enable the system's contribution to achieve or maintain the safe state of the item;

NOTE 4 This includes arbitration in the case of multiple control requests from safety mechanisms.

NOTE 5 Cybersecurity can be considered

Summer School IEEE-UNIPI Enabling Technologies for IoT - Riccardo Mariani - Intel 

A ZOOM ON H/W RANDOM FAILURES

34  HW random failures in a nutshell / 1

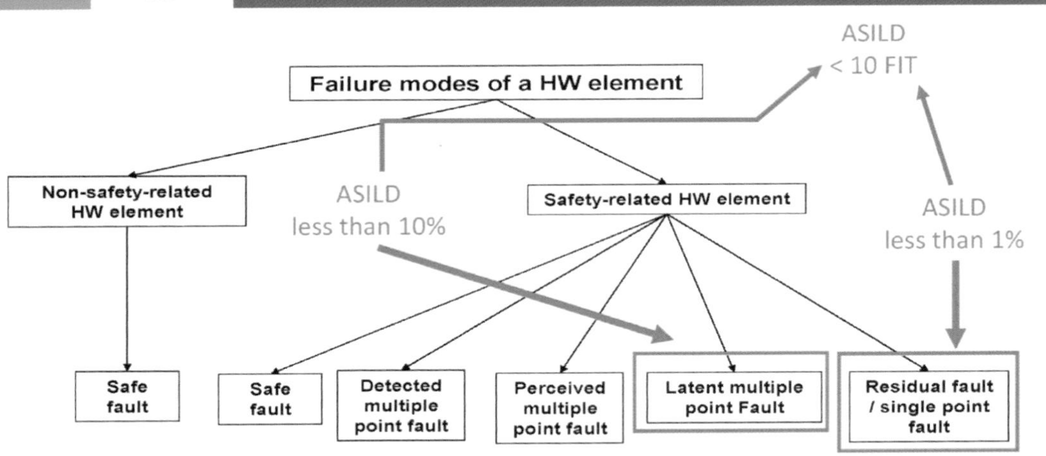

Multiple point fault = at least the combination of a fault in the mission logic and a fault in the corresponding safety mechanism

Summer School IEEE-UNIPI Enabling Technologies for IoT - Riccardo Mariani - Intel 

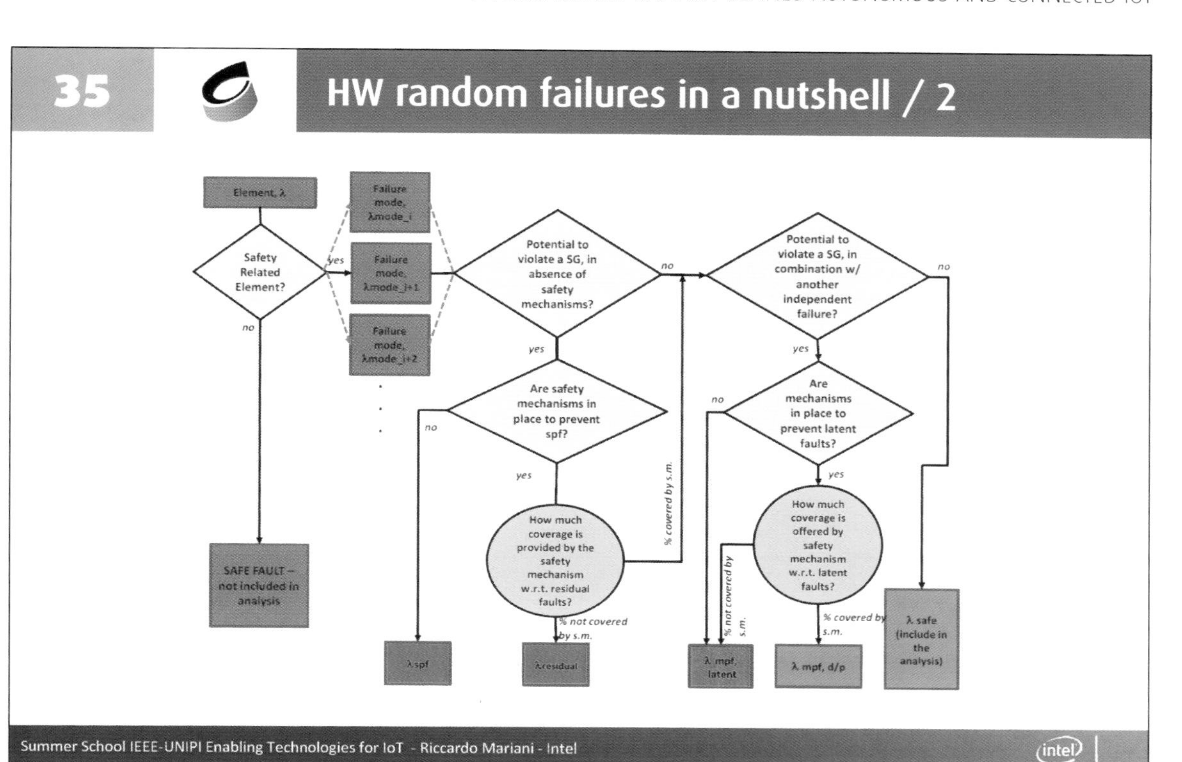

36

FuSa architectural and absolute metrics

$$\text{Single Point Fault metric} = \frac{\sum_{\text{Safety related HW elements}} (\lambda_{MPF} + \lambda_S)}{\sum_{\text{Safety related HW elements}} \lambda}$$

$$\text{Latent Fault metric} = \frac{\sum_{\text{safety related HW elements}} (\lambda_{MPF\ \text{perceived or detected}} + \lambda_S)}{\sum_{\text{safety related HW elements}} (\lambda - \lambda_{SPF} - \lambda_{RF})}$$

	ASIL B	ASIL C	ASIL D
Single point faults metric	≥ 90 %	≥ 97 %	≥ 99 %

	ASIL B	ASIL C	ASIL D
Latent faults metric	≥ 60 %	≥ 80 %	≥ 90 %

$$PMHF = \lambda_{SPF} + \lambda_{RF} + \lambda_{IF,\,DPF} \times 0.5 \times (\lambda_{SM,\,DPF,\,Latent} \times T_{lifetime}) + \lambda_{IF,\,DPF} \times 0.5 \times (\lambda_{SM,\,DPF,\,Detected} \times T_{service}) + \lambda_{SM,\,DPF} \times 0.5 \times (\lambda_{IF,\,DPF,\,Latent} \times T_{lifetime}) + \lambda_{SM,\,DPF} \times 0.5 \times (\lambda_{IF,\,DPF,\,Latent} \times T_{service})$$

ASIL	PMHF
D	$< 10^{-8}\ h^{-1}$ (10 FIT)
C	$< 10^{-7}\ h^{-1}$ (100 FIT)
B	$< 10^{-7}\ h^{-1}$ (100 FIT)

37 

Challenges to mitigate random failures

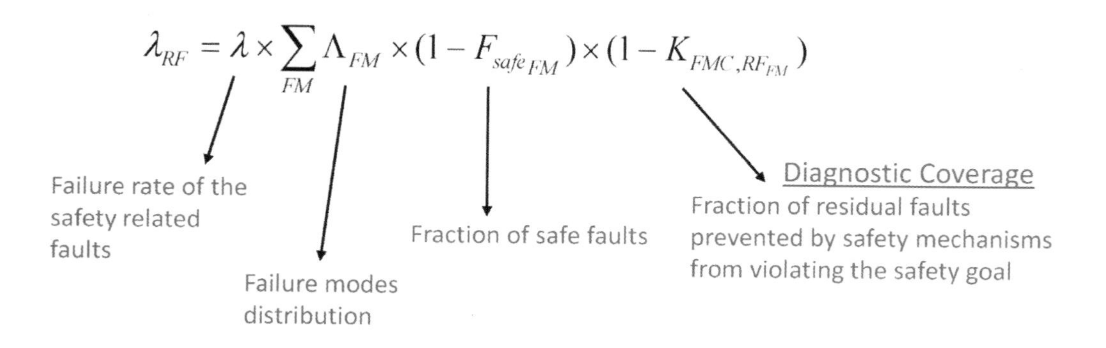

Summer School IEEE-UNIPI Enabling Technologies for IoT - Riccardo Mariani - Intel 

38 

Challenges to mitigate failure rates

$$\lambda_{RF} = \lambda \times \sum_{FM} \Lambda_{FM} \times (1 - F_{safe\,FM}) \times (1 - K_{FMC,RF_{FM}})$$

Failure rate of the safety related faults

Summer School IEEE-UNIPI Enabling Technologies for IoT - Riccardo Mariani - Intel 

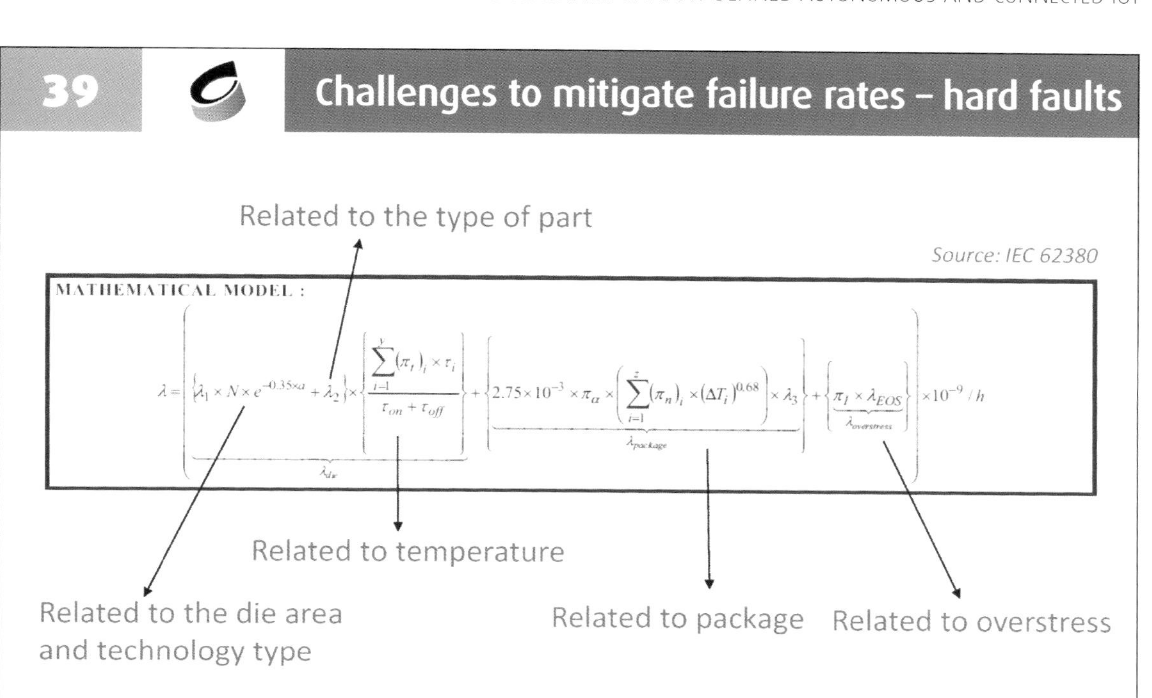

$$\lambda = \left\{ \underbrace{\left\{ \lambda_1 \times N \times e^{-0.35 \times a} + \lambda_2 \right\} \times \left\{ \frac{\sum_{i=1}^{y} (\pi_t)_i \times \tau_i}{\tau_{on} + \tau_{off}} \right\}}_{\lambda_{die}} + \left\{ \underbrace{2.75 \times 10^{-3} \times \pi_\alpha \times \left(\sum_{i=1}^{z} (\pi_n)_i \times (\Delta T_i)^{0.68} \right) \times \lambda_3}_{\lambda_{package}} \right\} + \left\{ \underbrace{\pi_I \times \lambda_{EOS}}_{\lambda_{overstress}} \right\} \right\} \times 10^{-9} / h$$

Challenges to mitigate failure rates – radiation effects

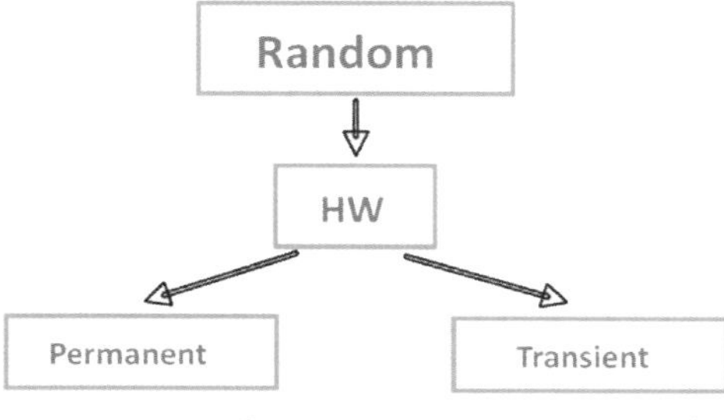

Example:

- *typical Soft Error Rate for a SRAM = 1000 FIT/Mbit*
- *If we consider 128KBytes RAM, the total raw failure rate is equal to λ = 1000 FIT !!!*

Hard errors

- *SEL = single event latch-up*

Soft errors

- SEU = single event upset
- MBU = multiple bit upset
- SET = single event transient

41 

Challenges to identify failure modes

$$\lambda_{RF} = \lambda \times \sum_{FM} \Lambda_{FM} \times (1 - F_{safe_{FM}}) \times (1 - K_{FMC,RF_{FM}})$$

Failure modes distribution

Summer School IEEE-UNIPI Enabling Technologies for IoT - Riccardo Mariani - Intel 

42 

Challenges to identify failure modes - ISO 26262 view

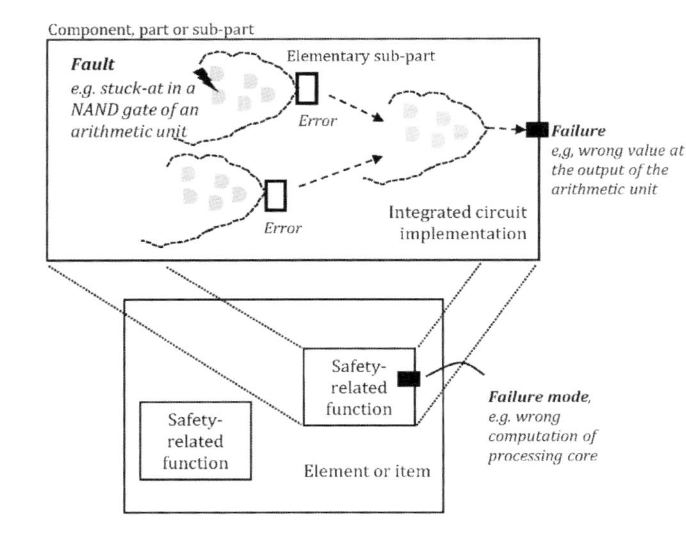

Faults need to be identified and connect them to error and failures

Source – ISO/DIS ISO 26262:2018

Summer School IEEE-UNIPI Enabling Technologies for IoT - Riccardo Mariani - Intel

43 Challenges to identify failure modes - examples

FM Name	Λ
Coprocessor Pipeline/datapath/regbank leading to wrong data computation	39.18%
Breakpoints leading to wrong program flow execution	10.37%
Instruction cache leading to wrong program flow execution	4.77%
Load/Store and MMU control leading to wrong program flow execution	4.60%
Bus Interface Unit leading to wrong data management	3.55%
Main Register Bank leading to wrong data computation	3.30%
Load/store queue & control - Watchpoints leading to wrong data computation	2.90%
Decoder unit (Dual) leading to wrong program execution	2.82%
Issue stage leading to wrong data computation	2.52%
Rename stage leading to wrong data computation	1.96%
MMU Buffers leading to wrong program flow execution	1.93%
Prefetch pipe and prediction logic leading to wrong program flow execution	1.89%
Store Buffer leading to wrong data management	1.65%
Instruction queue leading to wrong program flow execution	1.43%
Branch monitor and FIFO leading to wrong program flow execution	1.34%
MBIST interface leading to wrong program flow execution	1.26%
Dynamic Prediction queue leading to wrong program flow execution	1.23%
Execution unit (integer) 1/2 leading to wrong data computation	1.19%
MAC unit leading to wrong data computation	1.19%
....	....

Faults are NOT equi-distributed in failure modes !

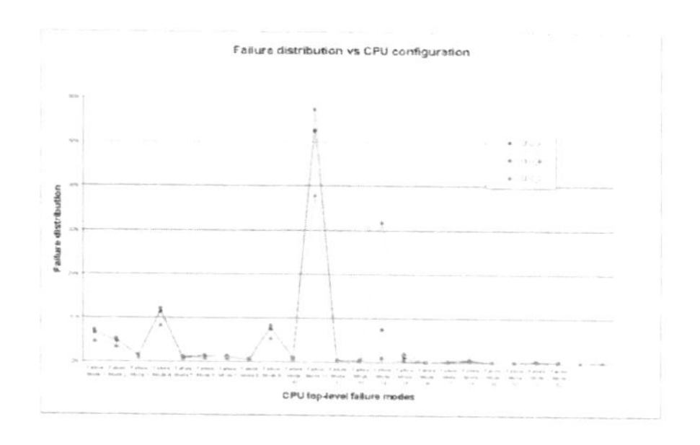

44 Challenges to measure safe faults

$$\lambda_{RF} = \lambda \times \sum_{FM} \Lambda_{FM} \times (1 - F_{safe_{FM}}) \times (1 - K_{FMC,RF_{FM}})$$

Fraction of safe faults

Challenges to measure safe faults - examples

$$F^{FM}_{SAFE\,tran} = F^{FM}_{SAFE\,tran,arch} \cup F^{FM}_{SAFE\,tran,sys}$$

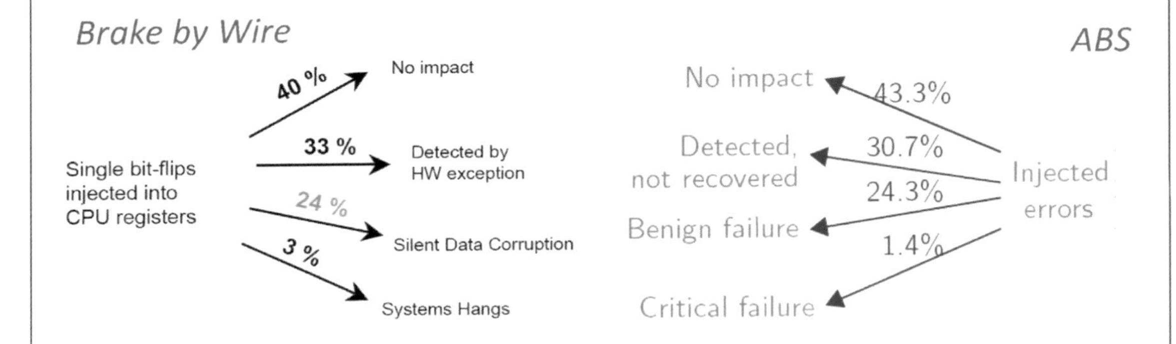

Summer School IEEE-UNIPI Enabling Technologies for IoT - Riccardo Mariani - Intel

46

Challenges to measure safe faults – vulnerability factors

AVF = the probability that a fault in a processor structure will result in a visible error in the final output of a program (introduced mainly by Shubhendu Sekhar Mukherjee, INTEL)

Table 2. AVF breakdown using Little's Law. # ACE inst = ACE IPC X ACE Latency. AVF ~= # ACE inst / # instruction queue entries.

Integer Benchmarks	ACE IPC	ACE Latency (cycles)	# ACE Inst	AVF	Floating Point Benchmarks	ACE IPC	ACE Latency (cycles)	#ACE Inst	AVF
bzip2-source	0.55	22	12	19%	ammp	0.23	92	21	33%
cc-200	0.57	18	10	16%	applu	0.82	21	18	27%
crafty	0.37	15	6	9%	apsi	0.31	31	9	15%
eon-kajiya	0.36	20	7	11%	art-110	0.68	37	25	40%
gap	0.78	17	13	21%	equake	0.26	12	3	5%
gzip-graphic	0.60	13	8	12%	facerec	0.41	7	3	5%
mcf	0.25	68	17	26%	fma3d	0.59	11	7	10%
parser	0.49	24	12	19%	galgel	1.10	21	23	35%
perlbmk-makerand	0.38	17	7	10%	lucas	1.23	17	21	33%
twolf	0.30	27	8	13%	mesa	0.47	16	8	12%
vortex_lendian3	0.42	22	9	15%	mgrid	1.28	10	13	21%
vpr-route	0.35	12	4	7%	sixtrack	0.66	20	13	21%
					swim	1.08	16	17	27%
					wupwise	1.60	13	20	31%
average	0.45	23	9	15%	average	0.77	23	14	23%

36th Annual International Symposium on Microarchitecture (MICRO), December 2003

Summer School IEEE-UNIPI Enabling Technologies for IoT - Riccardo Mariani - Intel

47 

Challenges to measure safe faults – AVF example

FM Name	Fsafe
Coprocessor pipeline/datapath/regbank leading to wrong data computation	99.80%
Breakpoints leading to wrong program flow execution	49.32%
Instruction cache leading to wrong program flow execution	48.37%
Load/Store and MMU control leading to wrong program flow execution	49.32%
Bus Interface Unit leading to wrong data management	49.32%
Main Register Bank leading to wrong data computation	49.32%
Load/store queue & control - Watchpoints leading to wrong data computation	49.32%
Decoder unit (Dual) leading to wrong program execution	48.82%
Issue stage leading to wrong data computation	49.32%
Rename stage leading to wrong data computation	49.32%
MMU Buffers leading to wrong program flow execution	49.32%
Prefetch pipe and prediction logic leading to wrong program flow execution	48.37%
Store Buffer leading to wrong data management	49.32%
Instruction queue leading to wrong program flow execution	48.37%
Branch monitor and FIFO leading to wrong program flow execution	72.52%
MBIST interface leading to wrong program flow execution	99.50%
Dynamic Prediction queue leading to wrong program flow execution	48.37%
Execution unit (integer) 1/2 leading to wrong data computation	82.48%
MAC unit leading to wrong data computation	99.97%
....	....

Faults are NOT all dangerous !

Summer School IEEE-UNIPI Enabling Technologies for IoT - Riccardo Mariani - Intel 

48 

Challenges to measure safe faults

IC supplier

End users (Tiers and OEM)

Soft error → Wrong line → Wrong object → Unintentional braking event

fault models | failure mode at image level | failure mode at algorithm level | hazard at item level

Summer School IEEE-UNIPI Enabling Technologies for IoT - Riccardo Mariani - Intel 

49 

Challenges to detect random failures

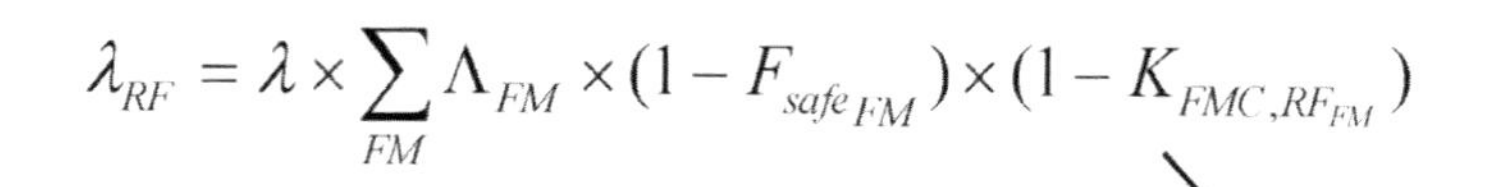

$$\lambda_{RF} = \lambda \times \sum_{FM} \Lambda_{FM} \times (1 - F_{safe_{FM}}) \times (1 - K_{FMC,RF_{FM}})$$

Diagnostic Coverage

Fraction of residual faults prevented by safety mechanisms from violating the safety goal

Summer School IEEE-UNIPI Enabling Technologies for IoT - Riccardo Mariani - Intel 

50 

Challenges to detect random failures - options

SW/system measures implemented by end user with or without reference implementation

FuSa-specific SW Test Libraries

FuSa-specific HW measures

«Instrinsic» HW measures (e.g. ECC)

Summer School IEEE-UNIPI Enabling Technologies for IoT - Riccardo Mariani - Intel 

51 Challenges to detect random failures – ECC pitfalls

How MBU can impact ECC diagnostic coverage....

$$k_{tran}^{RAM} = \frac{k_{SEC/DED\,tran}^{RAM} \times \lambda_{RAM} \cdot (1 - MBU\%) + k_{MBU\,tran}^{RAM} \times \lambda_{RAM} \cdot MBU\%}{\lambda_{RAM}}$$

- If we assume:

$k_{SEC/DED\,tran}^{RAM} = 100\%$

$k_{MBU\,tran}^{RAM} = 68\%$

MBU%	k_{tran}^{RAM}
10%	96,8%
1%	99,680%
0,1%	99,9680%
0,01%	99,99680%

52 Challenges to detect random failures – in field tests

- Availability of Memory and Logic in field tests is key:
  - Memory test = it shall be executed at power-on and at run-time on all safety critical memories and arrays, it shall cover different failure modes (including AF, ADF), it shall be quick (less than few ms)
  - Logic test = it shall be executed at least at power-on and possibly at run-time, at least on cores. It shall cover at least stuck-at faults.
  - Typical target coverage is at least 90%.
- Fault forecasting/prediction is an important feature (ref. ISO 26262-11)

53  Challenges to detect random failures – SW Tests

- SW Test Library (STL) executed periodically (e.g. each 100ms) to test a specific core
  - For permanent faults (ASILB, 90%)
- The diagnostic coverage of the STL shall be verified by means of fault injection at gate level
- STL shall be developed according to safety standards

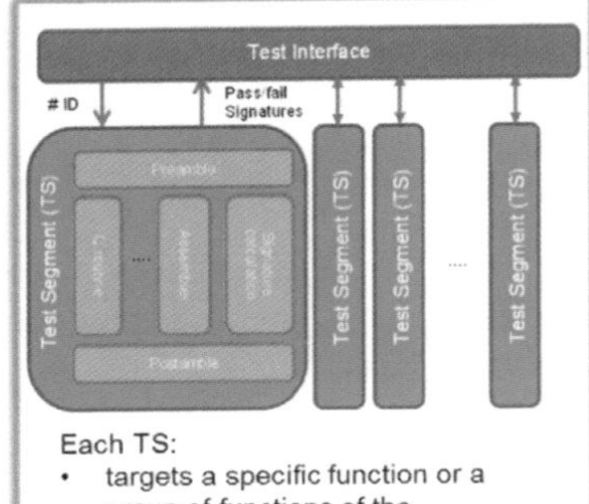

Summer School IEEE-UNIPI Enabling Technologies for IoT - Riccardo Mariani - Intel 

54  Challenges to detect random failures – "safe island"

«Safe island» or «safe MCU on chip», i.e. an/ASILD subsystem inside the SoC, typically implemented with two medium performance cores in HW lock-step, responsible of «FuSa housekeeping», interfacing with in vehicle networks and running SW to test the other portion of the SoC

- Smart Comparison
  - Fault discrimination
  - Functional downgrade
  - Latent faults detection
  - Timeout on system reset exiting
  - Common Cause Failures detection

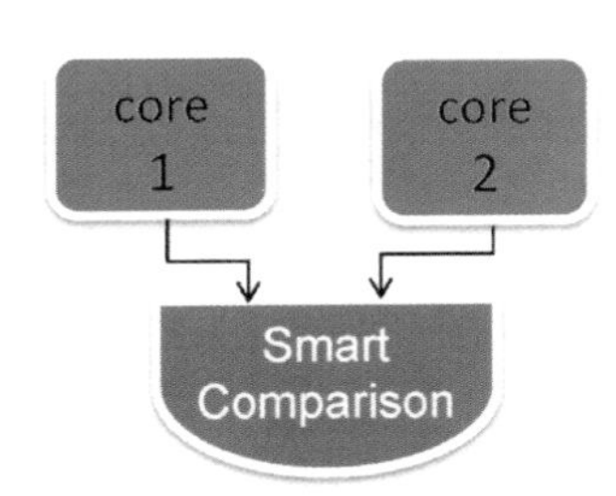

Summer School IEEE-UNIPI Enabling Technologies for IoT - Riccardo Mariani - Intel

55 

Challenges to detect random failures – LCLS

SW tasks fully replicated and running on two cores (global redundancy). Additional SW making the comparison task (usually done in an independent core).

Main ISO 26262 requirements
- Avoidance of common cause failures
- Verification of the diagnostic coverage achieved as a function of type/frequency of checks and type of algorithm

It seems an easy concept but it has many pitfalls (e.g. RTOS)

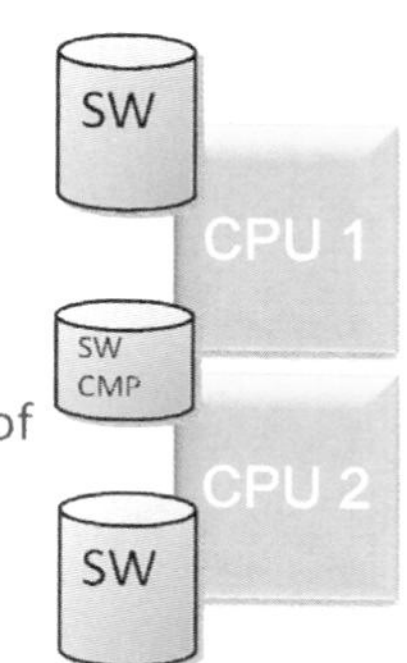

56 

Challenges to detect random failures – other redund

Proposed by the Association of the German Automotive Industry (VDA) as standardized monitoring concept for the engine control of gasoline and diesel engines.....

The max ASIL achievable by VDA E-Gas is typically limited to ASILC.

The coverage is highly application dependent – i.e. Level1/Level2 and even Level 3 shall be rewritten for each specific application

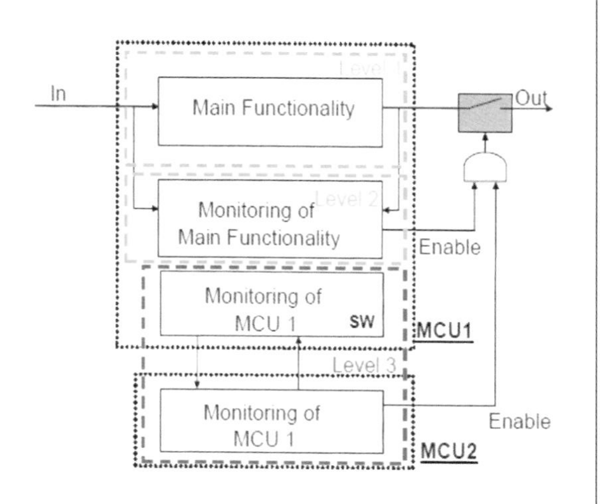

Source – VDA EGAS specification

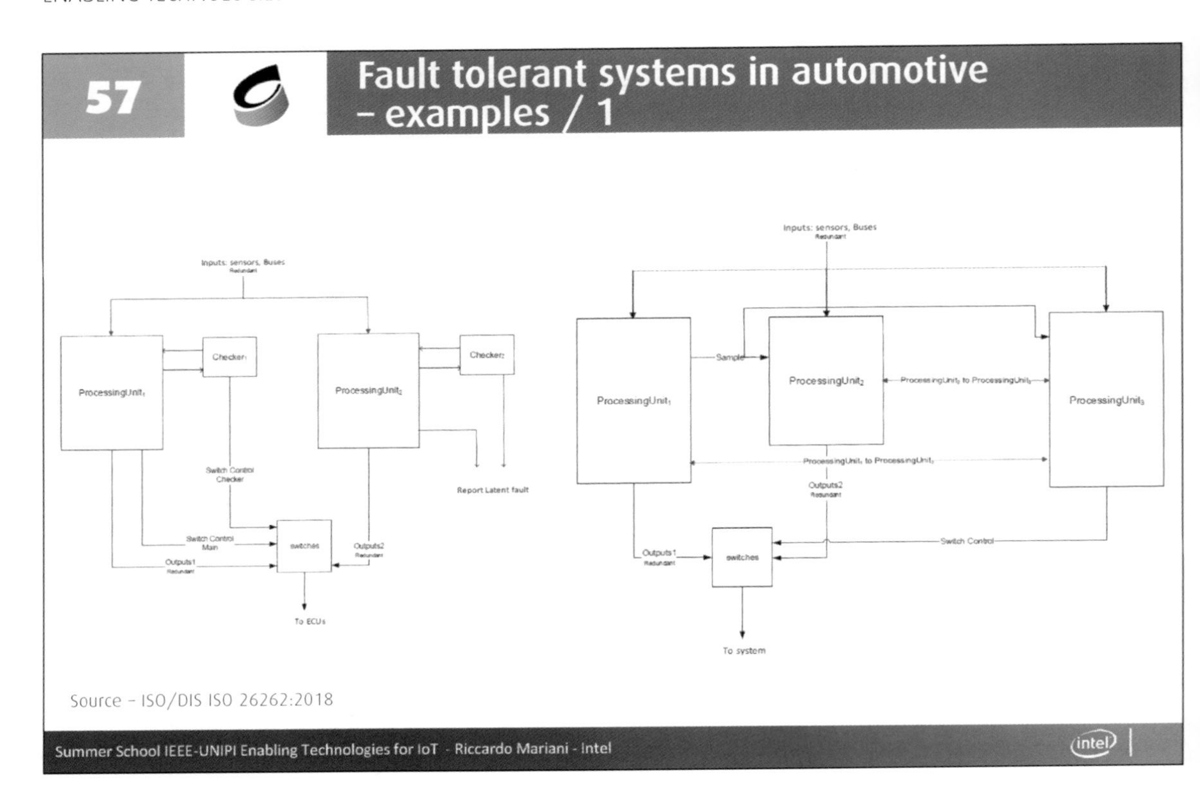
57
Fault tolerant systems in automotive – examples / 1
Inputs: sensors, Buses
Checker₁
ProcessingUnit₁
ProcessingUnit₂
Checker₂
Switch Control Checker
Report Latent fault
Switch Control Main
switches
Outputs2
Outputs1
To ECUs
Sample
ProcessingUnit₃
Switch Control
To system
Source – ISO/DIS ISO 26262:2018
Summer School IEEE-UNIPI Enabling Technologies for IoT - Riccardo Mariani - Intel
intel

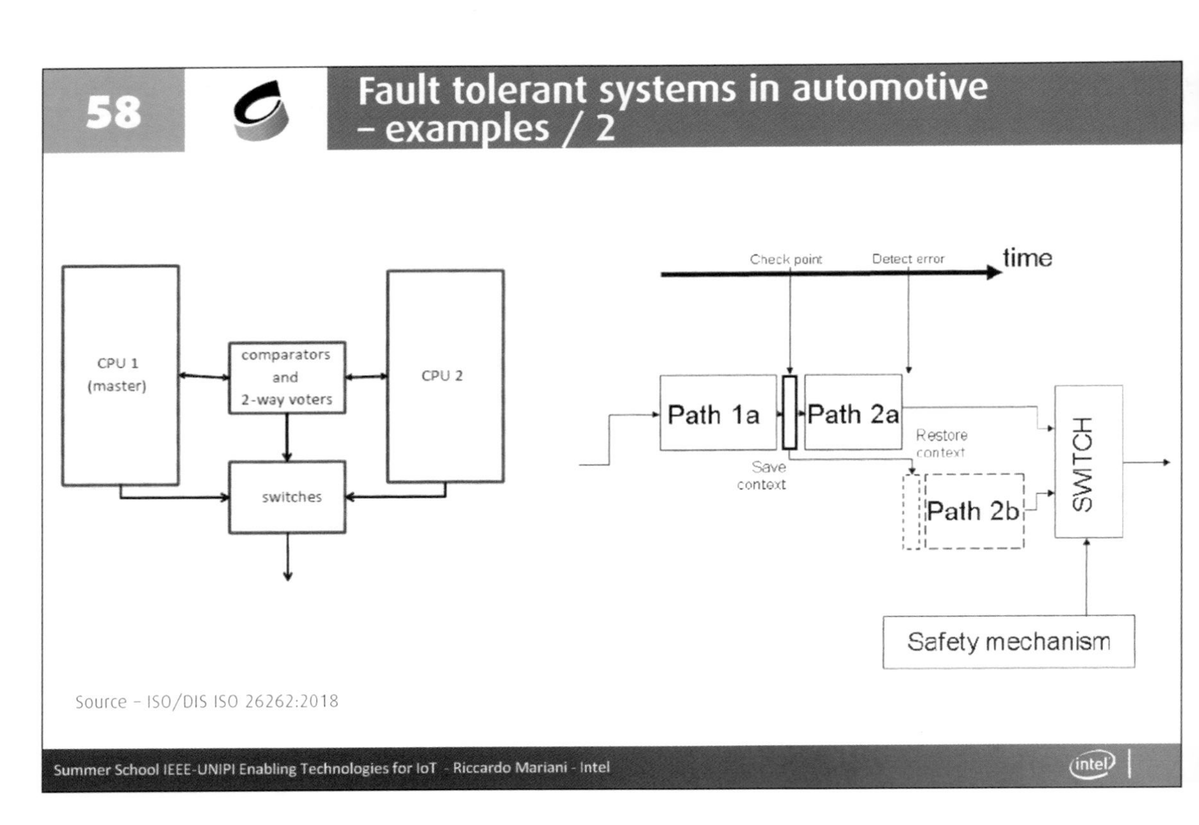
58
Fault tolerant systems in automotive – examples / 2
CPU 1 (master)
comparators and 2-way voters
CPU 2
switches
Check point
Detect error
time
Path 1a
Path 2a
Restore context
Save context
Path 2b
SWITCH
Safety mechanism
Source – ISO/DIS ISO 26262:2018
Summer School IEEE-UNIPI Enabling Technologies for IoT - Riccardo Mariani - Intel
intel

CONCLUSION

59  **Key takeaways**

How to make Functional Safety & Security effective in the new world of S/W defined autonomous things:

✓ Make them integrated: add H/W hooks, provide S/W libraries and safe&secure SDK, add safety and security «islands».

✓ Make them E2E: guarantee a solution from edge to data center, including connectivity.

✓ Make them configurable & self-adaptable: provide configurable systems, capable of analyzing failure and threats with machine learning and AI.

Summer School IEEE-UNIPI Enabling Technologies for IoT - Riccardo Mariani - Intel 

Bibliography

- ISO 26262:2017 DIS, *Road vehicles — Functional safety*, ISO, 2017.
- SAE J3016, *Taxonomy and Definitions for Terms Related to On-Road Motor Vehicle Automated Driving Systems*, SAE, 2014.
- SAE J3101, *Requirements for Hardware-Protected Security for Ground Vehicle Applications*, SAE, 2015.
- SAE J3061, *Cybersecurity Guidebook for Cyber-Physical Automotive Systems*, SAE, 2016.
- ISO/TC 22/SC 32 DPAS 21448, *Road vehicles — Safety of the Intended Functionality*, ISO.
- Shai Shalev-Shwartz, Shaked Shammah, and Amnon Shashua, "On a Formal Model of Safe and Scalable Autonomous vehicles", ARXIV, 2017.

About the Editor

Sergio Saponara

Sergio Saponara got the M.Sc. cum laude, and Ph.D. in Electronic Engineering from the University of Pisa. He is IEEE Senior Member and IEEE Distinguished Lecturer. In 2002, he was with IMEC, Leuven (B), as Marie Curie Research Fellow. He is Full Professor at University of Pisa in the field of electronic integrated circuits and systems, including wireless systems. He is also a Professor at the Italian Naval Academy. He co-authored about 300 scientific publications in peer-reviewed international journals and conferences, 17 patents, a book on radar design. He is also a research associate of CNIT and INFN and is an associate editor of IEEE Consumer Electronics Magazine, IEEE Vehicular Technology Magazine, IEEE Canadian Journal of Electrical and Computer Engineering, SpringerNature Journal of Real-time Image Processing, Electronics Letters, MDPI Designs, and guest editor of the SpringerNature Journal on Embedded Systems and IEEE Transactions on Industrial Informatics. He has been involved in the organization committee of about 130 conferences of IEEE and SPIE. He has been guest editor of several special issues, and invited/plenary speaker in several conferences, of IEEE, SPIE, Elsevier. He also got several best paper/presentation awards. He is the responsible of the summer school "Enabling technologies of IoT" and of the CrossLab "Industrial IoT" both at the University of Pisa. In 2014, he co-founded IngeniArs srl, a start-up company where he is Chief Technical Officer. He has been responsible for Pisa University or IngeniArs of several research projects funded under EU or ESA programs. He has been consultant for the R&D centers of industries such as STMicroelectronics, Renesas, AMS ag, Ericsson, SelexGalileo, Magna Closures, Rico, Flyby, PPC. He is co-founding member of the IEEE CASS IoT Special Interest Group and member of the World Forum on IoT plus other 3 IEEE standardization initiatives and 4 IEEE Technical Committees.

[Chapters 3 and 4]

About the Authors

Alice Buffi

Alice Buffi received the bachelor's and master's (summa cum laude) degrees in Telecommunications Engineering from the University of Pisa, Pisa, Italy, in 2006 and 2008, respectively, and Ph.D. in Applied Electro-magnetism in electrical and biomedical engineering, electronics, smart sensors, nanotechnologies from the University of Pisa in 2012, with *Doctor Europaeus* label. She was a Visiting Ph.D. Student with the Queen Mary University of London, London, U.K., in 2012. Since 2012, she has been with the University of Pisa, where she is currently an Assistant Professor with the Department of Energy, Systems, Territory and Construction Engineering. She co-authored several international journal papers and international conferences contributions and co-authored a European patent. Her research interests include design and characterization of far-field and near-field antennas for short-range communications and wireless power transfer. Her current research topics are on measurement methods for localization and classification with UHF-RFID systems. She is a member of the IEEE Instrumentation and Measurement Society and the Antennas and Propagation Society. Dr. Buffi was a recipient of the Young Scientist Award from the International Union of Radio Science, Commission B, in 2013 and 2016.

[Chapter 6]

Filippo Costa

Filippo Costa received the M.Sc. degree in Telecommunication Engineering and Ph.D. in Applied Electromagnetism from the University of Pisa, Italy, in 2006 and 2010, respectively. From March to August 2009, he was a Visiting Researcher at the Department of Radio Science and Engineering, Helsinki University of Technology, TKK, Finland. In July 2015, he was a Visiting Researcher at LCIS Lab - Systems Design and Integration del Grenoble Institute of Technology (INP-Grenoble). In February 2016, he was a Visiting Researcher at University Rovira I Virgili of Tarragona. He is currently an *Assistant Professor* at the University of Pisa. His research is focused on the analysis and modeling of Frequency Selective Surfaces and Artificial Impedance Surfaces with emphasis to applications in electromagnetic absorbing materials, leaky antennas, radomes, RFID, waveguide filters and methods for retrieving dielectric permittivity of materials. He was recipient of the Young Scientist Award of the URSI International Symposium on Electromagnetic Theory, URSI General Assembly, and URSI AT-RASC in 2013, 2014, and 2015, respectively. He was appointed as outstanding reviewer of IEEE Transactions on Antennas and Propagation in 2015 and 2016.

[Chapter 5]

Simone Genovesi

Simone Genovesi received the Laurea degree in Telecommunication Engineering and Ph.D. in Information Engineering from the University of Pisa, Italy, in 2003 and 2007, respectively. Since 2003, he has been collaborating with the Electromagnetic Communication Laboratory, Pennsylvania State University. From 2004 to 2006, he was a research associate at the ISTI institute of the National Research Council of Italy (ISTI-CNR) in Pisa. From 2010 to 2012, he was also a research associate at the CNR Institute for Microelectronics and Microsystems (IMM-CNR). He is currently an Assistant Professor at the Microwave and Radiation Laboratory, University of Pisa. His research topics focus on metamaterials, RFID systems, optimization algorithms, and reconfigurable antennas. He was the recipient of a grant from MIT in the framework of the MIT International Science and Technology Initiatives (MISTI) and he is involved in EMERGENT, an EU-funded project aiming at designing novel chipless RFID tags and sensors.

[Chapter 5]

David Girbau Sala

David Girbau Sala received the BS in Telecommunication Engineering, MS in Electronics Engineering, and PhD in Telecommunication from Universitat Politècnica de Catalunya (UPC), Barcelona, Spain, in 1998, 2002, and 2006, respectively. From February 2001 to September 2007, he was a Research Assistant with the UPC. From September 2005 to September 2007, he was a Part-Time Assistant Professor with the Universitat Autònoma de Barcelona (UAB). Since October 2007, he has been an Assistant Professor at Universitat Rovira i Virgili (URV). His research interests include microwave devices and systems, with emphasis on UWB, RFIDs, RF-MEMS, and wireless sensors.

[Chapter 5]

Ludger Klinkenbusch

Ludger Klinkenbusch received the Diploma, the Dr.-Ing., and the Habilitation degrees from the Ruhr-Universität Bochum (Germany) in 1986, 1991, and 1996, respectively. Since 1998, he has been with Kiel University (Germany) as Professor of Computational Electromagnetics in the Faculty of Engineering and Director of the Institute of Electrical Engineering and Information Technology. His research interests include theoretical aspects of scalar and electromagnetic fields, corresponding analytical and numerical methods, and their application in scattering and diffraction, bio-magnetic fields, and EMC-related problems. He is also active as a consultant for industry, e.g., in developing methods and software for antenna near-field measurements and for computational electromagnetics.

Dr. Klinkenbusch is a Fellow of IEEE 'for contributions to spherical-multipole analysis of electromagnetic fields', a member of German VDE, and an elected member of URSI Commission B. He currently serves as an Associate Editor of the IEEE Transactions on Antennas and Propagation. He co-organized the 2010 URSI Commission B International Symposium on Electromagnetic Theory in Berlin (Germany). He won the '2004 URSI Commission B International Electromagnetics Prize' paper competition with a contribution entitled 'Electromagnetic scattering by a plane angular sector.'

[Chapter 2]

Antonio Lazaro

Antonio Lazaro received M.S. degree and Ph.D. in Telecommunication Engineering from the Universitat Politècnica de Catalunya (UPC), Barcelona, Spain, in 1994 and 1998, respectively. He then joined the faculty of UPC, where he currently teaches a course on microwave circuits and antennas. Since July 2004, he has been a Full-Time Professor at the Department of Electronic Engineering, Universitat Rovira i Virgili (URV), Tarragona, Spain. His research interests are microwave device modeling, on-wafer noise measurements, monolithic microwave integrated circuits (MMICs), low phase noise oscillators, MEMS, RFID, UWB, and microwave systems.

[Chapter 5]

Massimo Macucci

Massimo Macucci graduated in Electrical Engineering in 1987 at the University of Pisa, he then obtained the "Perfezionamento" (Doctorate) degree from the Scuola S.Anna-Pisa (1990), and his master's degree (1991) and Ph.D. (1993) from the University of Illinois at Urbana-Champaign. Since 1992, he has been on the faculty of the Electrical and Computer Engineering Department of the University of Pisa, currently as Professor of Electronics. His research interests include novel nanoelectronic semiconductor devices, quantum effects in electron devices, and noise phenomena in electronic components and circuits, as well as some aspects of electromagnetic compatibility and of molecular electronics. He is also working on electronics for transportation applications, in particular safety systems for railways and solar-powered aircrafts. He has been the coordinator of the European research project QUADRANT, (QUAntum Devices foR Advanced Nano-electronic Technology) and the local principal investigator for the European Projects and Coordination and Support Actions ANSWERS, GRAND, PHANTOMS, MULTEU-SIM, NanoICT. He has authored about 150 peer-reviewed publications.

[Chapter 7]

Giuliano Manara

Giuliano Manara is a Professor at the College of Engineering of the University of Pisa, Italy. Since 1980, he has been collaborating with the Department of Electrical Engineering of the Ohio State University, Columbus, Ohio, USA, where, in the summer and fall of 1987, he was involved in research at the ElectroScience Laboratory. His research interests have centered mainly on the asymptotic solution of radiation and scattering problems to improve and extend the uniform geometrical theory of diffraction. He has also been engaged in research on numerical, analytical, and hybrid techniques (both in frequency and time domain), scattering from rough surfaces, frequency selective surfaces (FSS), and electromagnetic compatibility. More recently, his research has also been focused on the design of microwave antennas with application to broadband wireless networks, on the development and testing of new microwave materials (metamaterials), and on the analysis of antennas and propagation problems for Radio Frequency Identification (RFID) systems.

Prof. Manara was elected an *IEEE (Institute of Electrical and Electronic Engineers) Fellow* in 2004 for "contributions to the uniform geometrical theory of diffraction and its applications." In August 2008, he was elected the Vice-Chair of the *International Commission B "Fields and Waves" of URSI (International Radio Science Union)*. He served as the International Chair of *URSI Commission B* for the triennium 2011-2014. Prof. Manara has been elected a *URSI Fellow* in 2017.

[Chapter 6]

Riccardo Mariani

Riccardo Mariani is an Intel Fellow and the chief functional safety technologist in the Internet of Things Group at Intel Corporation. Based in Pisa, Italy, he is responsible for defining strategies, roadmaps and technologies for Internet of Things applications that require functional safety and high performance, including transportation and industrial systems. He is also the functional safety global domain lead for Intel's CISA Architecture Working Model Initiative. Mariani joined Intel in 2016 with the acquisition of Yogitech S.p.A., a leading provider of functional safety technologies. As Yogitech's chief technology officer and co-founder, he invented the company's flagship faultRobust technology and related products. Mariani led Yogitech to pioneer the certification of semiconductor intellectual property for the highest level of safety integrity, and to introduce electronic design automation tools specific to functional safety, known as the Design-for-Safety paradigm. Before founding Yogitech in 2000, Mariani was technical director at Aurelia Microelettronica S.p.A. His responsibilities in that position included leading high-reliability topics in projects with the CERN research center in Geneva. A recognized expert in functional safety and integrated circuit reliability, Mariani regularly contributes to industry standards efforts, including leading the ISO 26262-11 part specific to semiconductors. He also speaks frequently at industry conferences, lectures at universities, and coordinates functional safety topics for projects funded by Italy and the European Union. Mariani has co-authored a book and authored or co-authored more than 70 papers related to functional safety, high-reliability circuits, design for testability, advanced design techniques, and asynchronous circuits. He holds multiple patents in the field of functional safety, with additional patents pending. Mariani earned a bachelor's degree in Electronic Engineering and Ph.D. in Microelectronics, both from the University of Pisa in Italy. He won the SGS-Thomson Award and the Enrico Denoth Award for his engineering achievements.

[Chapter 8]

Andrea Michel

Andrea Michel received the B.E., M.E. (summa cum laude), and PhD degrees in Telecommunications Engineering from the University of Pisa, Pisa, Italy, in 2009, 2011, and 2015, respectively. Since 2015, he has been a PostDoc researcher in applied electromagnetism at the Microwave&Radiation Laboratory, Dept. of Information Engineering, University of Pisa. His research topics focus on design of integrated antenna for communication systems and smart antennas for near-field UHF-RFID readers. In 2014, he was a Visiting Scholar at the ElectroScience Laboratory, The Ohio State University, Columbus, OH, USA. During this period, he was involved in research on a theoretical analysis on the accuracy of a novel technique for deep tissue imaging. Recently, he has been working on the design of antennas for automotive applications, MIMO systems, and wearable communication systems, in collaboration with other research institutes and companies. Dr. Michel was a recipient of the Young Scientist Award from the International Union of Radio Science, Commission B, in 2014, 2015, and 2016. In 2016, he received the Best Paper Honorary Mention from the IEEE International Conference on RFID Technology and Applications, Shunde, Guangdong, China

[Chapter 6]

Paolo Nepa

Paolo Nepa received the Laurea (Doctor) degree in Electronics Engineering (summa cum laude) from the University of Pisa, Italy, in 1990. Since 1990, he has been with the Department of Information Engineering, University of Pisa, where he is currently an Associate Professor. In 1998, he was at the Electro Science Laboratory (ESL), The Ohio State University (OSU), Columbus, OH, as a Visiting Scholar supported by a grant of the Italian National Research Council. At the ESL, he was involved in research on efficient hybrid techniques for the analysis of large antenna arrays. His research interests include the extension of high-frequency techniques to electromagnetic scattering from material structures and its application to the development of radio propagation models for indoor and outdoor scenarios of wireless communication systems. He is also involved in the design of wideband and multiband antennas, mainly for base stations and mobile terminals of communication systems, as well as in the design of antennas optimized for near-field coupling and focusing. More recently, he has been working on channel characterization, wearable antenna design, and diversity scheme implementation, for body-centric communication systems. In the context of RFID systems, he is working on techniques and algorithms for UHF-tag localization and RFID-based smart shelves. He is an Associate Editor of the IEEE Antennas and Wireless Propagation Letters. Dr. Nepa received the Young Scientist Award from the International Union of Radio Science, Commission B, in 1998.

[Chapter 6]

Smail Tedjini

Doctor in Physics from Grenoble University in 1985. He was Assistant Professor at the Electronics Department of Grenoble Institute of Technology (Grenoble-inp) from 1981 to 1986, and Senior Researcher for the CNRS (Research French National Center) from 1986 to 1993. He became University Full Professor in 1993, and since 1996 he has been Professor at the esisar, Embedded Systems Dpt. of Grenoble-inp. His main teaching topics concern Applied Electromagnetism, RadioFrequency, Wireless Systems, and Optoelectronics. Now he has more than 30 years experience in academic education, Research and management of university affairs. He served as coordinator and staff member in numerous academic programs both for education and research. He was coordinator for Ph.D. program, Master and Bachelor Programs for the Universities at Grenoble, some of these programs are under collaboration with international universities from Europe, USA, Canada, Brazil, Vietnam, Egypt, Maghreb. He served as the Director of esisar, Dpt. of Grenoble-inp. He is involved in academic research supervision since 1982. His main topics in research are applied electromagnetism, modeling of devices and circuits at both RF and optoelectronic domains. Current research concerns wireless systems with specific attention to RFID technology and its applications. He is the founder in 1996 and past Director of the LCIS Lab. Now, he is project manager within ORSYS group that he leaded until 2014 and founded 15 years ago. He supervised more than 35 PhD and he has more than 300 publications. He serves as Examiner/reviewer for tens of Ph.D. degrees for universities in many countries. He supervised tens of research contracts with public administrations and industries. He is Member of several TPC and serves as expert/reviewer for national and international scientific committees and conferences including journals such Piers, IEEE (MTT, AP, Sensors, MGWL), URSI. He served as project expert for French and international research programs ISO, ANR, OSEO, FNQRT, CNPQ...

[Chapter 1]